Understanding Immunology

PEARSON

We work with leading authors to develop the strongest educational materials in biology bringing cutting-edge thinking and best learning practice to a global market.

Under a range of well-known imprints, including Prentice Hall, we craft high-quality print and electronic publications which help readers to understand and apply their content, whether studying or at work.

To find out more about the complete range of our publishing, please visit us on the World Wide Web at: www.pearsoned.co.uk.

Understanding Immunology

Third Edition

Peter Wood
University of Manchester

Prentice Hall
is an imprint of

Harlow, England • London • New York • Boston • San Francisco • Toronto • Sydney • Singapore • Hong Kong
Tokyo • Seoul • Taipei • New Delhi • Cape Town • Madrid • Mexico City • Amsterdam • Munich • Paris • Milan

Pearson Education Limited
Edinburgh Gate
Harlow
Essex CM20 2JE
England

and Associated Companies throughout the world

Visit us on the World Wide Web at:
www.pearsoned.co.uk

First published 2001
Second Edition 2006
Third Edition 2011

ISBN 978-0-273-73068-2

British Library Cataloguing-in-Publication Data
A catalogue record for this book is available from the British Library

Library of Congress Cataloging-in-Publication Data
Wood, Peter (Peter John), 1952-
Understanding immunology / Peter Wood. -- 3rd ed.
p. ; cm.
Includes bibliographical references and index.
ISBN 978-0-273-73068-2 (pbk.)
1. Immunology. I. Title.
[DNLM: 1. Immune System. 2. Immune System Processes. 3. Immunity. QW 504]
QR181.W735 2011
616.07'9--dc22

2011001686

10 9 8 7 6 5 4 3 2 1
15 14 13 12 11

Typeset in 9.5 Concorde BE by 30
Printed in Great Britain by Henry Ling Ltd., at the Dorset Press, Dorchester, Dorset

For my wife, Jude
And in memory of Jack and Kirsten. *Elsket og Savnet*

Contents

Preface

This book is aimed at introducing immunology to students who have never studied the subject before. It is intended primarily for undergraduate students in the biological sciences and biomedical fields and medical students, although it is hoped it will appeal to other groups. No previous knowledge of immunology is assumed, but people reading this book will benefit from a basic knowledge of biochemistry and cell biology.

The impetus to write this book came from experience of teaching introductory immunology courses to undergraduate science, medical and dental students. Often there is a stage of bewilderment at the complexity of the immune system, which seems to utilise convoluted mechanisms to perform what are, on the face of it, simple tasks. Therefore one of the aims of this book is to try to explain why, in many situations, a more complicated looking arrangement is actually much more efficient than an apparently simpler alternative.

The first part of the book is designed to take students step by step through the pathogenesis of infectious diseases and the molecules, cells and tissues of the immune system that provide protection against the wide variety of pathogens to which we are exposed. The latter part of the book deals more with the immune system operating in disease situations such as allergy, autoimmunity and transplantation. Inevitably, with an introductory book, some topics receive little or no attention. Therefore immunodeficiencies are referred to briefly and other topics such as xenotransplantation and mucosal tolerance are not covered.

The book is organised so that it is easiest to understand if the chapters are read sequentially. This has led to two basic rules in design. The first is that, wherever possible, new words and concepts are explained when they are first introduced, although inevitably reference is made to later chapters in some instances. Second, it is assumed in the later chapters that previous chapters have been read and an understanding of the previous concepts acquired. Where it is felt important to refer the reader to previous chapters this is done. Another concept of the book is the use of boxes. The material in the boxes is not essential for understanding of the rest of the text. The boxes are provided (i) to describe in more detail material covered in the text and (ii) to describe key observations or experiments that led to important advances in immunological knowledge (e.g. the discovery of the thymus), but they can be bypassed if time or inclination warrant this.

Chapter 1 begins with a description of the variety of pathogens, illustrating their diversity, and emphasises that no simple defence mechanism could deal with this variety of threats. The immediate response to infection is described in Chapter 2 and the requirement to distinguish between pathogens and host tissue introduced. Recognition by the innate immune system is covered. Chapter 3 describes the antibody molecule as part of the solution to having a better system of recognising pathogens than that provided by innate recognition and also introduces one of the cells of the specific immune system – the B lymphocyte. Other important types of lymphocyte, T cells, how they recognise antigen and the structure of the major histocompatibility complex (MHC) are described in Chapter 4. Chapter 5 describes the development of lymphocytes and explains the mechanism that allows the generation of a huge repertoire of antigen specificities from a limited number of genes. Chapter 6 describes the anatomy of the immune system and how it promotes the cellular interactions required in immune responses. The anatomical and cellular aspects of antibody production are expanded in Chapter 7. How antibody contributes to the elimination of pathogens or neutralises their effects is covered in Chapter 8. Chapter 9 is concerned with cellular immunity, which involves different types of immune responses than antibody production and describes the different types of CD4 helper T cells. One of the major successes of manipulating the immune system, vaccination, and its basis in immunological memory are described in Chapter 10. Chapter 10 also incorporates new sections on how antibodies, including monoclonal antibodies, are used as experimental and clinical tools. Chapter 11 deals with immunological tolerance and the mechanisms that are in place to try to prevent the immune system from attacking our own cells and tissues. The remaining chapters are devoted to the immune system in disease. Chapters 12 and 13 cover autoimmunity and allergy respectively and suggest reasons for the severe increase in incidence of these diseases in developed countries, and Chapter 14 is about another immunological disease that has attracted attention – AIDS. The final chapter describes attempts, clinical and experimental, to manipulate the immune system in the fields of transplantation and tumour therapy. To reiterate, throughout the book attempts have been made wherever space has allowed to explain why the immune system is organised the way it is, so that its complexity can be understood.

Peter Wood

Acknowledgements

We are grateful to the following for permission to reproduce copyright material:

Figures
Figure 1.2: The changing incidence of infectious disease. From 'The Effect of Infections on Susceptibility to Autoimmune and Allergic Diseases', *The New England Journal of Medicine*, 347, 911–920 (Bach, JF., M.D.,D. Sc. 2002); Figure 10.11a: Western Blotting. Adapted from *Experimental Biochemistry*, W. H. Freeman and company/ Worth Publishers (Switzer, R. L, Garrity, L. F. 1999) p. 296; Figure 15.4: The effect of HLA-DR matching on graft survival. Reprinted with permission of Dr. Philip A. Dyer.

Plates
Plate 1: Elephantiasis. Richard Suswillo, Department of Infectious Disease Epidemiology, Imperial College School of Medicine, London UK; Plate 2: Cells of the phagocyte lineage; Plate 3: Mast cells and basophils; Plate 4: Lymphocyte and plasma cell; Plate 9: Histological appearance of thymus; Plate 10: Histological section of part of the spleen and lymph node; Plate 11: Histological appearance of mucosal associated lymphoid tissue (MALT). All courtesy of Mike Mahon, John S Dixon and Philip F Harris, University of Manchester; Plate 5: Antibody structure. From Saphire E. O., et al (2001) Crystal structure of a neutralizing human IgG against HIV-1: A template for vaccine design. Reprinted with permission from Science Vol 293, copyright © 2001 American Association of the Advancement of Science (AAAS); Plate 6: Antibody–antigen interaction; Plate 8: Interaction between the T cell receptor (TcR) and antigen/ class I MHC. Reprinted with permission of Dr Annemarie Honegger and Prof. Andeas Pluechthun, Biochemisches Institut Der Universitat Zurich, Winterthurerstrasse 190, CH-8057 Zurich, Switzerland; Plate 7: C1 I and C1 II MHC. Copyright © 1994 Elsevier. Reprinted with permission; Plate

12: Langerhans cells and dendritic cells. Courtesy of Dr Pedram Hamrah and Dr Reza Dana, Schepends Eye Research Institute, Harvard Medical School, Boston MA, USA.

In some instances we have been unable to trace the owners of copyright material, and we would appreciate any information that would enable us to do so.

CHAPTER ONE

The threat to the body: the role and requirements of the immune system

Learning objectives

To be acquainted with the role of the immune system. To learn about the huge variety of threats posed to the body by infectious organisms. To appreciate that this requires a complex defence system.

Key topics

- The role of the immune system
- Types of pathogen
- Disease production by pathogens
    - Infection
    - Replication
    - Spread
    - Pathogenesis
- Barriers to infection

1.1 The role and complexity of the immune system

Immunology is the study of the immune system which evolved primarily to provide defence against infectious disease caused by bacteria, viruses, fungi and parasites. We are under constant attack from these infectious organisms that can invade the body and potentially cause disease and sometimes death. There have been a number of plagues that have ravaged civilisations since humans began to live in towns and cities, which encouraged the rapid spread of infection. Smallpox and bubonic plague were two of the major culprits and it is estimated that during the major plagues (starting in 540, 1346 and 1665 BCE) over a third of the population were killed in affected areas. Another consequence of city living was poor sanitation with open sewers and human waste thrown in streets. This meant that the average lifespan of 40 years in 1900 had hardly risen from the estimated 35-year lifespan of our hunter gatherer ancestors some 40 000 years before! What is more, 40% of children in cities still died before the age of five. It was only during the twentieth century, with better sanitation and the development of vaccination and antibiotics, that child mortality dropped dramatically and average lifespan rose to nearly 80 years by the end of the century. However billions of people are still exposed to major infections in many parts of the world. Worldwide, infection is the second leading cause of death after cardiovascular disease and the major cause of death in the under 50s! In the last thirty years the AIDS (acquired immunodeficiency syndrome) epidemic, which is still ongoing, has brought the consequences of having a poorly functioning immune system to the attention of the public at large. AIDS patients usually die from infections, such as the yeasts *Pneumocystis carinii*, which causes pneumonia, or *Cryptococcus neo-formans*, a cause of meningitis. The immune system normally controls these infections with little or no damage to the host.

The primary role of the immune system is to provide defence against the threat of disease posed by infectious organisms. Like any other physiological system, the immune system consists of proteins, cells and organs that are shown in Figure 1.1.

Immunology today consists of much more than the study of defence against infection. As the incidence of infection dropped in the twentieth century, the incidence of immunologically related diseases, such as **allergies**, and **autoimmune diseases**, including rheumatoid arthritis, multiple sclerosis and some types of diabetes, has risen equally dramatically (Figure 1.2). Both allergies and autoimmune diseases are caused by malfunctioning of the immune system. Allergies are caused by an inappropriate immune response against generally harmless material such as pollen or food and autoimmune diseases occur when the immune system attacks the body's own tissue.

There are other clinically relevant situations where the immune system plays a role. Evidence is accumulating that the immune system can provide protection against some **tumours**. Exciting new developments suggest that immune responses can be induced against tumours that normally do not provoke an immune response. These so-called tumour vaccines offer hope of additional weapons in the armoury against cancer. The immune system is also responsible for the rejection of **transplants**. In this instance the immune system is acting normally in trying to defend the body against a foreign invader, even though the 'invader' is beneficial. Rejection of transplants is a major cause of graft loss and much effort is being devoted to try to prevent transplant rejection. Finally there is increasing evidence that some of the chronic diseases associated with increased lifespan, such as atherosclerosis (causing heart disease and stroke) and Alzheimer's disease, also involve the immune system although the exact nature of the involvement is not known.

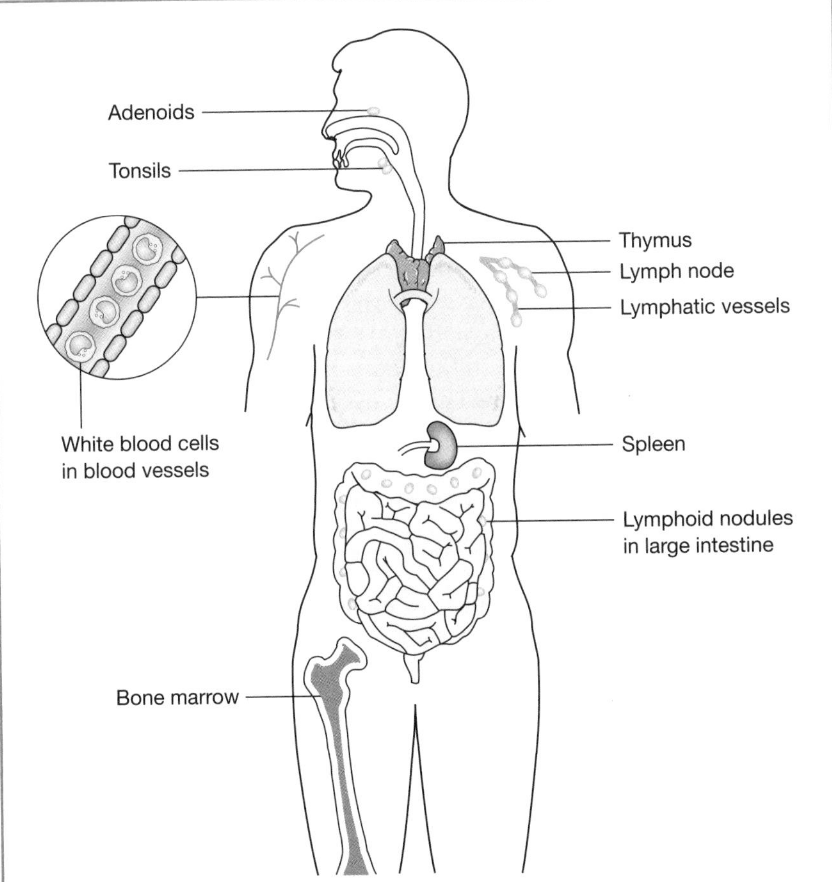

Figure 1.1 Some components of the immune system. The bone marrow and thymus are sites of production of lymphocytes, white blood cells involved in immune responses. Adenoids, tonsils, lymph nodes, spleen and lymphoid nodules are sites where immune responses are generated. Lymphatic vessels are similar to blood vessels and allow cells of the immune system to travel throughout the body.

So although the immune system is capable of causing harm, the threat is not as great as that posed by infection and on balance a properly functioning immune system is essential for life in the microbe-strewn world in which we live.

1.1.1 So what exactly is the threat from infectious organisms?

Infectious organisms that cause disease are called **pathogens** and the individual (person or animal) infected by a pathogen is called the **host**. Not all infectious organisms cause disease and some are actually beneficial, for example bacteria living in the gut help to digest certain foods. Infectious organisms that help the host are called **commensal organisms**. However, many viruses, bacteria, fungi, yeasts and parasites are pathogenic and we

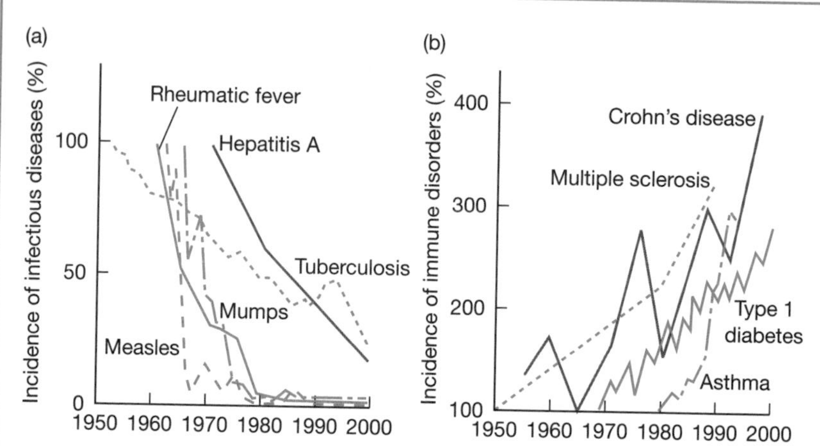

Figure 1.2 The changing incidence of infectious disease (a) and allergic and autoimmune diseases from 1950 to 2000. (Source: Bach, JF. (2002), *New England Journal of Medicine* Vol. 347, pp. 911–920.)

are constantly in danger of infection and disease caused by them. It is not known how many different pathogens there are but there are at least 350 diseases are listed as being caused by infection and some of these, such as diarrhoea, can be caused by many different bacteria, viruses and parasites. It is therefore likely that the number of pathogens runs into the thousands. Table 1.1 shows some examples of infections caused by various pathogens.

Unfortunately the number of human pathogens may be increasing because of the newly emerging infectious diseases. These are diseases that have not been identified before; in the 1980s AIDS was a newly emerging infectious disease. More recent examples are West Nile virus, Nipah virus and sudden acute respiratory syndrome (SARS). Most of these diseases are **zoonoses**, that is they are caused by pathogens that have jumped species from their normal animal host to man. The same thing happened with AIDS where the causative agent, the human immunodeficiency virus (HIV), jumped from chimpanzee to man. Many factors are contributing to the emergence of these infectious diseases including: increased migration, increased travel, increased incursion into remote wild habitats and the keeping of exotic animals as pets.

1.1.2 Why is immunology so complicated?

There are thousands of components to the immune system, and during the course of learning about some of these it can appear that the immune system is far more complicated than necessary for achieving what is, on the surface, the simple task of eliminating an infectious organism. There are a number of reasons why the immune system is complex. The first of these is the desirability of eliminating pathogens without causing damage

Table 1.1 Examples of infectious diseases caused by different families of pathogens

Organism	*Disease*
Viruses	
Hepadnavirus	Hepatitis B
Herpesvirus	Chickenpox
Poxvirus	Smallpox
Picornovirus	Polio, common cold
Myxovirus	Measles, mumps
Retrovirus	AIDS
Bacteria	
Streptococcus	Pneumonia
Staphylococcus, MRSA	Boils, diarrhoea
Clostridium	Tetanus, botulism
Neisseria	Gonorrhoea
Salmonella	Food poisoning
Vibrio	Cholera
Mycobacterium	Tuberculosis, leprosy
Fungi (yeasts and moulds)	
Trychophyton	Athlete's foot, jock itch
Candida	Thrush
Cryptococcus	Meningitis
Protozoan parasites	
Plasmodium	Malaria
Giardia	Diarrhoea
Cryptosporidium	Diarrhoea
Trypanosomes	Chagas' disease, sleeping sickness
Helminth parasites	
Taenia	Tapeworm
Schistosoma	Schistosomiasis (flukes)
Ascaris	Roundworm
Onchocera	River blindness

to the host. Getting rid of a pathogen is theoretically easy. If you had an infection in your liver you could produce a nasty toxin that would kill the pathogen; unfortunately it would also destroy your liver. Killing pathogens is not difficult, but getting rid of pathogens without damaging the host is much more complicated. Imagine if a city in your country was infiltrated by soldiers from another country; you could get rid of the foreign invaders by dropping a nuclear bomb on the city but this would also kill a lot of your compatriots. To go into the city to eliminate or capture the foreign soldiers without causing harm to your compatriots is much more difficult. This analogy raises a major issue concerning the immune system – that of **recognition**. To eliminate foreign soldiers without killing your own requires that you can tell the two apart. In the same way the immune system must be able to distinguish between pathogens and host cells so that it can direct its destructive powers towards the pathogens. Many of the specialised features of the immune system are involved in recognition of foreign pathogens.

In additional to the large number of different pathogens, additional problems facing the immune system are that pathogens come in all shapes and sizes, with different lifestyles and different ways of causing disease. To understand fully the complexity that the immune system must deal with, it is necessary to have some understanding of infectious organisms and the ways in which they cause disease. The rest of this chapter describes how pathogens differ, so that hopefully it will be possible to get an appreciation of the problems faced by the immune system.

1.2 Pathogens differ in size, lifestyle and how they cause disease

The types of pathogen that can cause disease include many groups of single-celled microorganisms and larger multicellular parasites. Viruses, bacteria, some yeasts and protozoan parasites are examples of single-celled pathogens. Fungi and helminths (parasitic worms) are the major multi-cellular pathogens (Table 1.1). These pathogens come from very different parts of the biological kingdom and vary considerably in many aspects. Pathogens differ enormously in their size. They also have very different lifestyles and cause disease in a variety of ways (Table 1.2).

1.2.1 Size of pathogens: from tiny viruses to very big worms

One feature of the range of pathogenic organisms listed in Table 1.2 is the enormous variation in **size**. Viruses are the smallest infectious organisms, being 20–400 nm in size. At the other end of the scale some parasitic worms, such as the tapeworm, can be up to 7 m (20 ft) in length. This represents a difference in scale of a factor of 10^9. To put that into some sort of perspective, if a virus were the size of a tennis ball, a fully developed tapeworm would reach from London to Los Angeles. It does not stretch the imagination too far to appreciate that the problems posed to the immune system by these two organisms would require very different solutions.

Table 1.2 Size and lifestyle of pathogens

Organism	*Size*	*Habitat*	*Mode of multiplication*	*Multiplication rate (doubling time)*
Viruses	20–400 nm			
Poliovirus		Intracellular: pharynx, intestine, nervous system	Intracellular synthesis of viral components	<1 hour
Poxvirus		Intracellular: upper respiratory tract, lymph nodes, skin	Intracellular synthesis of viral components	<1 hour
Bacteria	1–5μm			
Streptococcus Pyogenes		Extracellular: pharynx	Cell fission	3 hours
Mycobacterium Leprae		Intracellular: macrophages, endothelial cells, Schwann cells	Cell fission	2 weeks
Fungi	2–20 μm			
Candida Albicans		Extracellular: mucosal surfaces	Asexual budding	Hours
Histoplasma Capsulatum		Intracellular: macrophages	Asexual budding	Hours
Protozoan parasites	1–50 mm			
Trypanosomes		Extracellular: bloodstream	Binary fission	6.5 hours
Plasmodium		Intracellular: red blood cells, hepatocytes	Asexually in hepatocytes (cell fission)	8 hours
Metazoan parasites (worms)	3 mm to 7 m			
Ascaris lumbricoides		Intestine	Lays eggs	200 000 eggs/day
Taenia solium (tapeworm)		Gut	Releases body segments containing eggs	800 000 eggs/day

1.2.2 Stages of disease production by pathogens

Size is not the only way in which infectious organisms vary. They also vary enormously with respect to how they enter and live within the body and actually cause disease. Infection and disease production by pathogenic organisms can be divided into four stages:

1. Invasion.
2. Multiplication.
3. Spread.
4. Production of disease (pathogenesis).

Although infection usually involves all of these steps, there are many exceptions in terms of both the steps involved and their order. Some pathogens do not spread significantly or even technically gain entry to the body. The bacterium causing cholera, *Vibrio cholerae*, enters the gut and attaches to the luminal side of epithelial cells; as such it does not technically enter the body but secretes powerful toxins that affect the epithelial cells causing vast volumes of watery diarrhoea. Organisms may replicate locally before spreading or may spread through the body before beginning significant replication. Pathogens show considerable variation at each of these stages of infection, as will be described below.

1.3 How do pathogens cause disease and what protection is there?

The first stage of disease production by pathogens is infection, or entry of the pathogen into the body. Nearly all pathogens must gain entry into the body before they can begin to replicate or spread. A few pathogens can exist on the skin (e.g. viruses causing warts) or in the gut (e.g. bacteria causing cholera) without technically entering the body. However, infection is not made easy for pathogens because the body has many physical and chemical barriers to try to prevent pathogens entering the body.

1.3.1 Barriers to infection

The body has many physical, chemical and biochemical barriers that make it much more difficult for pathogens to gain entry into the body (Figure 1.3). The **physical barriers** to infection are as follows:

- **Skin and mucosa.** Intact skin and mucosa provide a physical barrier to prevent entry of organisms.
- **Cilia.** The respiratory tract is lined with little hair-like structures that beat in such a way as to propel particles towards the throat, where they can be expelled by coughing or swallowing and excretion.
- **Mucus.** Mucus is secreted by epithelial cells of the gut, respiratory tract and genito-urinary (GU) tract. It has the unusual properties of being

sticky and slimy at the same time and is able to entrap microorganisms so they can be expelled. In the respiratory tract, cilia and mucus combine to provide an effective way of trapping and eliminating microbes.

- **Coughing and sneezing.** Coughing and sneezing provide a way of expelling microorganisms. It is estimated that the expelled air in a sneeze can travel at an impressive 200mph. However there are some arguments that coughing and sneezing have been exploited by some microbes to promote their spread from the infected person to others.

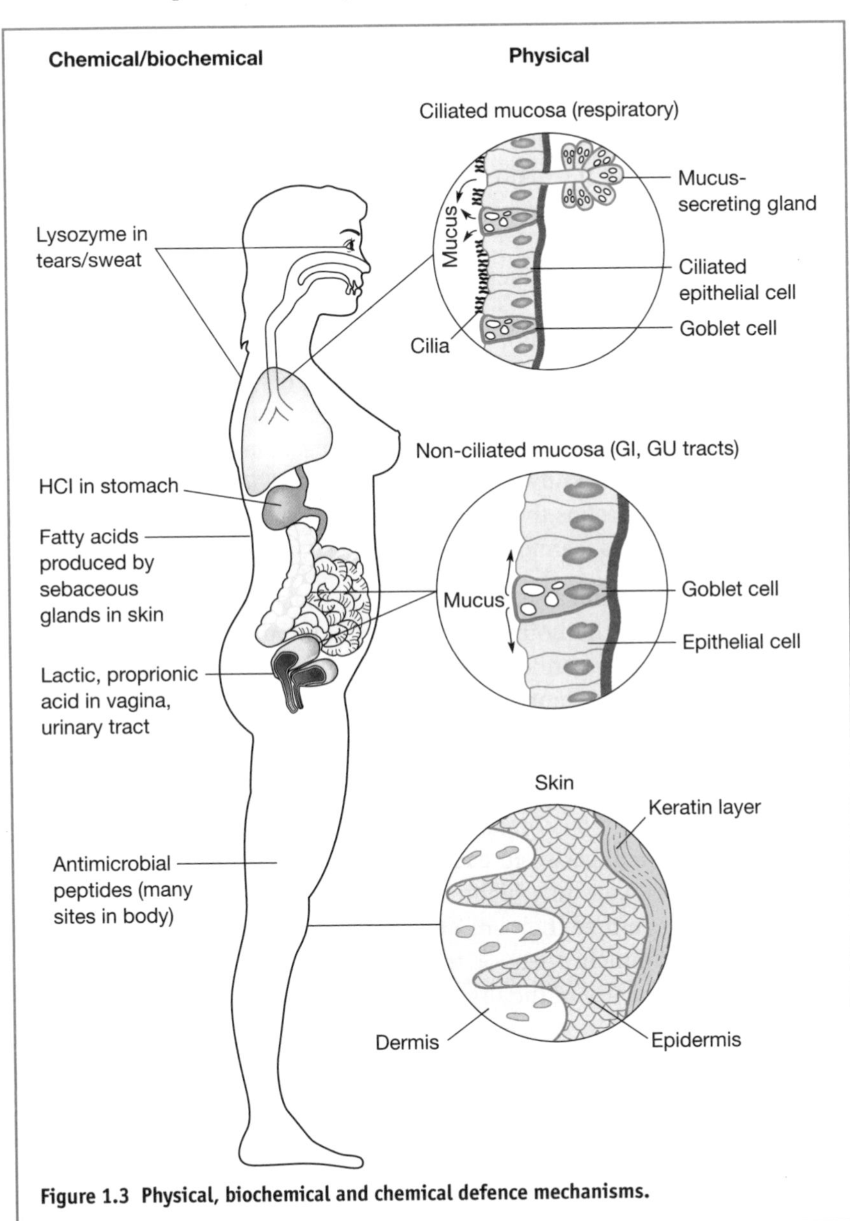

Figure 1.3 Physical, biochemical and chemical defence mechanisms.

The **chemical and biochemical** defences are as follows:

- **Acids.** Hydrochloric acid secreted by the stomach is lethal to many (though not all) bacteria. Commensal bacteria in the vagina produce lactic and proprionic acid resulting in a low pH, which is inhibitory to the division of many bacteria.
- **Fatty acids.** Sebaceous glands in the skin produce fatty acids that have antimicrobial properties.
- **Lysozyme.** This is present in sweat, tears and many other secretions. It breaks down peptidoglycans in bacterial cell walls, thus damaging and killing the bacteria.
- **Anti microbial peptides.** Over the last few years it has become apparent that we produce over 1000 different small peptides with anti microbial properties. Some of the main types are:
  - **Defensins** are antimicrobial peptides that are found in the secretions of mucosa and skin.
  - **Cathelicidins** are antibacterial peptides that were originally discovered as insect defence peptides. Other members of the cathelicidin family have been found in mucosal secretions.
  - **Collectins** are proteins that can bind sugars on microbial surfaces and promote the elimination of microbes. Proteins that bind sugars are known as lectins; because collectins bind sugars in a calcium-dependent manner, they are known as C-type lectins. The A and D lung surfactants are collectins that provide protection at the lung surface; other collectins, such as mannose-binding protein, are found in serum.

The physical and chemical barriers are very effective at preventing pathogens from entering the body and they exclude more than 99.9% of the infectious organisms we are exposed to. However, organisms do infect the body. This can occur in a number of ways.

1.3.2 Invasion – entry of pathogens into the body

Routes by which infectious organisms gain entry into the body include the skin, respiratory tract, gastro-intestinal (GI) tract and GU tract. There are fundamentally two ways in which infectious agents cross the physical and chemical barriers: either they are able to penetrate the intact barriers at one or more anatomical sites, or the physical barriers are damaged and breached, allowing entry of the organism (Figure 1.4).

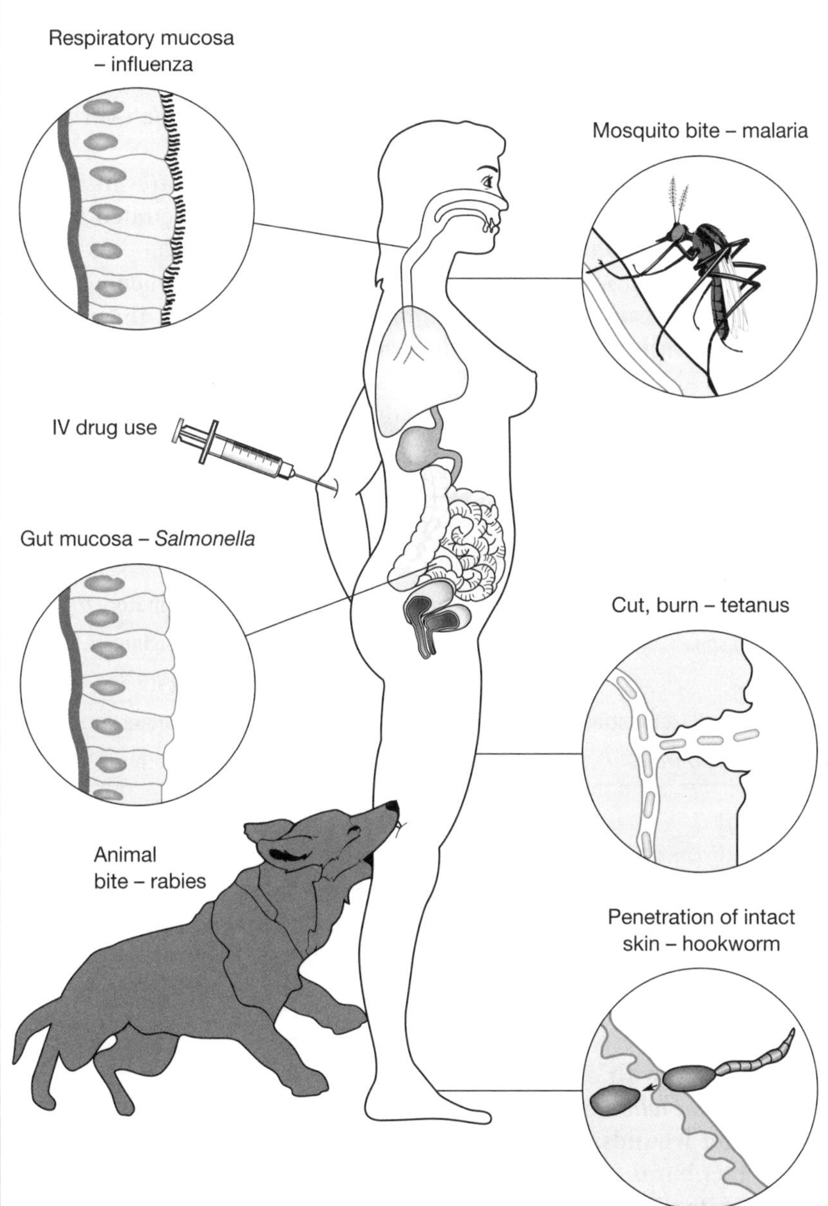

Figure 1.4 Entry of pathogens into the body. Insect bites, cuts, burns and animal bites breach the skin barrier, allowing entry of pathogens. Some parasites can penetrate intact skin while many pathogens penetrate intact mucosa of the respiratory, intestinal and genito-urinary tracts.

Penetration of intact skin or mucosa

- **Skin.** Few organisms are able to penetrate intact skin. However, some parasites (e.g. hookworm) or their larvae (e.g. schistosoma) can do this. Other agents, such as wart viruses, set up infection in the skin and do not enter further into the body.
- **Mucosa.** Mucosa, being softer and damper than skin, are much more frequent sites of entry and all intact mucosa can be penetrated by some organisms. Examples are shown in Table 1.3. Pathogens can cross epithelia by passing through epithelial cells, as in the case of the meningococcus (a bacterium causing meningitis), or by passing between the epithelial cells, seen with *Haemophilus influenzae*.

Table 1.3 Mucosal sites of entry for pathogens

Pathogen	*Disease*	*Mucosal site of entry*
Rhinovirus	Common cold	Nasal epithelium
Influenza virus	Influenza	Upper respiratory tract
Bordetella pertussis	Whooping cough	Lower respiratory tract
Salmonella spp.	Food poisoning	Small intestine
Rotavirus	Diarrhoea	Small intestine
Escherichia coli (some strains)	Urinary tract infection	Bladder, ureter
Neisseria gonorrhea	Gonorrhoea	Vagina, urethra

Penetration of damaged skin or mucosa

There are many ways in which skin or mucosa can be damaged, allowing entry of infectious organisms that could not cross intact skin or mucosa. Damage to skin is a particularly important route of infection and can occur in a number of ways:

- **Burns.** Burns, especially severe ones, pose a major risk for infection, particularly with *Staphylococcus, Streptococcus, Pseudomonas* and *Clostridium tetanus*.
- **Cuts and wounds.** These can allow entry of similar organisms to those seen after burns.
- **Insect bites.** Numerous infections are transmitted via insect bites. These include malaria, typhus and plague.
- **Animal bites.** Animal bites can provide direct transmission of infection, such as in rabies. Because they cause significant damage to the skin, bites can allow the entry of the same environmental pathogens as burns, cuts and wounds (see above).
- **Human behaviour.** Various aspects of uniquely human behaviour can result in the skin being penetrated. Sharing of syringes by intravenous (IV) drug users exposes them to risk of hepatitis and human

immunodeficiency virus (HIV). A number of viral infections (hepatitis, HIV) have been transmitted by blood transfusion and blood products (e.g. factor VIII for haemophiliacs) before appropriate screening procedures were developed. Transplantation has also resulted in transmission of infection before the introduction of appropriate donor screening.

Damage to mucosa may not increase the likelihood of infection to the same extent as damage to the skin. However, physical or chemical damage may allow entry of some organisms, e.g. smoking increases the risk of respiratory bacterial infections. Furthermore, infection of the mucosa with a virus may cause damage and facilitate the entry of bacterial pathogens.

1.3.3 Multiplication of pathogens

Most initial infections are local, i.e. the infectious agent gains entry to the body at a single site, e.g. via an insect bite or infection of a particular mucosal surface. The next stages of infection involve multiplication and spread of the pathogen. These can be considered part of the lifestyle of the pathogen, and infectious organisms vary enormously in lifestyle.

Multiplication of pathogens provides variety at three levels: the **mode** of multiplication, the **site** of replication and the **rate** of multiplication.

Mode of multiplication

Different pathogens multiply in very different ways (Figure 1.5). Many single-celled organisms, including bacteria, yeasts and protozoan parasites, divide by simple cell division. Viruses, however, have a completely different mode of multiplication called replication. Following infection of a cell, viral particles disassemble and, under direction of viral nucleic acid (DNA or RNA), new viral proteins and genetic material are synthesised. Eventually new viral particles are assembled and leave the cell. This can occur by the cell bursting open and releasing viral particles to infect other cells, resulting in cell lysis and death of the cell. Alternatively the cell can shed viral particles in a more gradual manner, a process known as budding, which does not result in the death of the cell. Finally many parasitic worms do not multiply directly but lay eggs, which provide additional sources of infection for other organisms.

Site of replication

Pathogens can live and multiply inside host cells or outside the cells. Many bacteria, yeasts and parasites multiply extracellularly. Viruses by their nature have to replicate intracellularly because they lack enzymes and other cofactors necessary for synthesising viral proteins. Many bacteria and protozoan parasites also replicate intracellularly. Some organisms can live in either an intracellular or an extracellular environment (e.g. *Mycobacterium tuberculosis, Neisseria gonorrhoeae*). Parasites (e.g. trypanosomes) have the most complicated life cycles, which can often involve both an intracellular and extracellular stage.

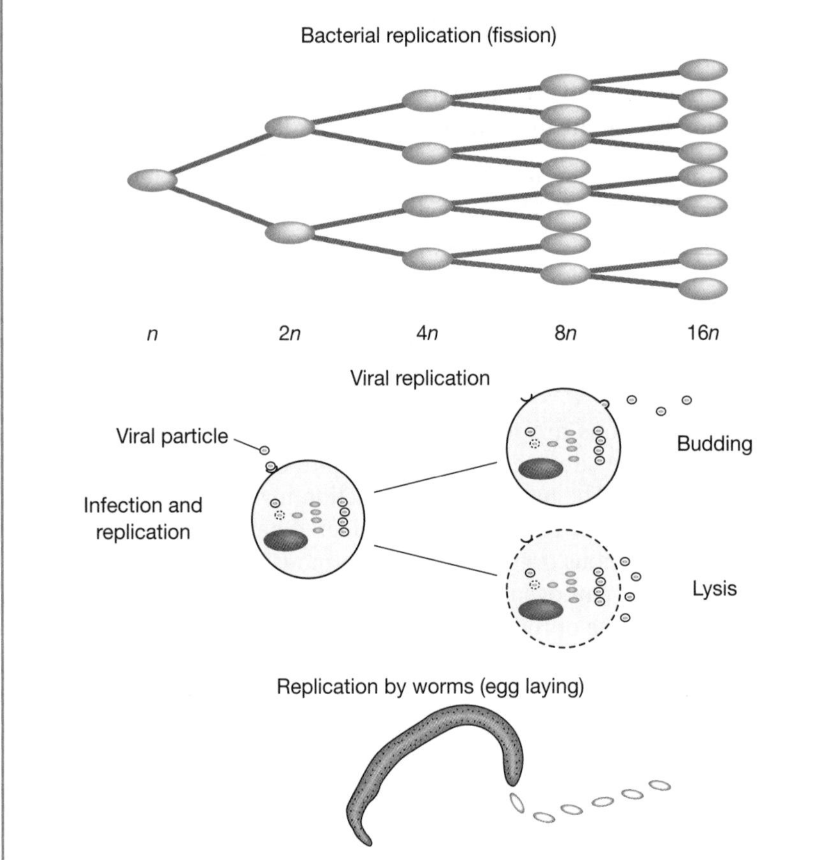

Figure 1.5 Multiplication of pathogens. Many bacteria divide by simple cell fission. Viruses must infect host cells to replicate. Parasitic worms (helminths) often lay eggs, which are transmitted to new hosts before developing into worms.

The site in which pathogens live and multiply poses different problems for the immune system. The most important of these is whether the pathogen has an intracellular stage, because during this stage the organism may be partially hidden from the immune system. However, as we shall see in Chapter 7, the immune system has even evolved ways of detecting whether infected host cells are harbouring hidden pathogens.

Rates of multiplication

The time taken for pathogens to reproduce themselves varies enormously. Some bacteria under optimal conditions *in vitro* can divide every 20 minutes. At this rate of division a single bacterium would produce over 10^{21} progeny in a day! Obviously this rate of replication is unsustainable for long, even under optimal *in vitro* conditions, and it is debatable whether it is ever reached *in vivo*. Viral replication can result in hundreds or thousands of progeny being produced from a single virion in hours. Other pathogens have low rates of replication. Not all bacteria have the capacity to divide rapidly and some, such as the mycobacteria, the causes of tuberculosis and leprosy, have a doubling time of many days. Some parasitic worms never replicate within the host, although they may lay eggs, thereby increasing the number of organisms that can infect other hosts. However, again the rate of egg laying can vary enormously: *Schistosoma mansoni*, the cause of the disease schistosomiasis, lays only 200 eggs a day while *Ascaris lumbricoides*, a roundworm, may lay over 200 000.

1.3.4 Pathogens spread through the body in many different ways

The way in which organisms spread through the body is influenced to some extent by whether they live intracellularly, extracellularly or both. Organisms that live extracellularly are able to spread via body fluids such as blood. However, even organisms that replicate intracellularly may be able to leave the cell and spread via an extracellular route. Organisms can spread in the following ways:

- **Cell to cell contact.** Many organisms, especially viruses, spread directly from cell to cell with essentially no extracellular component to their lifestyle. These pathogens tend to cause localised infections such as seen in influenza, where only the respiratory tract is infected. However, localised infections can still cause widespread symptoms, so that 'flu causes headache, fever and muscle-ache.
- **Via blood and lymphatic vessels.** The commonest, and fastest, way in which pathogens can spread through the body is via the bloodstream. Since all organs and tissues require a blood supply, microorganisms in the blood have the potential to spread to all sites. However, individual pathogens show a preference to localise in particular organs or tissues that may differ from pathogen to pathogen (see Table 1.2).

  The lymphatic vessels form a circulatory system that parallels that of the blood (Figure 1.1; see also Chapter 6). There are important differences between the two systems, however. The circulation of the lymphatic fluid is maintained not by the heart but by the movement of the muscles surrounding the lymphatic vessels; thus lymphatic fluid flows at a much more sluggish rate than blood. Moreover, tissue fluid can drain directly into lymphatic vessels. Organisms can easily enter lymphatic vessels draining the site of infection, where they will be conveyed to the local lymph nodes.
- **Spread via body cavities.** Microorganisms that have infected one organ in a body cavity such as the peritoneum may occasionally spread via the cavity to other organs located within it.
- **Spread via nervous system.** This is a particularly important route of spread for certain viruses. Viruses can spread via peripheral nerves to

the central nervous system (CNS) or vice versa. In some instances this route of spread allows the virus to become more widespread within the nervous system where it resides and causes disease (e.g. herpes simplex virus). In other cases the virus travels via nerves to infect other organs. The rabies virus infects the salivary glands in this way, enabling the virus present in the saliva to be transmitted via a bite.

1.3.5 Pathogens cause disease in many different ways

The final stage of the disease process (although it may not be the final stage of the infection) is the actual production of disease. Many microorganisms live in or on the body without causing disease. These organisms are called **commensal** organisms and may be beneficial to the host: the production of lactic and proprionic acids by lactobacilli in the vagina inhibits the growth of many other bacteria and many commensal organisms compete with pathogens for 'living space' in the gut. Pathogens differ in that they cause disease by one or more mechanisms (Figure 1.6). These include the following:

- **Secretion of toxins.** Many organisms, especially bacteria, secrete toxins that either directly or indirectly account for most of the pathology caused by the organism. These include the powerful neurotoxins secreted by the *Clostridium* family of bacteria responsible for tetanus or botulism food poisoning, toxins of the bacteria *Shigella dysenteriae* and *Vibrio cholerae* that cause dysentery and cholera respectively, and toxins secreted by *Streptococcus pyogenes*, which can cause scarlet fever (Box 1.1). Some protozoa and fungi also secrete exotoxins.
- **Endotoxins.** Endotoxins, rather than being secreted, are components of the cell wall of pathogens. They are particularly prevalent in Gram-negative bacteria (e.g. *Salmonella*) but are also found in other bacteria, some yeasts and protozoa (Box 1.1). Unlike exotoxins, which have direct, very specific toxic effects, endotoxins act by causing cells of the host to produce factors that cause fever, a fall in blood pressure and other symptoms.
- **Direct killing of host cells.** Some intracellular dwelling pathogens replicate within cells and leave the cells (usually by budding from the cell surface) with relatively little damage to the cell. This results in the continuous production of infectious particles by an infected cell. Other pathogens replicate within the cell and kill the cell, which bursts open (a process called cell lysis), thereby releasing many infectious particles (see Section 1.3.3). Many viruses and protozoa lyse host cells in this way; if this lysis is extensive enough, it will result in disease.
- **Physical blockage.** Larger pathogens may cause pathology purely by their physical presence. Probably the most dramatic example of this is elephantiasis caused by the filarial worms. By blocking lymphatic drainage these organisms can cause massive swelling of the breasts, testes and legs (Plate 1).

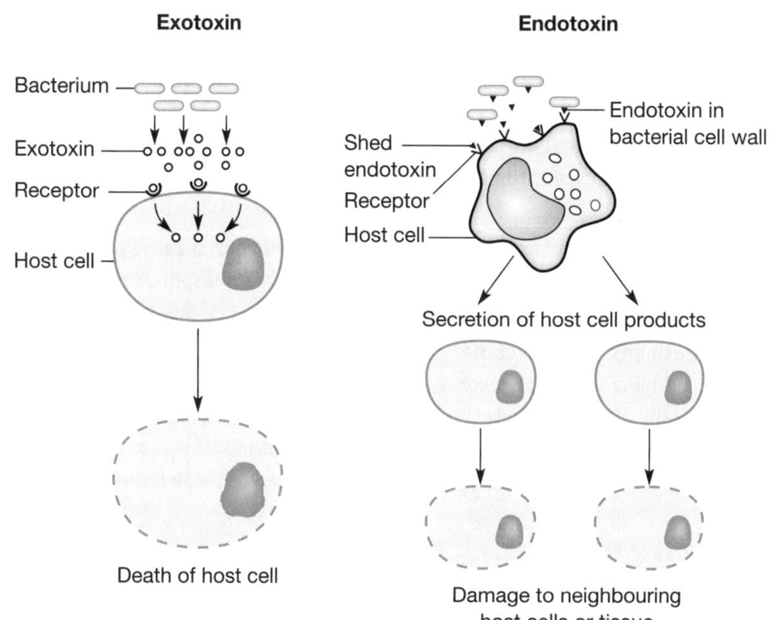

Figure 1.6 Damage caused by microbial exotoxins and endotoxins. Many pathogens secrete exotoxins, which bind to molecules on host cells, enter the host cell and kill it. Endotoxins are components of the cell wall of some pathogens. Endotoxins in the cell wall, or shed endotoxin, bind to receptors on certain host cells (such as macrophages) and stimulate the host cells to secrete products that damage neighbouring host cells and tissue.

BOX 1.1: TOXINS

Exotoxins are secreted products, usually of bacteria but sometimes protozoa and fungi. They can act in a number of ways:

- **Inhibition of protein synthesis.** *Corynebacterium diphtheriae*, the cause of diphtheria, produces a toxin that causes ADP-ribosylation of elongation factor-2, thereby stopping protein synthesis. It is extremely potent and one molecule of toxin is capable of killing a cell. *Escherichia coli, Vibrio cholerae* and *Bordatella pertussis* (the cause of whooping cough) also produce toxins that cause ADP-ribosylation of proteins.

  Toxins from *Shigella dysenteriae* and *E. coli* strain 0157:H7 (a cause of dangerous food poisoning) inhibit protein synthesis by removing adenine from 28s rRNA.
- **Increase in cAMP.** A number of bacteria produce toxins that raise cAMP levels. These include *V. cholerae, Bacillus anthracis, B. pertussis* and some strains of *E. coli*. The consequence of increased cAMP levels is a change in ion transport and hence fluid movement, often resulting in severe oedema.

- **Neurotoxins.** Members of the *Clostridium* family produce particularly potent neurotoxins. *Clostridium tetani* produces a toxin that prevents the release of glycine, an inhibitory neurotransmitter. This results in overactivity and muscle spasm including the typical lockjaw. *Clostridium botulinum* produces a neurotoxin that stimulates release of acetyl choline, leading to paralysis. It is one of the most potent toxins known and it is estimated that less than 1 μg can kill a person.
- **Enzymes that disrupt cell walls.** *Clostridium perfringens*, a cause of gas gangrene, produces a toxin called α-toxin, which is a phospholipase that hydrolyses lecithin in the cell membrane, resulting in cell death.
- **Superantigens.** Some bacteria, particularly *Staphylococcus* and *Streptococcus*, produce toxins that cause excessive stimulation of the immune system (specifically of T lymphocytes; see Chapter 6). This leads to the production of factors by the immune system that cause the symptoms of shock. Toxic shock syndrome and food poisoning are two consequences of these toxins.

1.4 Conclusion

From the above description of the variety of pathogens and the way they live and cause disease, it can be appreciated that the immune system is faced with an enormous variety of problems when trying to protect the body from disease caused by all the different types of pathogens. Box 1.2 summarises the lifestyles of two pathogenic organisms. There is one additional factor that further challenges the immune system and increases the complexity of the immune responses required. Pathogens have co-evolved with the immune system and have developed survival strategies to counter attempts to eliminate them. It is obviously in the pathogen's best interests to survive in the host, and natural selection occurs so that pathogens with an improved ability to survive and multiply within hosts will have a selective advantage and become more common. The evolution of some pathogens seems to have been strongly influenced by the need to evade the immune response; for instance, cytomegalovirus (a cause of pneumonia) has devoted 30% of its genome to subverting the immune response against it. The immune system has accordingly had to evolve an equally complex variety of mechanisms to deal with the wide range of threats posed by different pathogens.

BOX 1.2: EXAMPLES OF THE LIFESTYLE OF PATHOGENIC ORGANISMS

Measles

The measles virus enters the body through the respiratory tract. It then travels to local lymph nodes and lymphoid tissue located in the mucosa. After a few days the virus spreads to other lymphoid tissue, including the spleen, where it begins to replicate. After a week or so, large quantities of the virus spread via the bloodstream to epithelial sites throughout the body. The presence of large amounts of virus at these sites gives rise to the various symptoms seen in measles. Virus in the respiratory tract causes runny nose and coughing. There is inflammation of the conjunctiva, and the presence of the virus in the skin causes the characteristic rash seen in measles.

Typhoid

If the *Salmonella typhi* bacterium is ingested and the dose is big enough, some bacteria will survive the acid environment of the stomach and enter the intestine. Bacteria penetrate the gut mucosa through specialised lymphoid structures known as Peyer's patches (see Chapter 6) and spread to the intestinal lymph nodes, where they proliferate in macrophages. Eventually the organisms reach the bloodstream, where they spread mainly to the liver, bone marrow and spleen, where they continue to multiply. This results in a further large increase in bacterial numbers and subsequent spread of the organism to other tissues such as the kidney and the gall bladder via blood or the biliary tract. The bacteria can also spread to the brain, heart and skin. The bacteria then invade the intestinal tract in much larger numbers than seen with the original infection and cause inflammatory lesions in the Peyer's patches, which may result in ulceration of the intestinal wall. The presence of the bacteria in other sites may cause meningitis, osteomyelitis, endocarditis and rashes.

1.5 Summary

- The body is continually exposed to infectious organisms that have the potential to cause disease (pathogens). Most pathogens are prevented from entering the body by a combination of physical, chemical and biochemical defence mechanisms. However, some pathogens can breach the barriers and in some cases the barriers are breached by injury or other causes.

- Pathogens vary enormously in terms of size, ways in which they enter the body, how they multiply, whether they replicate intra- or extracellularly, replication rates, mechanisms by which they spread through the body and ways in which they actually cause disease.
- The variety of pathogenic lifestyles means that the immune system must have an equally varied repertoire of mechanisms for dealing with the diversity of threats.

1.6 Questions

1) What are the four major families of pathogens?
2) Zoonoses are:

   A) Infections that affect the nasal cavity
   B) Infections that are more prevalent in zoos
   C) Infections that jump from one species to another
   D) Diseases that mainly affect animals
   E) Infections that occur mainly in the jungle

3) In the figure opposite, which letters correspond to the physical, chemical and biochemical barriers to infection from the following: (i) cilia, (ii) hydrochloric acid, (iii) lactic acid, (iv) lysozyme, (v) mucus, (vi) skin.
4) Explain the main differences between endotoxins and exotoxins.
5) Why does damage to the skin increase the likelihood of infection?

The answers to these questions can be found on page 333.

1.7 Further reading

1) Pitt, TL. (2007) Classification, identification and typing of micro-organisms. In *Medical Microbiology* (17th Ed.) Greenwood D, Slack R, Peutherer J, Barer M. Churchill Livingstone Elsevier. Edinburgh. p24–37.
2) Brooker S, Hotez PJ, Bundy DAP. (2010) The Global Atlas of Helminth Infection: Mapping the Way Forward in Neglected Tropical Disease Control. *PLOS Neglected Tropical Diseases* 4:e779
3) **www.medic8.com/infectious-diseases/index.html**
   Website listing infectious diseases alphabetically

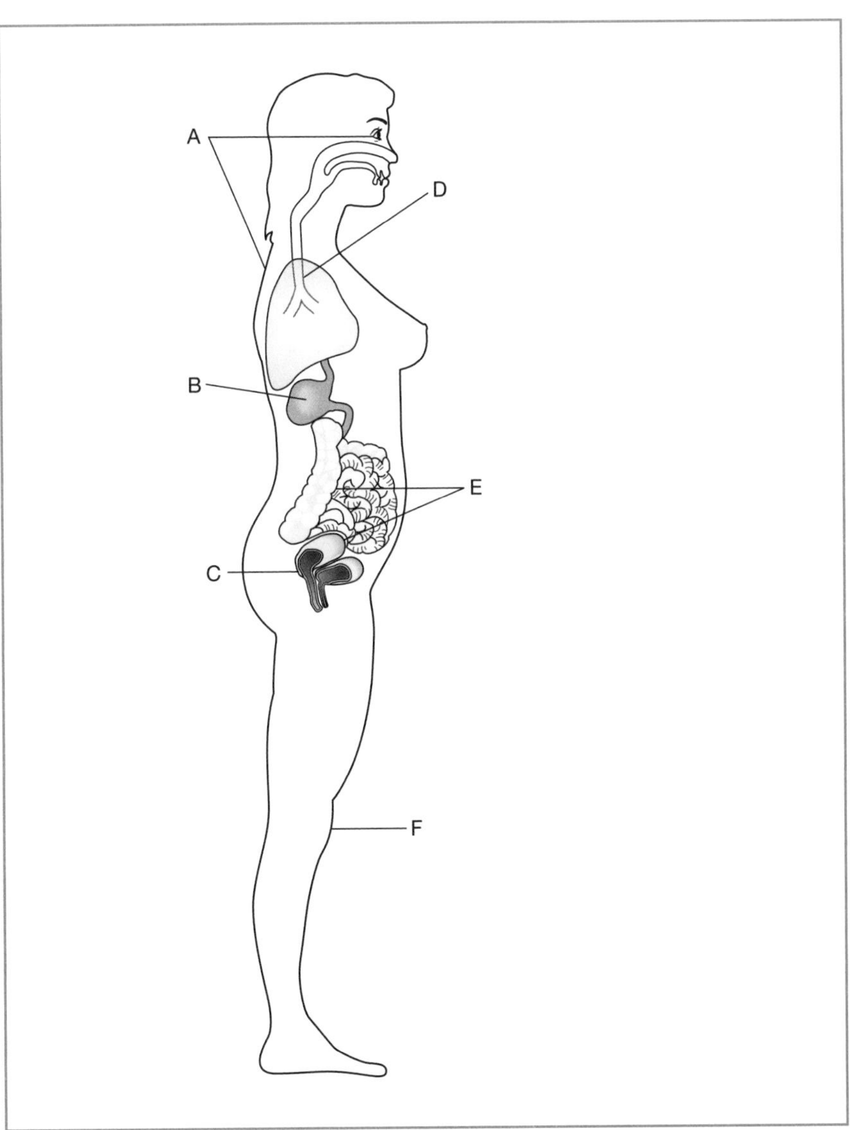
A
D
B
E
C
F

CHAPTER TWO

The immediate response to infection: innate immunity and the inflammatory response

Learning objectives

To be aware of the stages of the immune response to infection. To understand the concept and molecular aspects of recognition of foreign organisms by the innate immune system. To learn about the inflammatory and acute phase responses and how leukocytes move around the body.

Key topics

- The response to infection
- Recognition of pathogens by the innate immune system
- Phagocytosis
- Cytokines
- Cell migration
- The inflammatory response
  - Cellular components
  - Activation of the complement, kinin and clotting systems
- The acute phase response
  - Effect on the brain
  - Effect on the liver
  - Phagocytosis and opsonins
- Natural killer cells and interferons

2.1 The response to infection

Chapter 1 covered pathogenic organisms and the stages of infection leading to disease, illustrating the tremendous variety of organisms and disease processes to which the body is exposed. Although the physical and chemical barriers are effective at preventing most pathogenic organisms from entering the body, many pathogens are able to infect the body and potentially cause disease. So how does the body attempt to counter the threats posed by the enormous number of different infectious organisms?

The response to infection can be divided into five stages:

1. **Awareness** of infection. Obviously the body cannot begin to mount a defensive response against a pathogen until it is aware of the presence of the pathogen.
2. The **immediate** response to infection. This involves the activity of cells and other factors that are present at the time of infection but may require their recruitment to the site of infection and activation once there.
3. The **delayed** response to infection. Some infections can be resolved by the innate system but usually infection is accompanied by a new immune response that involves the generation of new cells and factors to deal with the infection. This involves the second component of the immune system, the specific immune system, which is introduced in Chapter 3.
4. **Destruction** or **elimination** of the pathogen or **neutralisation** of the threat posed by pathogens. The optimal way of dealing with pathogens is to kill them or eliminate them from the body. However, in some cases where pathogens are producing a powerful toxin it may be more beneficial to neutralise the toxin first before attempting to destroy or eliminate the pathogen.
5. Provision of **immunity** so that you do not get ill if you are infected again with the same pathogen.

2.2 The immediate response to infection – the innate immune system

The term 'innate immune system' is used to describe pre-existing defence mechanisms that are designed to prevent infection by pathogens or to mount an immediate defence against the infectious agent. The physical, chemical and biochemical barriers to infection described in Chapter 1 are part of the innate immune system. There are also many cells and proteins found throughout the body that are part of the innate immune system and are involved in defence against pathogens. They are called 'innate' or 'natural' because they are present before infection, although the amount of some components may increase following infection. Collectively, these cells and proteins perform two important functions: they are able to recognise the presence of a foreign body, for example a bacterial infection, and they provide an immediate response to the presence of a pathogen.

2.2.1 Cells of the innate immune system

Some cells of the innate immune system reside in tissues and organs of the body ready to respond when a pathogen infects that tissue. All tissues contain cells called **macrophages**, which are bone marrow-derived cells (Plate 2). Macrophages are particularly abundant in the liver, lungs and spleen but are found everywhere and even the brain has a resident population of macrophages called microglial cells. Because of their location, macrophages are often the first cells to encounter foreign pathogens.

Although macrophages may be the first cell to encounter pathogens, they usually are not enough to deal with the pathogens on their own. Therefore it is often necessary to recruit other cells of the innate immune system to sites of infection. The best way to get cells to a particular site in the body is via the blood, so many of the cells of the innate immune recirculate in the bloodstream. When a tissue becomes infected, chemical messages are sent to the cells in the blood so that they leave the bloodstream and enter the infected tissue to combat the invading pathogen. One such type of blood cell is the monocyte (Plate 2), which is also derived from bone marrow. When monocytes leave the bloodstream and enter tissue, they can differentiate into macrophages. Because of their relationship, macrophages and monocytes are known as cells of the monocyte/macrophage lineage.

Another very important white blood cell type of the innate immune system is the neutrophil (Plate 2), which is also called a polymorphonuclear leukocyte (PMN). Neutrophils can also leave the bloodstream and enter damaged or infected tissue where they try to destroy the invading pathogen (see Chapter 8). About 80% of all neutrophils are actually in the bone marrow and they can be mobilised after infection to enter the bloodstream and travel to the infected area.

Many other white blood cells contribute to fight against infection and Figure 2.1 is a simplified depiction of their production in the bone marrow. The specific functions of these cells will be described in the appropriate sections of the book.

2.2.2 Recognition of pathogens by cells of the innate immune system

When we say that cells of the innate immune system can 'recognise' pathogens, what do we actually mean? By 'recognition' we mean that molecules, or receptors, on cells of the innate immune system bind to other molecules that are present on pathogens but not present on our own healthy cells. In this way the immune system can distinguish foreign objects and respond to a foreign pathogen but not respond against our own healthy tissue or cells. It is a crucial feature of the immune system that it can target foreign objects. There are many receptors present on the various cells of the innate immune system that can recognise molecules on pathogens. These receptors have been given the general name of **pattern recognition receptors**

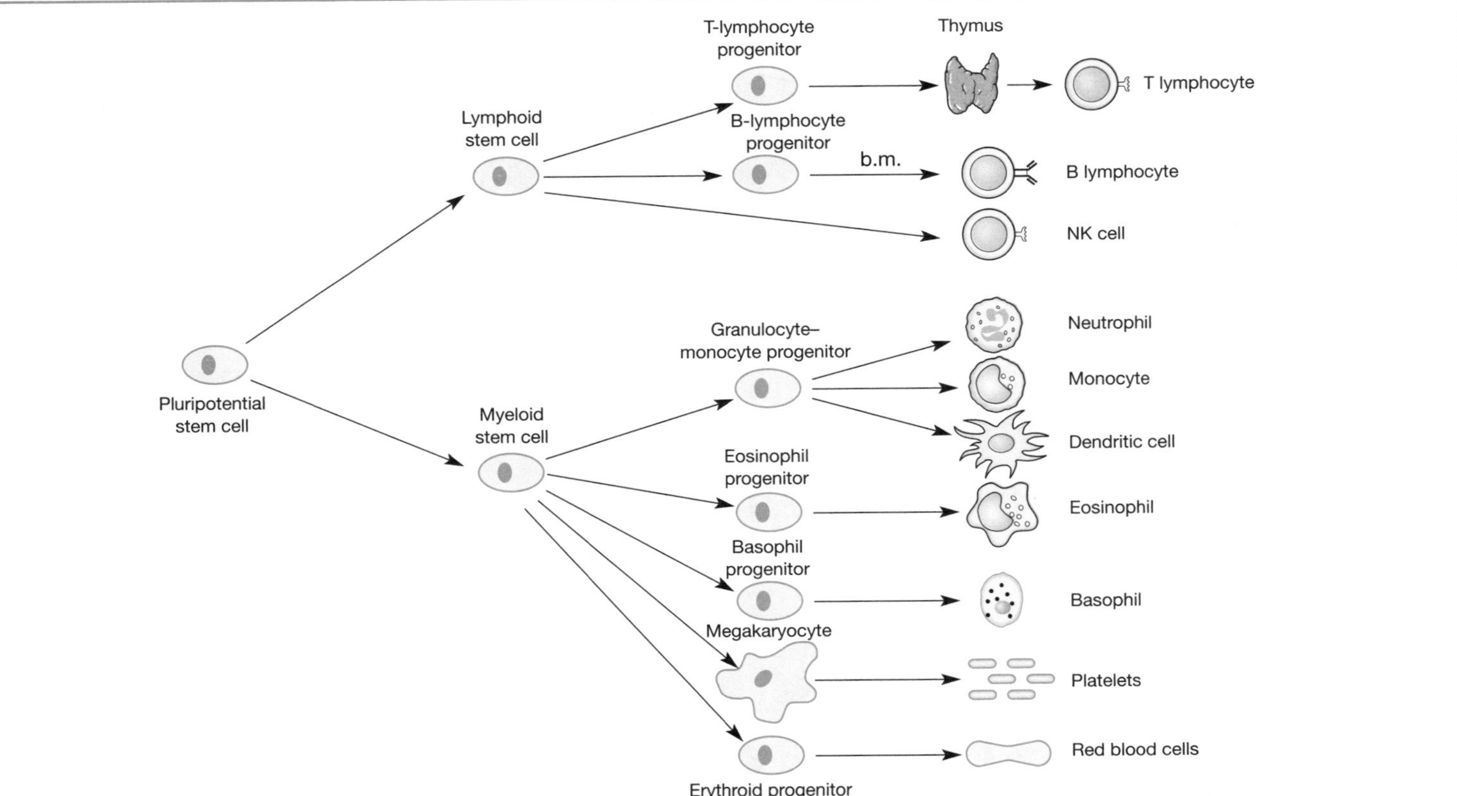

Figure 2.1 Summary of haematopoiesis. Except for the final development of T cells in the thymus, all other haematopoietic cells, including B cells, are produced in the bone marrow (b.m.).

(PRRs) and the molecules on pathogens recognised by them are called **pathogen-associated recognition patterns (PAMPs)**. Some of the PRRs and the PAMPs recognised by them are shown in Tables 2.1 and 2.2 and Figure 2.2. Not all cells of the immune system express all of the different PRRs and many of the PRRs are also on other cells, such as mucosal epithelial cells, so that these cells can also contribute to sensing the presence of pathogens. Pathogens and components derived from them can be located in extra-cellular fluid, or in various locations inside cells such as the cytoplasm or endosomes. As can be seen from Figure 2.2, PRRs are expressed in all these locations so that no matter where the pathogen, or its products, go there is a good chance that one or more PRRs will be there to recognise the pathogen or some of its components. This makes it hard, although not impossible, for the pathogen to evade recognition.

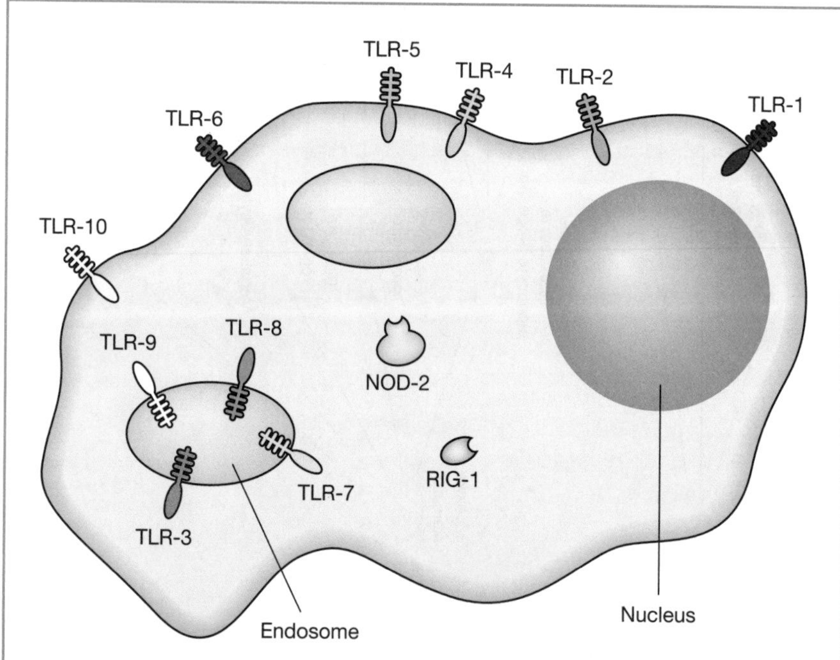

Figure 2.2 Pattern recognition receptors. Pattern recognition receptors can be expressed on the cell surface, in endosomes or in the cytoplasm. TLR – toll-like receptor. NOD – nucleotide-binding oligomerisation domain. RIG - Retinoic acid-inducible gene.

It is now becoming clear that many PRRs also recognise endogenous ligands, that is, molecules from our own bodies. These ligands are usually molecules that are released from, or exposed on, dead, damaged or stressed cells and are a sign of tissue damage. They have been called damage-associated molecular patterns (DAMPs) and include some sugars, heat shock proteins (HSP), high mobility group box 1 (HMGB-1) protein, uric acid, adenosine triphosphate (ATP) and fragments of hyaluronate, a component of the extra cellular matrix. In this case the stimulation of PRRs on macrophages stimulates the macrophages to phagocytose and remove dead or damaged cells or bits of damaged tissue and contribute to the wound healing process. Overall, then, the innate immune system is designed to distinguish between microbial pathogens and our own damaged/dead cells, which must be removed, and our own healthy cells which are not to be 'attacked'.

2.2.3 Different PRRs stimulate different types of cellular responses

All of the PRRs stimulate a response in the cell upon binding of their particular PAMP. Some of the receptor families, such as TLR and NLR, stimulate cells to respond by differentiating and/or secreting various factors that contribute to the response against the pathogen. Other receptor families, the C-type lectins, scavenger receptors and complement receptors stimulate a particular response – phagocytosis, which is described below.

Table 2.1 Pattern recognition receptors and their ligands – PAMPS

Receptor	*Cellular distribution (phagocytes)*	*Pathogen molecules recognised*	*Pathogen distribution*
Mannose receptor	Macrophages, neutrophils	Mannose-containing carbohydrates (polysaccharides)	Many bacteria
Scavenger receptor	Macrophages	Sialic acid	Bacteria and yeast
CD14	Macrophages, neutrophils	Lipopolysaccharide (LPS) component of bacterial cell walls	Gram-negative bacteria
Complement receptors CR3 and CR4	Macrophages	LPS, lipophosphoglycans	Bacteria and yeast

Table 2.2 Recognition by Toll-like receptors

Toll-like receptor	*Ligands*
TLR1	Bacterial lipopeptide
TLR2	Peptidoglycans, lipopeptides, zymosan (from yeast)
TLR3	Double-stranded RNA (viral)
TLR4	LPS (Gram-negative bacterial cell wall)
TLR5	Flagellin
TLR6	Lipopeptide, zymosan
TLR7	Single-stranded RNA
TLR8	Single-stranded RNA
TLR9	Unmethylated DNA (bacterial)
TLR10	? Unidentified molecules on some bacteria

2.2.4 The cellular response to infection – phagocytosis

Phagocytosis is the ingestion and destruction of microbes by cells called phagocytes. The two main types of phagocytes are the macrophages and neutrophils described above. The way in which macrophages and neutrophils phagocytose particles is essentially the same and can be divided into four stages (Figure 2.3):

1. **Attachment** of the phagocyte to the particle being phagocytosed, which may be a pathogen, a dead or damaged host cell or a piece of tissue.
2. **Ingestion.** By extending membrane protrusions called pseudopodia around the particle, the phagocyte is able to engulf the particle, which is taken into the cell in a phagocytic vacuole.
3. **Killing.** If the ingested particle is a live cell of a pathogen (e.g. a bacterium) the phagocyte will normally kill the cell by one of a number of mechanisms (described in more detail in Chapter 8).
4. **Degradation.** The phagocytosed particle, whether it is a dead cell or a piece of tissue, is broken down by enzymes in the phagocytic vacuole.

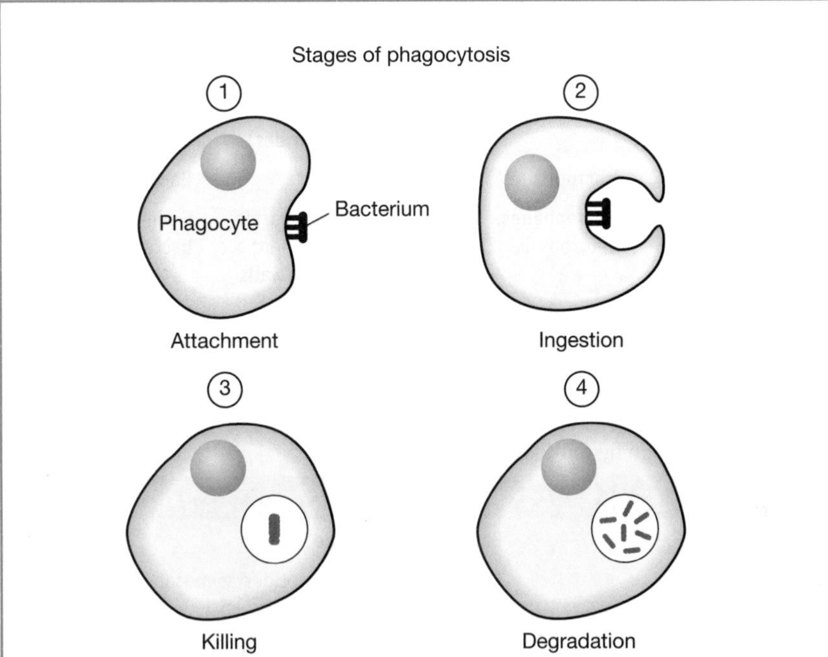

Figure 2.3 Phagocytosis. Phagocytes can take up and remove bacteria and dead host cells or tissue debris. The figure shows phagocytosis of a bacterium: ① The phagocyte binds to the bacterium. ② The phagocyte extends projections around the bacterium and engulfs it in a phagocytic vacuole. ③ The phagocyte kills the engulfed bacterium. ④ The bacterium is degraded by proteolytic enzymes.

Although the basic process of phagocytosis is similar in neutrophils and macrophages, there is an important difference. While neutrophils are only able to phagocytose small organisms such as bacteria and viruses, macrophages are able to phagocytose larger particles such as dead host cells and tissue debris in addition to microorganisms. Therefore macrophages are involved in eliminating pathogens from tissues and also in cleaning up damaged tissue by removing dead or damaged host cells. Macrophages are able to distinguish between healthy host cells and dead/damaged cells because some of the PRRs on macrophages recognise molecules that are exposed by dead or damaged host cells as described in Section 2.2.2 (see Figure 2.4).

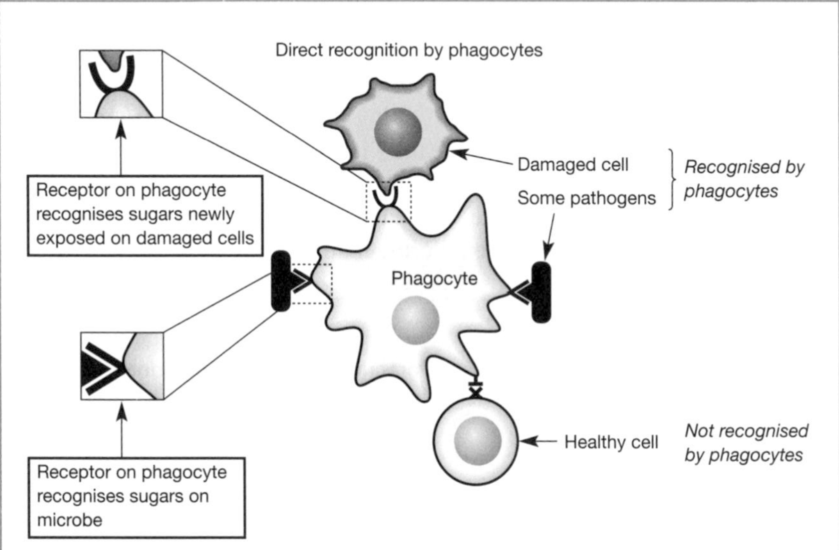

Figure 2.4 Recognition by phagocytes. Phagocytes must distinguish microbes and dead host cells from healthy host cells so that healthy host cells are not phagocytosed. Phagocytes have PRRs on their surface that recognise molecules present on microbes or molecules that are newly expressed on dead or damaged host cells. These sugars are not present on healthy host cells and therefore the host cells are not phagocytosed.

2.2.5 The cellular response to infection – production of new factors

Another way in which cells respond to the recognition of pathogen is by the synthesis and/or secretion of many new products. Some of these products are directly involved in killing pathogens and others are involved indirectly in recruiting other cell types to try to eliminate the pathogen. An important group of proteins that can be secreted in response to pathogenic stimuli are known as cytokines.

2.3 Cytokines – hormones of the immune system

The term cytokine covers a large number of smallish proteins (usually less than 20 kDa) that serve a hormone-like function in enabling cells to communicate with each other. Most people are familiar with hormones such as insulin and growth hormone, which are produced in one organ or tissue and travel through the bloodstream to other organs where they bind to receptors on the cells of that organ and stimulate a particular response. Hormones that are produced in one organ and act on a distant tissue are said to be acting in an **endocrine** manner (Figure 2.5). Cytokines do not usually act in an endocrine manner; rather, they act locally. They are produced by cells in a particular tissue and act on 'cells' in that tissue. Cytokines therefore act in a **paracrine** or **autocrine** manner (Figure 2.5). Paracrine action means that the cytokine binds to receptors on cells close to those producing the cytokine; by 'close' we are probably talking about a microenvironment of a few microns to 1 mm. Autocrine means that the cytokine actually binds to receptors on the cell that produced the cytokine. Thus the role of cytokines is to enable cells to communicate with each other in a local environment. A few cytokines can also act in an endocrine manner.

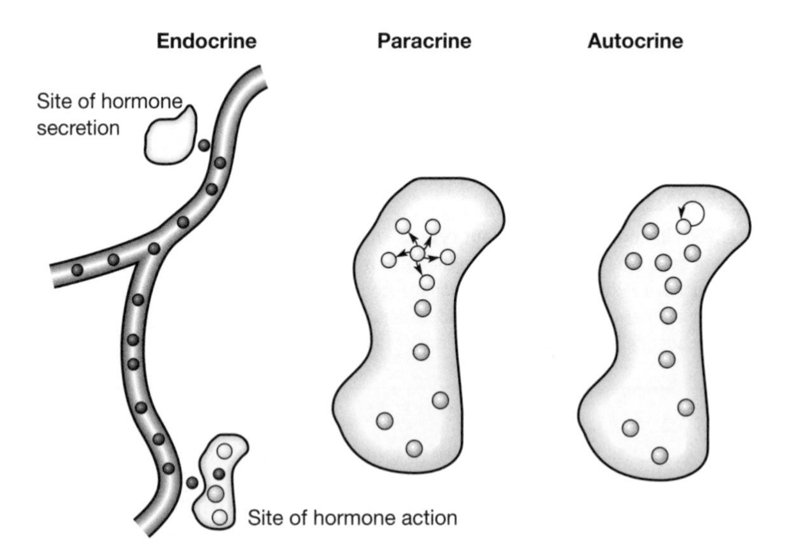

Figure 2.5 Action of hormones. Endocrine: the hormone is secreted at one site of the body and travels through the bloodstream. The hormone will bind to receptors on cells at a distant site (blue cells) and cause a response in those cells. **Paracrine:** hormones produced by cells in a tissue bind to receptors on other cells in the immediate vicinity (blue cells). Cells in other parts of the same tissue are not affected by the hormone (white cells). **Autocrine:** the secreted hormone binds to receptors on the cell that produced the hormone (blue cell).

There are many cytokines and they can be divided into families (Table 2.3). The main families of cytokines are the interleukins (ILs), colony-stimulating factors (CSFs), interferons (IFNs), tumour necrosis factors (TNFs), chemokines (CCs) and growth factors (GFs). Growth factors, such as transforming growth factor-β and epidermal growth factor, are now included in the cytokine list although they were not initially identified as having a role in the immune system. Cytokines control many aspects of cell behaviour, including proliferation, differentiation, cell function and leukocyte migration.

Table 2.3 Cytokine families

Family	*Members*	*Comments*
Interleukin (IL)	IL-1 to IL-35	Different IL have different functions and are secreted by different cells.
Interferon (IFN)	IFNα IFNβ IFNγ	Leukocyte IFN. Inhibits viral replication. Fibroblast IFN. Inhibits viral replication. Secreted by lymphocytes. Many immunoregulatory functions.
Tumour necrosis factor (TNF)	TNFα	Secreted by monocytes and other cells. Factor activates macrophages and endothelium.
	TNFβ	Secreted by T cells. Similar activity to TNFα.
Colony-stimulating factor (CSF)	G-CSF, M-CSF, GM-CSF and others	Originally identified by ability to make bone marrow cells differentiate into particular cell type, e.g. neutrophil. Also have effects on mature cells of same lineage, e.g. monocytes, macrophages, neutrophils.
Chemokine	MCP, Eotaxin and many others	Very important in controlling the migration of cells between and within tissues. Also influence function of many cells.
Growth factor	TGF, IGF and many others	Originally identified because of non-immune-related function but may have effects on immune cells.

G-CSF, granulocyte-CSF; M-CSF, macrophage-CSF; GM-CSF, granulocyte/monocyte-CSF; MCP, macrophage chemotactic protein; TGF, transforming growth factor; IGF, insulin-like growth factor.

The functions of cytokines will be described in detail at the appropriate times when particular mechanisms are being explained. It is important to realise that in the body, cells are never exposed to a single cytokine – they will be exposed to a number of different cytokines, probably produced by a number of different cell types. Different cytokines can either act cooperatively in promoting a response or act antagonistically in inhibiting each other's actions. It is the combination of cytokines to which a cell is exposed that determines the behaviour of the cell.

2.4 The inflammatory response and cell migration

If a pathogen has successfully invaded a tissue, the macrophages in the tissue may recognise the pathogens with one or more of the PRRs described in Section 2.2.2 and attempt to phagocytose and kill the pathogens. Often there are not enough macrophages present in a tissue to phagocytose and remove all the pathogens and therefore the tissue macrophages initiate a response that will bring additional phagocytes, together with a variety of proteins, to the site of infection from the blood. These cells and proteins then help to remove the pathogen. This response is known as the **inflammatory** response. The aim of the inflammatory response is to recruit cells and other factors from the bloodstream into tissues to aid in the removal of pathogens and dead cells or tissue. Leukocytes (white blood cells) are unique in their ability to move throughout the body. They travel through the bloodstream and also have the ability to leave the bloodstream and enter tissue or organs. This ability to move around the body is also referred to as 'cell migration'.

2.5 Cell migration – through blood and into tissue

The movement of cells around the body must be carefully controlled so that the cells go only to where they are required. If you have an infection in your big toe there is no point sending a lot of cells to your ear. The control of cell movement is at two levels: one level is controlling where leukocytes leave the bloodstream and enter a particular tissue or organ; the second level is controlling where the cells go within tissues and organs once they have left the bloodstream. For a single cell, most organs are pretty big places and the cell must go to the right location within the organ or tissue.

Two important factors play an important role in controlling the movement of cells to and within specific tissue sites. **Adhesion molecules** are present on leukocytes and endothelial cells, and interactions between adhesion molecules allow leukocytes to bind to endothelium as part of the process of migrating across the endothelium. **Chemotactic** factors, especially the chemokines, are also important in controlling cell migration. They can act directly on cells and cause them to move in a particular direction or they can act indirectly by altering the expression or binding activity of adhesion molecules.

2.5.1 Adhesion molecules – controlling cellular interactions

There are four families of adhesion molecules called **selectins, integrins, mucin-like vascular addressins** and members of the **immunoglobulin superfamily** (Figure 2.6) and each family contains many members.

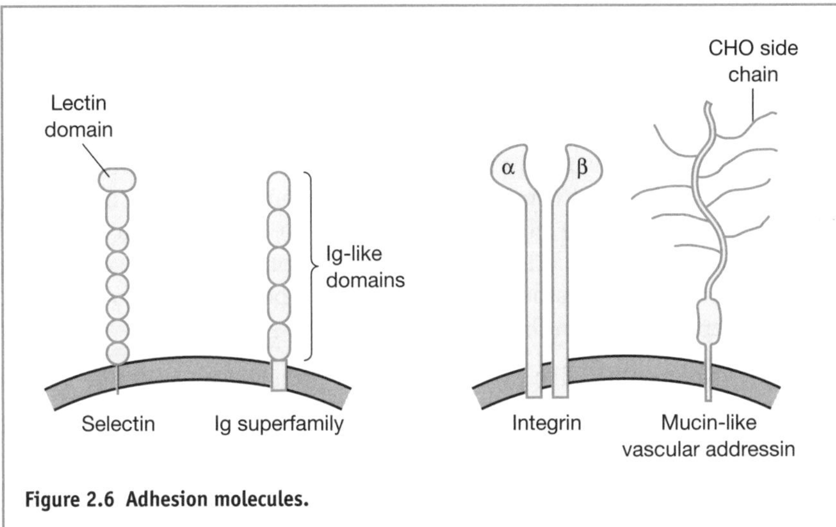

Figure 2.6 Adhesion molecules.

Different adhesion molecules bind to each other in a specific manner and enable cells to interact with each other. Cell–cell adhesion is controlled both by the expression of particular adhesion molecules and in some cases by the activation status, or actual binding capacity, of the adhesion molecules. Different adhesion molecules are expressed on different cell types; some are expressed constantly on the cell surface and others are induced by cell activation, e.g. by cytokines.

By altering cell-adhesion molecule expression or activity on endothelial cells or leukocytes, it is possible to control whether particular leukocytes bind to endothelium at a particular tissue site and, hence, the entry of the leukocytes into the tissue.

Selectins are glycoproteins that are lectins, i.e. sugar-binding molecules, some of which are expressed on leukocytes and some on endothelial cells.

Mucin-like vascular addressins are heavily glycosylated proteins and therefore can bind to the selectins. Some are expressed on leukocytes and some on endothelial cells.

Integrins are heterodimeric proteins consisting of an α-chain and a β-chain and are expressed on leukocytes. There are many α- and β-chains and they can pair to give many combinations of integrins with different expression and binding specificity. Some integrins will bind to target molecules only following activation of the leukocyte by various factors.

Immunoglobulin superfamily: these molecules contain immunoglobulin (Ig)-like domains (110 amino acids flanked by an intra chain disulphide bond) and are the binding target for the integrins. They are expressed on endothelial cells.

2.5.2 Migration of cells from the blood into tissue

The process by which cells leave the bloodstream and cross the endothelium to enter into various tissues is called **extravasation**. Although the particular molecules involved may differ in different situations, the fundamental process is the same. Extravasation can be divided into three stages – rolling, activation and firm attachment, and transendothelial migration. Once cells have left the bloodstream they must be guided to the right location within the tissue. The entrance of neutrophils into a site of inflammation is the best understood example and will be described to illustrate the basic steps involved in these processes (Figure 2.7).

1. **Rolling.** Neutrophils, like other leukocytes, normally travel in the centre of the blood flow away from the endothelium. At a site of inflammation vasodilation occurs, slowing down and disturbing the blood flow so that the neutrophils can 'bump' along the endothelium, a process known as rolling. Due to the action of inflammatory mediators, especially TNFα, the endothelial cells are activated to express P-selectin and E-selectin on their surface. These selectins can bind to sialyl-Lewis^x on the surface of the neutrophil, slowing down the neutrophil so that it rolls along the endothelium.
2. **Activation and firm attachment.** The binding of the selectins to the sialyl-Lewis^x is not strong enough for the neutrophil to adhere strongly to the endothelium. Strong attachment requires the binding of the integrin LFA-1 on the neutrophil to ICAM-1 on the endothelium. Before it can bind to ICAM-1, the LFA-1 must change conformation. One of the factors produced in an inflammatory response is interleukin-8 (IL-8), which is a **chemokine.** Chemokines are a group of cytokines with chemotactic and other functions (Table 2.4). Some of the IL-8 produced is held in the extracellular matrix on the endothelial cell surface and can bind to IL-8 receptors on the neutrophil surface. The binding of IL-8 to the neutrophil activates the neutrophil and LFA-1 changes conformation and binds firmly to ICAM-1 on the endothelium.
3. **Transendothelial migration.** Once the neutrophil is firmly attached to the endothelium it squeezes between the endothelial cells, making contact with the basement membrane underneath. This process is poorly understood but involves additional adhesion molecules. Finally enzymes digest the basement membrane, allowing the leukocyte to pass through into the tissue space.

2.5.3 Movement inside tissues

The movement of cells within tissues is controlled by chemotactic factors. In the inflamed tissue there will be a gradient of the chemokine interleukin-8 (IL-8), with maximum levels at the centre of infection. Neutrophils that have left the bloodstream and entered the tissue will travel along the IL-8 gradient, moving towards increasing concentration of the chemokine so that they will accumulate at the centre of infection.

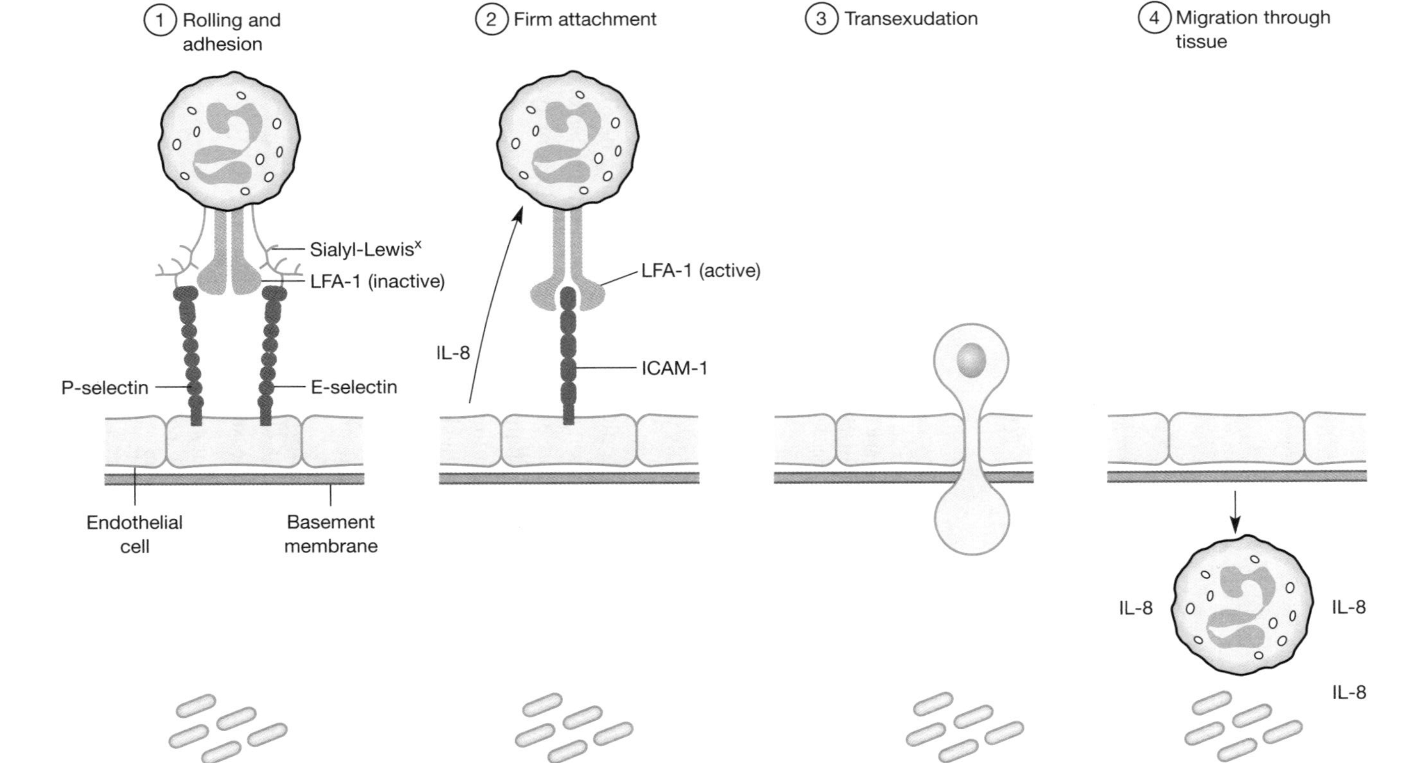

Figure 2.7 The stages of neutrophil migration into sites of inflammation. The first stage involves sialyl-Lewis^x on the neutrophil binding to E-selectin and P-selectin on the endothelial cell. Activation of the neutrophil by IL-8 results in a change of conformation of LFA-1 so that it binds firmly to ICAM-1 on the endothelial cell. The neutrophil then squeezes between the endothelial cells and under the influence of chemokines such as IL-8 migrates through the tissue to the site of infection.

Table 2.4 Chemokines: families, receptors and cellular expression

Chemokine	*Receptors*	*Cells affected*
CXC family		
IL-8 (CXCL8)*	CXCR1, CXCR2	Neutrophils
GRO-α (CXCL1)	CXCR2	Neutrophils
IP-10 (CXCL10)	CXCR3	T cells
CC family		
MIP-1α (CCL3)	CCR-1, CCR-5	T cells, monocytes, DCs
MIP-1β (CCL4)	CCR-5	T cells, monocytes, DCs
MCP-1 (CCL2)	CCR2	T cells, monocytes
RANTES (CCL5)	CCR-1, CCR-3, CCR-5	Eosinophils, monocytes, DCs, T cells
Eotaxin (CCL11)	CCR-3	Eosinophils, monocytes, T cells

The chemokine families are based on the number and pattern of conserved cysteines near the -NH_2 terminal of the protein. The CXC family has two cysteines separated by an amino acid. In the CC family the cysteines are adjacent. There are two other families (C) and (CXXXC), which contain two and one members respectively. DC - dendritic cell.

* Recently a systematic basis for chemokine nomenclature has been adopted and the names under this system are shown in parentheses.

The way in which other leukocytes cross endothelia, leave the bloodstream and migrate through tissues is essentially the same as for neutrophils, although the adhesion molecules and chemokines may be different for different cell types. Many adhesion molecules and chemokines exist to control adhesion, integrin activation and movement of different types of cells in various tissues. In sites of inflammation other factors such as complement components and prostaglandins can also act as chemoattractants. Tables 2.5 and 6.1 show the main adhesion molecules involved in leukocyte recirculation and migration to sites of inflammation.

Table 2.5 Adhesion molecules involved in migration of leukocytes to sites of inflammation

Adhesion molecule	*Cellular distribution*	*Endothelial ligand*
L-selectin	All types of leukocyte	CD34
$\alpha_L\beta_2$ integrin (LFA-1)	T cells, monocytes, macrophages, neutrophils, dendritic cells	ICAM-1, -2, -3
$\alpha_4\beta_1$ integrin (VLA-4)	T cells, monocytes, neutrophils	VCAM-1, fibronectin
CR3	Monocytes, neutrophils, macrophages	ICAM-1
PSGL-1	Neutrophils	E- and P-selectin

2.6 The inflammatory response

As mentioned above, one of the main aims of an inflammatory response is to recruit cells and soluble factors from the bloodstream to help fight off pathogens that have infected a particular tissue site. Although in this instance we are concerned with inflammatory responses in the setting of infection, inflammatory responses can also be triggered by physical, chemical and physical trauma. Four important events occur during an inflammatory response to promote these aims (Figure 2.8):

- **Vasodilation** causes increased blood flow to the area, increasing the supply of cells and factors.
- **Activation of endothelial** cells lining the blood vessels causes increased expression of adhesion molecules, making the endothelium more 'sticky' to white blood cells so that blood cells can adhere more strongly to the endothelium, thereby promoting the migration of leukocytes from the blood into the tissue.
- **Increased vascular permeability** makes it easier for cells and proteins to pass through the blood vessel walls and enter the tissue.
- **Chemotactic factors** are produced that attract cells into the tissue from the bloodstream.

All of these events are controlled by factors that either are produced by cells involved in the inflammatory response or enter the site of inflammation from the blood.

2.6.1 Activation of macrophages

The first stage in the inflammatory response following infection is recognition of the pathogen and activation of tissue macrophages. The activated macrophages produce a number of factors including prostaglandins, platelet activating factor (PAF) and cytokines. Prostaglandins are a group of small biologically active lipid molecules derived from arachidonic acid. Three of the cytokines produced by the macrophages, interleukin-1 (IL-1), interleukin-8 (IL-8) and tumour necrosis factor-α (TNFα), are important in the inflammatory response. These factors have a number of effects.

TNFα, PAF and the prostaglandins act directly on the endothelium to increase vascular permeability. PAF also causes platelets to release histamine, which is another potent agent for increasing vascular permeability.

IL-1 and TNFα activate endothelial cells lining the blood vessels at the site of infection. This causes the endothelial cells to express on their cell surface,molecules that neutrophils in the bloodstream can bind to, enabling the neutrophils to leave the bloodstream and enter the tissue. Neutrophils and macrophages ingest and kill bacteria and other microorganisms. The recruitment of neutrophils is also promoted by IL-8, which is chemotactic for neutrophils.

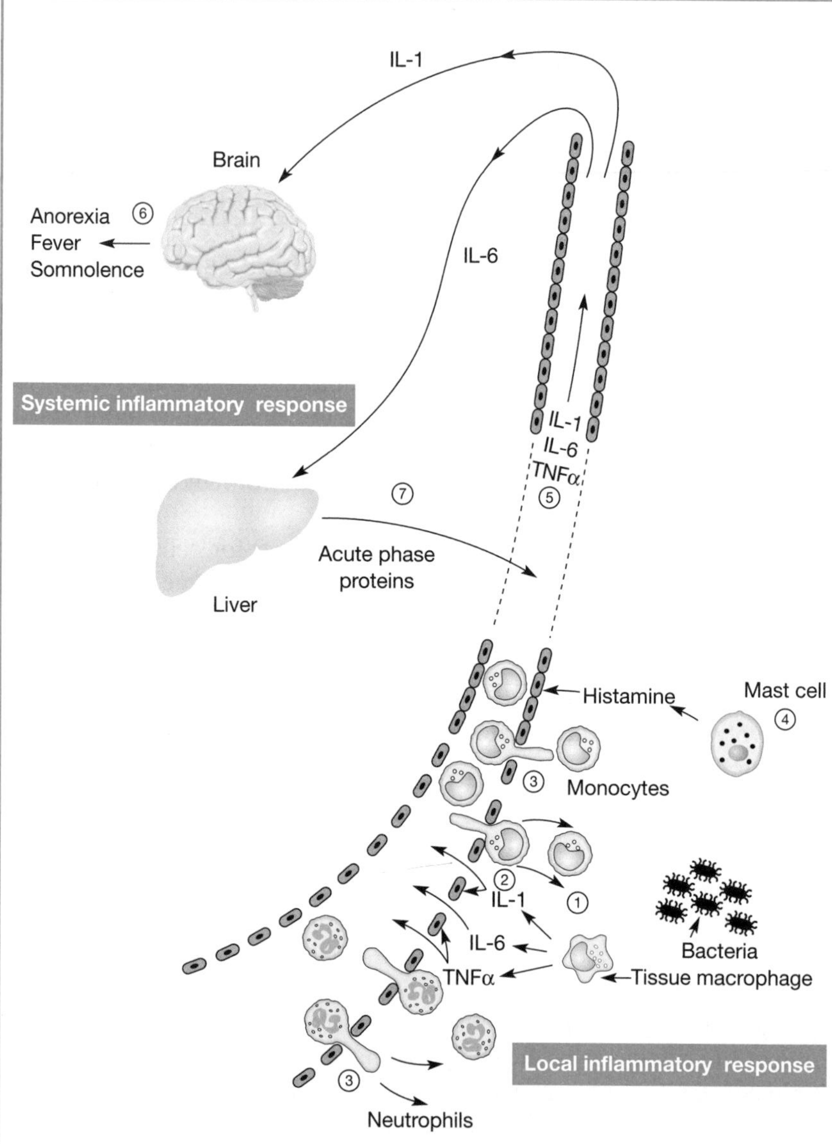

Figure 2.8 Inflammatory responses. Inflammatory responses can be local or systemic. ① Tissue macrophages recognise microbial products. ② The macrophages release cytokines and other inflammatory mediators (IL-1, TNFα, IL-6) that cause vasodilation and increased vascular permeability and have chemotactic effects on monocytes and neutrophils. ③ Monocytes and neutrophils are recruited to the site and there is accumulation of plasma fluid and proteins at the site, causing oedema. ④ Inflammatory mediators can activate mast cells to release further mediators that amplify the response. ⑤ If the local production of cytokines is high enough, the cytokines travel in the blood and affect other organs. ⑥ IL-1 affects the brain, causing fever, anorexia and somnolence. ⑦ IL-6 stimulates hepatocytes to produce acute phase proteins.

2.6.2 Activation of other pathways during inflammatory responses

A number of other cell types and biochemical pathways can also be activated during an inflammatory response.

Mast cells

Mast cells are distributed throughout the body. They contain many large granules and have similar properties to basophils, which are a type of white blood cell (Plate 3). There are two types of mast cells, mucosal mast cells and connective tissue mast cells, which, although sharing most properties, do have some differences (see Chapter 13 for more details on mast cells).

Mast cells express some of the TLRs and other receptors of the innate immune system and can therefore be activated by the presence of pathogens through PAMPs. When they are activated, mast cells release the contents of their granules in a process known as mast-cell degranulation. The contents of the granules include histamine, heparin and proteolytic enzymes. These factors result in vasodilation and increased vascular permeability. Activated mast cells also start to synthesise new products, especially prostaglandins and leukotrienes, which are products of the arachidonic acid pathway (Box 2.1). These new products also cause vasodilation and increased vascular permeability and attract neutrophils to the site.

Clotting system

Activation of the clotting system leads to the cleavage of fibrinogen to generate fibrin threads, which form blood clots, and **fibrinopeptides**, which are chemotactic for phagocytes. Formation of a blood clot is important if there has been damage to blood vessels, because the clot can limit the entry of pathogens into the bloodstream and therefore the spread of pathogenic organisms.

Complement system

The complement system is made up of a number of different plasma proteins that play many roles in resistance to infection. It is described in more detail in Chapter 8. In a manner analogous to the clotting cascade, it consists of a series of pro-enzymes and related factors that sequentially activate each other resulting in the production of a variety of biologically active proteins. During inflammatory responses, a complement component called C5a is produced that causes increased vascular permeability. Other complement components, C3a and C5a, can cause mast-cell degranulation, thereby amplifying the inflammatory process.

Kinin system

Kinins are small polypeptides of 9–11 amino acids. They are cleaved from larger plasma proteins called kininogens by specific esterases called kallikreins. The most important kinin in inflammation is **bradykinin**, which causes pain and vasodilation and increases vascular permeability.

BOX 2.1: PROSTAGLANDINS AND LEUKOTRIENES

Prostaglandins (PGs) and leukotrienes (LTs) are small molecules derived from membrane phospholipids. The precursor of PGs and LTs is arachidonic acid, which is produced after cleavage of phospholipids by the enzyme phospholipase. Arachidonic acid can enter the cyclooxygenase pathway, resulting in the formation of PG-G_2, which is converted to other PGs and thromboxane B_2. Alternatively, arachidonic acid can enter the lipoxygenase pathway, where it is converted to LTA_2, which is converted to other LTs. Both prostaglandins and leukotrienes are large families of structurally similar molecules containing 20 carbon atoms. They are chemical messengers and have a number of different functions depending on the particular PG or LT. They are involved in many physiological processes, including control of inflammation, ovulation, parturition, gastric secretion, steroidogenesis and blood pressure.

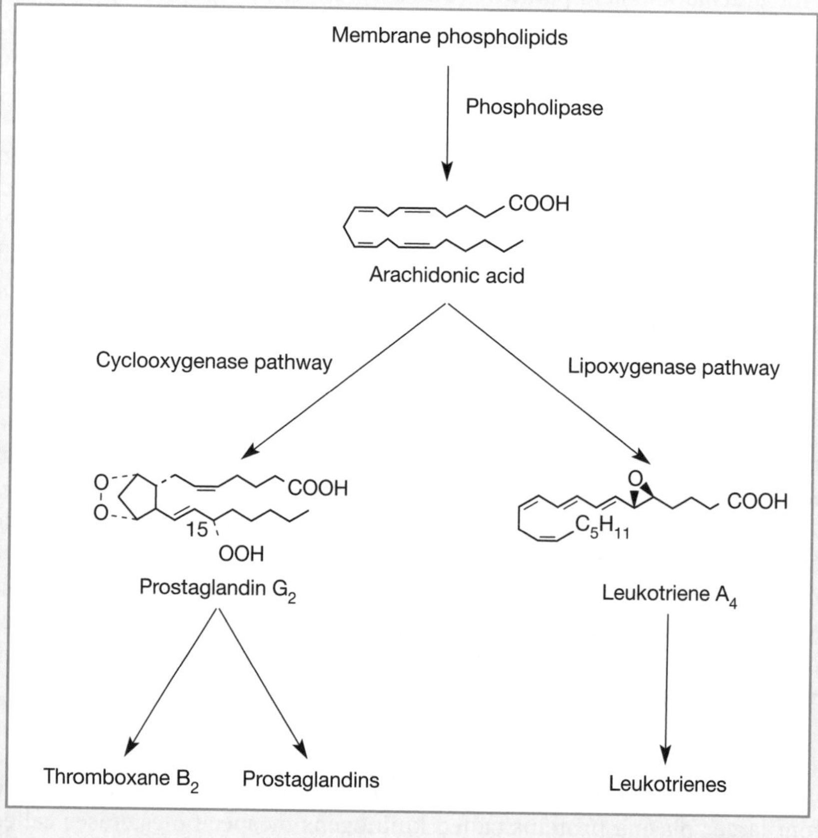

All the inflammatory mediators described above cause an increase in blood flow, increase in vascular permeability and chemotactic activity that results in the accumulation of granulocytes and monocytes at the site of inflammation and their activation. The activated macrophages and granulocytes can then begin to remove the pathogenic organisms by the process of phagocytosis.

2.7 Systemic inflammation – involvement of the brain and liver

In some cases the inflammatory response will succeed in eliminating the pathogen. In this case the response will be acute (short-lived) and confined to the area of tissue damage. If the pathogen is not eliminated the continued recruitment and stimulation of macrophages will result in a rise in the concentration of the macrophage-derived cytokines in the plasma. These cytokines can affect other organs, particularly the brain and the liver, leading to a systemic inflammatory response.

2.7.1 Cytokines and the brain – modifying behaviour to fight infection

IL-1 affects the brain, causing fever, somnolence (sleepiness) and anorexia (loss of appetite). Many of the symptoms you feel when you are ill with an infection are due to the actions of cytokines in the brain. Fever is known to have a protective effect in infection, and the replication of some pathogens is inhibited at higher temperatures. Somnolence reduces physical activity and hence energy consumption. Anorexia, by limiting the desire to engage in food-gathering activity, also reduces physical activity. Basically the body is saying, through the actions of these cytokines on the brain, 'rest and concentrate your energies on overcoming this infection'.

2.7.2 Cytokines and the liver – the acute phase response

IL-6 has a potent effect on hepatocytes, stimulating them to produce a series of proteins called **acute phase proteins** (APPs). Acute phase proteins are found in the serum at basal (background) levels in healthy normal individuals but rise in concentration following stimulation of the liver. They can be divided into two categories based on the degree to which they increase. The concentration of some acute phase proteins increases only 1.5- to 10-fold while that of others increases 10- to 1000-fold.

APPs that increase 1.5- to 10-fold

- **Fibrinogen.** As mentioned in Section 2.6.2 this is involved in clotting and the generation of fibrinopeptides.
- **Haptoglobulin.** This protein binds to iron-containing haemoglobin and reduces the concentration of iron that many bacteria require for their metabolism, thereby reducing bacterial growth.

- **Complement component C3.** This can be cleaved to generate C3a, which activates mast cells, and C3b, which helps phagocytes recognise pathogens (see Section 2.6.2).
- **Mannose-binding protein (MBP).** MBP is able to bind to mannose-containing sugars on the surface of pathogens and also helps phagocytes recognise pathogens.

APPs that increase 10- to 1000-fold

- **Serum amyloid A (SAA).** This protein inhibits fever and platelet activation. As such, it provides an important negative feedback control loop typical of that seen in many physiological systems.
- **C-reactive protein (CRP).** This protein binds to phosphoryl choline, which is found on the surface of a variety of bacteria, fungi and parasites and is exposed in damaged cells. As described below, it helps phagocytes to recognise bacteria or damaged cells.

The increased serum concentration of the acute phase proteins results in their increased accumulation at the site of inflammation; this is also aided by the increases in blood flow and vascular permeability caused by mediators of the inflammatory response. The acute phase proteins provide additional factors that help in the elimination of infectious agents, especially extracellular dwelling bacteria, yeasts and parasites. Three important APPs are CRP, C3b and MBP, which can act as opsonins to help phagocytes recognise pathogens. The production of acute-phase proteins by the liver is known as an **acute phase response**.

2.8 Opsonins can promote phagocytosis

The name opsonin derives from Greek, meaning ‘prepare for the table’. Victorian biologists likened phagocytosis to eating and hence something that prepared the phagocytes’ ‘food’ for ‘eating’ was termed an opsonin. The concept of an opsonin is very simple (Figure 2.9): one end of the opsonin recognises and binds to a molecule on the surface of a foreign organism and the other end of the opsonin binds to a receptor on the phagocyte. Engagement of the receptor on the phagocyte then stimulates the phagocytic process.

2.9 Interferons and natural killer cells

The cells and proteins involved in the inflammatory and acute phase responses are part of the **innate** or **natural** immune system. There are two other components to the innate immune system that contribute to the early response to infection; these are the interferons and natural killer cells.

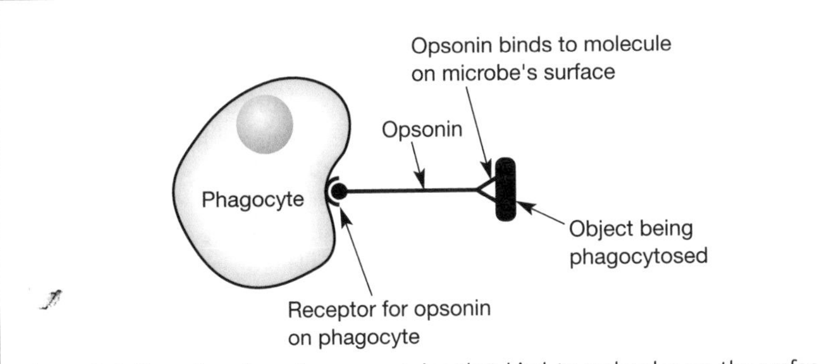

Figure 2.9 Opsonins. Opsonins are proteins that bind to molecules on the surface of microbes and to specific receptors on phagocytes. The binding of the opsonin to the phagocyte receptor activates phagocytosis.

2.9.1 Interferons are produced in response to viral infection of cells

Interferons are cytokines that inhibit viral replication in infected cells. They can inhibit viral replication in the cells that produce them, or they can be secreted by the cells and bind to specific receptors on other cells, making them resistant to viral infection (Figure 2.10). The way in which they inhibit viral replication is shown in Box 2.2.

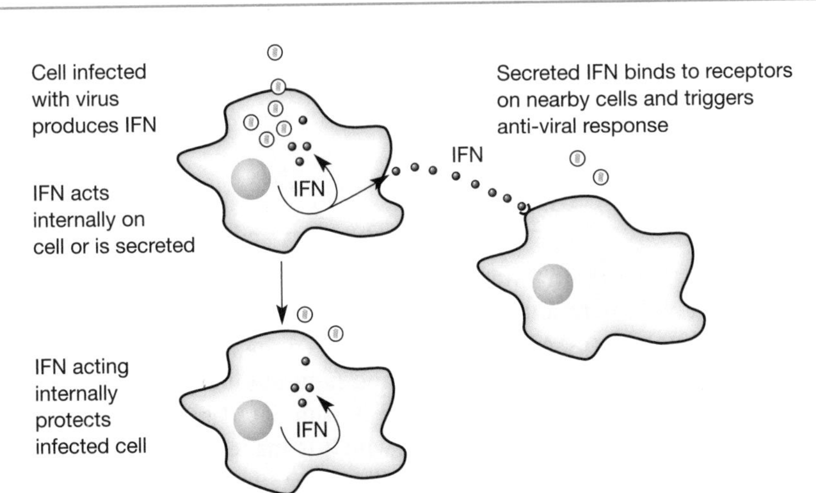

Figure 2.10 Interferons. Interferons activate anti-viral responses within the cells that produce them or are secreted and bind to receptors on neighbouring cells, inducing protection against viral infection.

BOX 2.2: ANTI-VIRAL ACTION OF INTERFERONS

The action of interferon α/β involves the induction and activation of two main pathways. An enzyme called 2′-5′-oligoadenylate synthase (2–5(A)-synthase) catalyses the production of 2′-5′-adenylate from ATP. 2′-5′-Adenylate activates another enzyme, RNAse, which degrades viral RNA. The second pathway involves the activation of a serine/threonine kinase, called P1 kinase, which phosphorylates the protein synthesis initiating factor elF2, resulting in inactivation of elF2 and inhibition of protein synthesis.

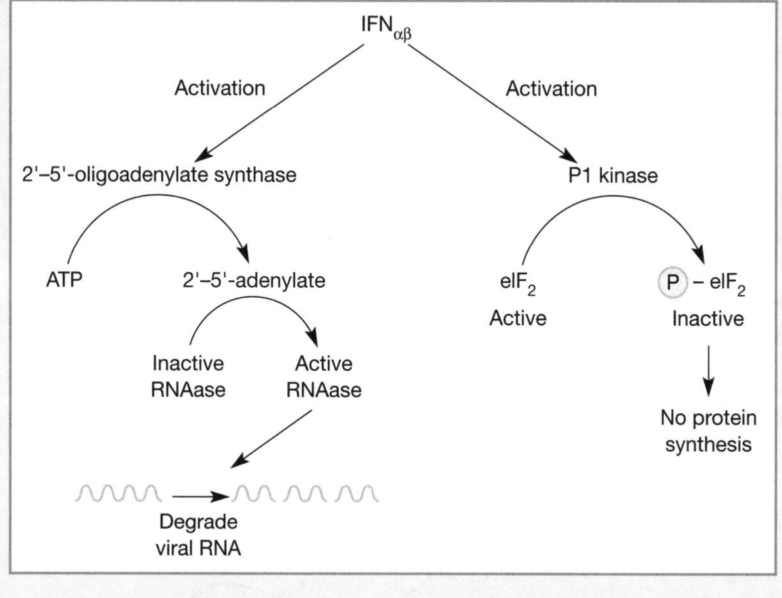

There are three main interferons – IFN-α, IFN-β and IFN-γ. Both IFN-α and IFN-β are produced by many cell types, including macrophages, fibroblasts, lymphocytes, endothelial cells and epithelial cells. IFN-γ is produced by lymphocytes (see Chapter 7) and natural killer cells (see Section 2.9.2). Production of IFN-α and IFN-β is stimulated by microbial products, especially double-stranded RNA, which is made only by viruses and therefore indicates viral infection of a cell. Other stimulators of IFN-α and IFN-β production include products of bacteria, fungi and parasites, and a number of cytokines.

IFNs have many other functions in addition to inhibiting viral replication, two of which are to activate macrophages and natural killer cells (see below). They can therefore serve to amplify the innate immune response against the virus and also against other pathogens that stimulate interferon production.

2.9.2 Natural killer cells

Another population of cells that form part of the innate immune system are **natural killer** (NK) cells (Figure 2.11). These cells make up 1–5% of white blood cells. As their name implies, they were first described by their ability to kill other cells, especially tumour cells. It has now been demonstrated that they play an important role in resistance to certain viral, bacterial and protozoan infections. There have been very few people born with no functional NK cells but these individuals appear to suffer from various viral infections, one suffered from repeated fungal infections and one died of infection with *Mycobacterium avium*, which, as its name implies, normally affects birds and only causes problems in severely immunosuppressed humans, for example those with AIDS. You can see from this that NK cells play a vital role in defence against some pathogens.

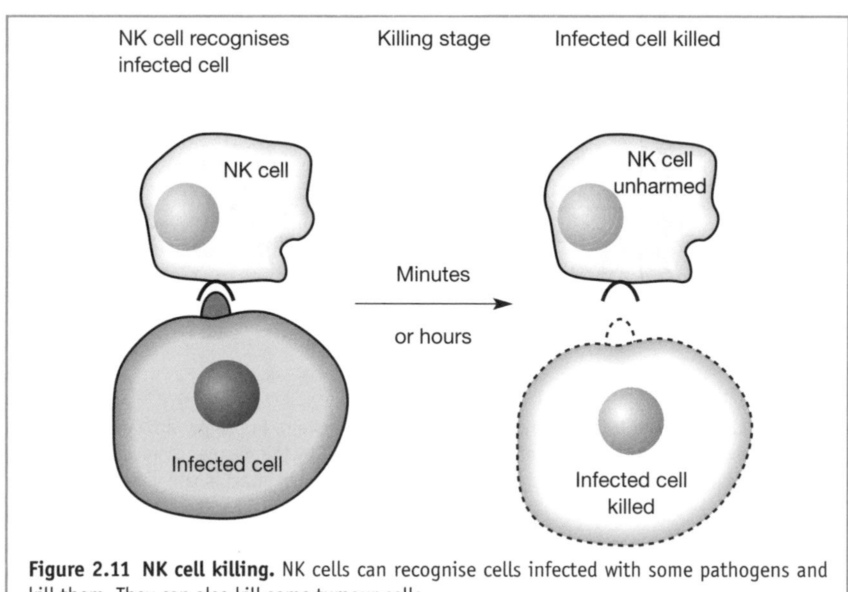

Figure 2.11 NK cell killing. NK cells can recognise cells infected with some pathogens and kill them. They can also kill some tumour cells.

NK cells enter sites of infection, where they can be stimulated by a cytokine called IL-12, which is produced by activated macrophages. The NK cells are stimulated by IL-12 to produce IFN-γ, which is a powerful activator of macrophages. This provides an amplification loop to maintain macrophage and NK activation. Although NK cells are able to recognise and kill cells infected with certain viruses, it is not clear whether the killing or cytokine-secreting ability of NK cells is most important in resistance to infection.

2.10 The innate immune response limits the early replication of pathogens

Although the innate immune response may not be enough on its own to destroy all the infectious particles of a bacterium, virus or other pathogen, innate immunity is very important in limiting the size of the infection in its early stages. The timing of the response of macrophages, neutrophils and NK cells is shown in Figure 2.12. There are rare individuals who have mutations in genes of the innate immune system and they are at increased susceptibility to various infections (see Figure 2.12).

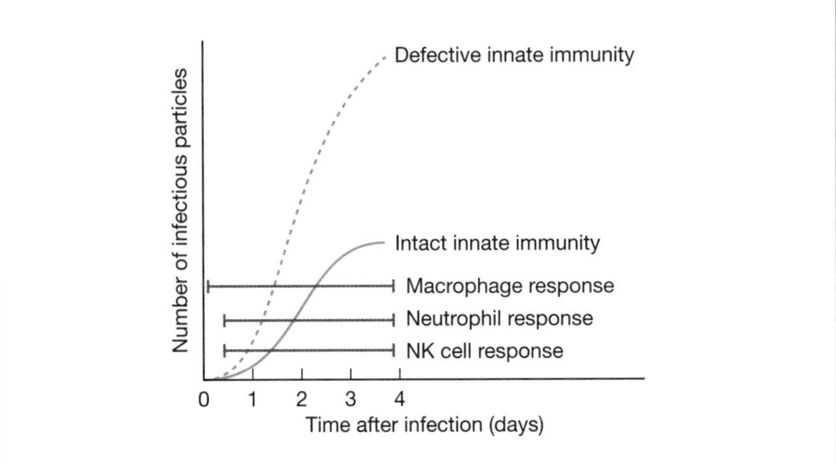

Figure 2.12 Control of infection by the innate immune system. In people with defects in the innate immune system the spread of infection, as measured by the number of infectious particles, is much more rapid and extensive than with a fully functional intact innate immune system.

2.11 Summary

- The first step in responding to infection is to be aware that infection has occurred. Tissue macrophages have receptors that are able to recognise molecules on some pathogens.
- Tissue macrophages can respond to the presence of a pathogen by stimulating an inflammatory response that results in the activation of many cells and protein pathways and the recruitment of phagocytes to the site of inflammation. These phagocytes can remove pathogens and damaged tissue.
- Adhesion molecules and chemokines also control the movement of many cell types into sites of inflammation.

- If the inflammatory response is severe enough, it can also affect the brain, leading to behavioural changes, and the liver, triggering an acute phase response.
- Some of the acute phase proteins can act as opsonins, which help phagocytes phagocytose pathogens.
- Natural killer cells and interferons are other important components of the innate immune system and contribute to the inflammatory response and to protection against pathogens.

2.12 Questions

1) What is the relationship between a monocyte and a macrophage?

2) PRR stands for:

   A) Pathogen recognition receptor
   B) Pattern resembling receptor
   C) Pattern recombination receptor
   D) Pathogen resembling receptor
   E) Pattern recognition receptor

3) Which of the following is NOT a PRR?

   A) Mannose receptor
   B) Polyamine receptor
   C) Scavenger receptor
   D) TLR
   E) NOD

4) What is meant by endocrine, paracrine and autocrine?

5) The four stages of phagocytosis are, in alphabetical order: (i) attachment, (ii) degradation, (iii) ingestion, (iv) killing. In what order do they occur during the phagocytic process?

6) How do adhesion molecules, cytokines and chemokines contribute to the migration of leukocytes from the bloodstream to a site of infection in a tissue?

7) Redness and swelling are typical of inflammatory responses. (i) What are the mechanisms for these? (ii) What are the benefits of redness and swelling?

8) The diagram on page 48 depicts the opsonisation of a bacterium. Match the following labels to the appropriate arrows: A) phagocyte, B) bacterium, C) opsonin, D) receptor for opsonin, E) molecule being recognised by opsonin.

The answers to these questions can be found on page 333.

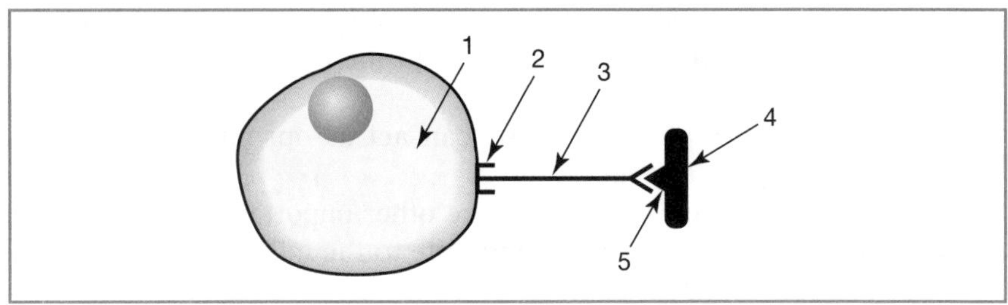

2.13 Further reading

1) Akira S, Uematsu S, Takeuchi O. (2006) Pathogen recognition and innate immunity. *Cell* 124:783–801.

2) Muller WA. (209) Mechanisms of transendothelial migration of leukocytes. *Circulation Research* 105:223–230.

3) Medzhitov R. (2008) Origin and physiological roles of inflammation. *Nature* 454:428–435.

CHAPTER THREE

Specific immune recognition: B lymphocytes and the antibody molecule

Learning objectives

To be introduced to the cells and molecules of the specific immune system. To learn that antibodies are molecules of the specific immune system. To understand how antibodies recognise molecules on pathogens and other structures.

Key topics

- Basic antibody structure
- The nature of binding of antibodies to foreign molecules (antigens)
- Antibody classes: IgG, IgM, IgA, IgE, IgD
- Antibody as a soluble protein and cell-surface receptor

3.1 Introduction to the specific immune system

The ultimate aim of an immune response is to eliminate or neutralise threats to the body posed by an infectious agent (pathogen). Following infection, the first encounter between the pathogen and the infected individual is through the innate immune system, often resulting in an inflammatory response (see Chapter 2). However there are many reasons why pathogens are able to avoid elimination by the innate immune system. One reason is that pathogens evolve ways of avoiding being recognised by the cells and opsonins of the innate immune system. A good illustration of this is streptococcal bacteria. Non-virulent streptococcal bacteria are recognised and killed by phagocytes and do not cause disease. Virulent streptococcal bacteria synthesise a waxy polysaccharide (sugar) coat that surrounds them and prevents them being recognised by phagocytes or opsonins of the innate immune system. They cannot be eliminated by the innate immune system and go on to cause disease. Other pathogens have evolved a variety of ways to avoid being recognised and/or destroyed by the innate immune system.

So how do we deal with a pathogen that the innate immune system alone cannot eliminate? The answer is that all but the most primitive of mammals possess a second type of immune system in addition to the innate immune system. This is known as the **specific immune system** and consists of a collection of specialised organs, tissues and cells. The cells of the specific immune system are able to recognise the presence of a pathogen in a much more efficient way than the innate immune system. Following recognition of pathogen the cells of the specific immune system are stimulated to generate a **specific immune response**. This results in the production of new types of cells which have different functions that contribute to fight against pathogens.

3.1.1 Lymphocytes – the cells of the specific immune system

The main cells of the specific immune system are the lymphocytes, which are a type of white blood cell (Plate 4). There are many different types of lymphocytes with different functions, although morphologically they look the same. Two important types of lymphocytes are B lymphocytes and T lymphocytes. Like all blood cells, the lymphocytes are produced in the bone marrow (see Figure 2.1). However, while B lymphocytes complete their maturation in the bone marrow and are then released into the bloodstream, T lymphocytes start off their maturation in the bone marrow but then travel to an organ called the **thymus** where they finish off their maturation and are then released into the blood (this is described in Section 5.3). There are two important sub-types of T lymphocytes called CD4 and CD8 T lymphocytes. Although they are morphologically identical, the CD4 and CD8 lymphocytes have different functions as described in Chapters 7 and 9.

The lymphocytes travel through the blood and can enter tissues ‘looking’ for the presence of pathogens. When the lymphocytes identify the presence of a pathogen they are stimulated to respond and mount a specific immune response. As mentioned above, a vital feature of B and T lymphocytes is their ability to recognise molecules on pathogens (or indeed any object foreign to the body). They are able to recognise foreign proteins that differ from our own by one, or a few, amino acids. The receptors of the innate immune system cannot do this; they recognise foreign molecules that are usually quite different from our own. Lymphocytes recognise foreign objects through the expression of receptors on their cell surface or through the secretion of proteins. The receptors that B and T lymphocytes use are different but share the ability to recognise small differences between us and foreign objects. The nature of this recognition is simplest to explain by looking at B lymphoctes. B lymphocytes are responsible for the production of important glycoproteins that are part of the specific immune system. These glycoproteins are called **antibodies** and they have the ability to recognise foreign molecules on the surface of pathogens. Antibodies are also called **immunoglobulins** and the terms antibody (Ab) and immunoglobulin (Ig) mean the same thing. Antibodies are vital for human life and people who cannot make antibodies die of overwhelming infection unless treated with pooled Ig from healthy people. The way in which antibodies recognise molecules derived from pathogens illustrates some of the differences between the innate and specific immune systems and the additional demands that are required of the specific immune system.

3.2 Antibody structure

The basic antibody molecule is depicted as a Y-shaped structure consisting of four protein subunits (Figure 3.1a). The two longer subunits are called **heavy (H) chains** and are identical to each other; they have a molecular mass of 50–75 kDa. The two shorter subunits are also identical to each other; they are called **light (L) chains** and have a molecular mass of about 25 kDa. There are two types of light chain called κ and λ; they are very similar in structure but are coded for by different genes. An individual antibody molecule will contain two H chains and two κ-chains *or* two λ-chains; κ- and λ-chains are never seen together in the same antibody molecule. The heavy chains are linked to each other, and to the light chains, by disulphide bridges.

Further analysis of antibody structure shows that both the heavy and the light chains have repeating substructures called **domains** (Figure 3.1b,c). These domains are regions of approximately 110 amino acids within the heavy and light chains and are flanked by intrachain disulphide bridges. The heavy chain of the immunoglobulin molecule has four or five domains and the light chain has two. The domains at the *N*-terminal ends of the heavy and light chains are called the variable domains (V_H and V_L, respectively) because the amino acid sequences of these domains were found to differ from antibody to antibody. The other domains are called constant region domains because they do not differ to the same extent from antibody to antibody. The heavy chain constant domains are called C_H1, C_H2, C_H3 and C_H4; the light chain domain is called C_L. Between the C_H1 and C_H2 domains of the heavy chain is a region containing several prolines; this makes this part of the molecule quite flexible and it is therefore known as the hinge region. X-ray crystallography has revealed that the heavy and light chains fold together to form a molecule that is globular in structure (Plate 5).

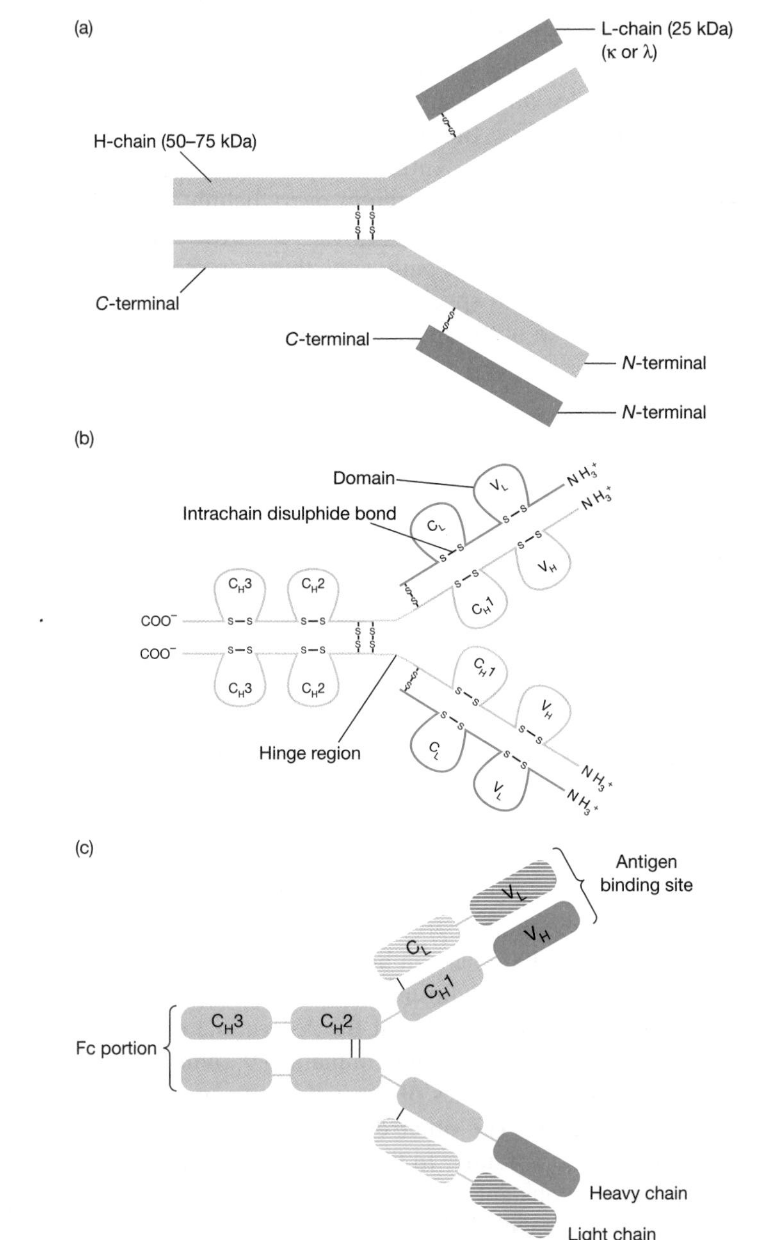

Figure 3.1 Antibody structure. (a) Each antibody molecule contains two identical larger (heavy) chains and two identical smaller (light) chains. (b) The heavy and light chains contain domains that are structural parts of the proteins, about 110 amino acids in size, flanked by intrachain disulphide bonds. The hinge region contains several prolines giving that part of the molecule flexibility. (c) The *N*-terminal domains are responsible for binding antigen, and the *C*-terminal domains of the heavy chain are called the Fc portion of the molecule.

3.3 Recognition by antibody – antigens and epitopes

Antibodies bind to molecules that are 'foreign' to the body. These molecules may be on the surface of a pathogen or they may be soluble products such as toxins secreted by pathogens. The molecules that antibodies bind to are called **antigens**. Antigens are nearly always macromolecules; they are usually proteins but they may also be polysaccharides (sugars) or, less commonly, lipids or nucleic acids. An antibody molecule does not bind to the whole of an antigen; it binds to a part of the antigen that is called an **antigenic epitope** (see Figure 3.2). For a protein antigen, an antigenic epitope will be a structural conformation within the protein ranging from 8 to 22 amino acids in size. Epitopes recognised by antibody can be linear or conformational (Figure 3.2). A linear epitope is a conformation on the antigen formed by a continuous sequence of amino acids. By contrast, a conformational epitope is formed by the folding of the protein (Figure 3.2). Given that some proteins have an M_r of over one million and contain thousands of amino acids, a single antigen can potentially have a number – possibly several hundred – of different epitopes. Although it might be expected that a large protein could theoretically form a large number of

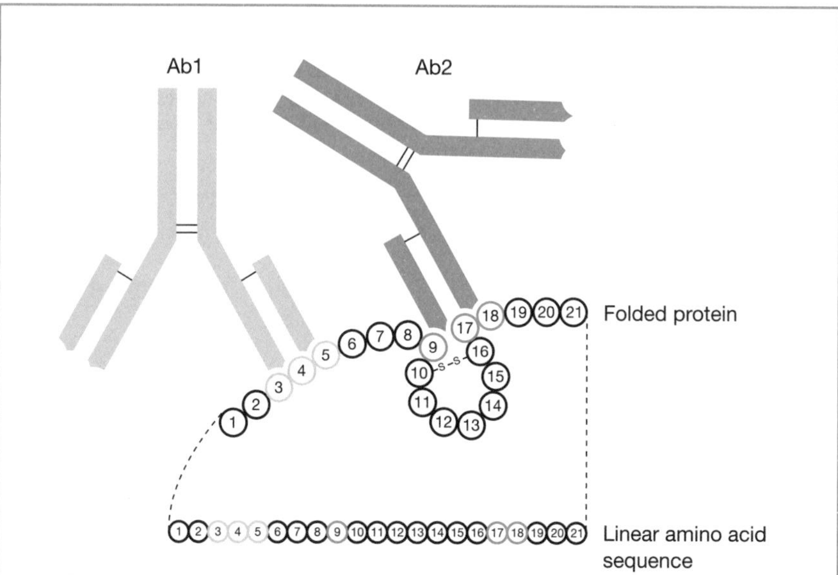

Figure 3.2 Linear and conformational epitopes. Ab1 is binding to a linear epitope that is composed of amino acids in a continuous sequence in the protein amino acid sequence (amino acids ③, ④ and ⑤). Ab2 is binding a conformational epitope formed by the folding of the protein antigen. The amino acids forming the structural epitope (⑨, ⑰ and ⑱) do not occur in a linear sequence in the amino acid sequence of the protein antigen. Note that in reality antigenic epitopes are formed by 5–15 amino acids and not the three depicted here.

epitopes, in reality this is not the case. An experimental antigen, hen-egg lysozyme, is a protein of M_r 14 000 and appears to have eight different antigenic epitopes. Most antigens are thought to have a similar number of epitopes in relation to their M_r. For most antigens each epitope on one molecule of antigen will be different, although some polysaccharides have many repeats of the same epitope (Figure 3.3).

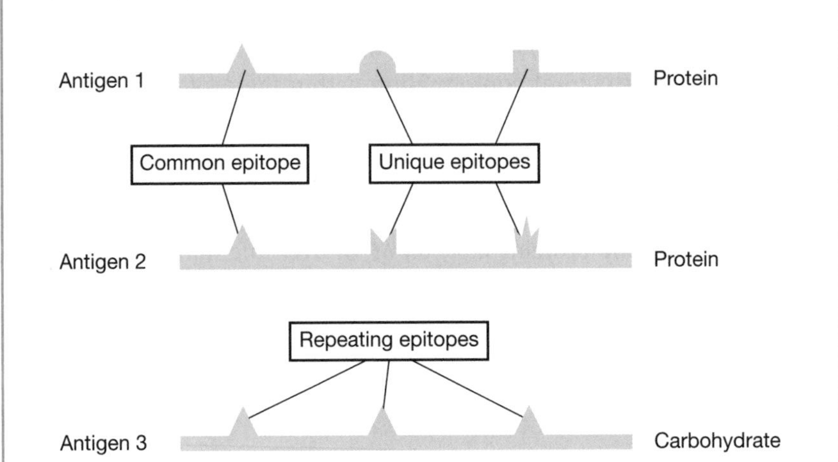

Figure 3.3 Repeating and unique epitopes. Antigen 1 has three different epitopes depicted by the triangle, circle and square shaped protrusions. Antigen 2 has one epitope (triangle) that is shared by antigen 1 and two unique epitopes. Antigen 3 has repeats of the same epitope; this is typical of polysaccharide antigens.

3.3.1 Antibodies have special regions that are involved in binding antigen

The parts of the antibody molecule that bind to antigenic epitopes are called the antigen-binding sites and are located in the variable domains of the H- and L-chains (Figures 3.2 and 3.4). Because they are responsible for antigen binding they have also been called the antigen-binding fragments, or FAbs, of the molecule (Box 3.1). The antigen-binding sites are formed by the folding of the variable domains of the heavy and light chains (Figure 3.4). Antibodies against different antigens have different amino acid sequences in their variable regions. More detailed analysis shows that the variability is confined to three regions, each of 5–15 amino acids in length, within each heavy and light chain (Figure 3.4); these regions are therefore called the **hyper-variable** regions of the antibody. The parts of the variable regions between the hyper-variable regions do not differ so much between antibodies and have been called the **framework** regions.

They were so named because they contribute to the overall structure of the antibody and provide a framework on which the hyper-variable regions sit. Although the hyper-variable regions are equally spaced in the linear amino acid sequence of the variable region, when protein folding is taken into account it can be seen that the three hyper-variable regions are brought together at the end of the molecule (Figure 3.4). Furthermore, the heavy and light chains fold together so that their hyper-variable regions form a single surface. This surface forms the antigen-binding site and interacts with the antigenic epitope (Figure 3.4 and Plate 6). Different antibodies will have antigen-binding regions with different shapes, as depicted in Figure 3.5. These shapes determine whether an antigenic epitope will bind to the antibody.

BOX 3.1: FRAGMENTS OF ANTIBODIES

Much of the terminology about the different parts of antibodies comes from early (1950s and 1960s) experiments to try to elucidate the structure of antibodies. It was known that antibody (the IgG fraction of serum) had a molecular weight of 150 000 but it was not known whether Ig consisted of one protein chain of M_r 150 000 or was made up of smaller units.

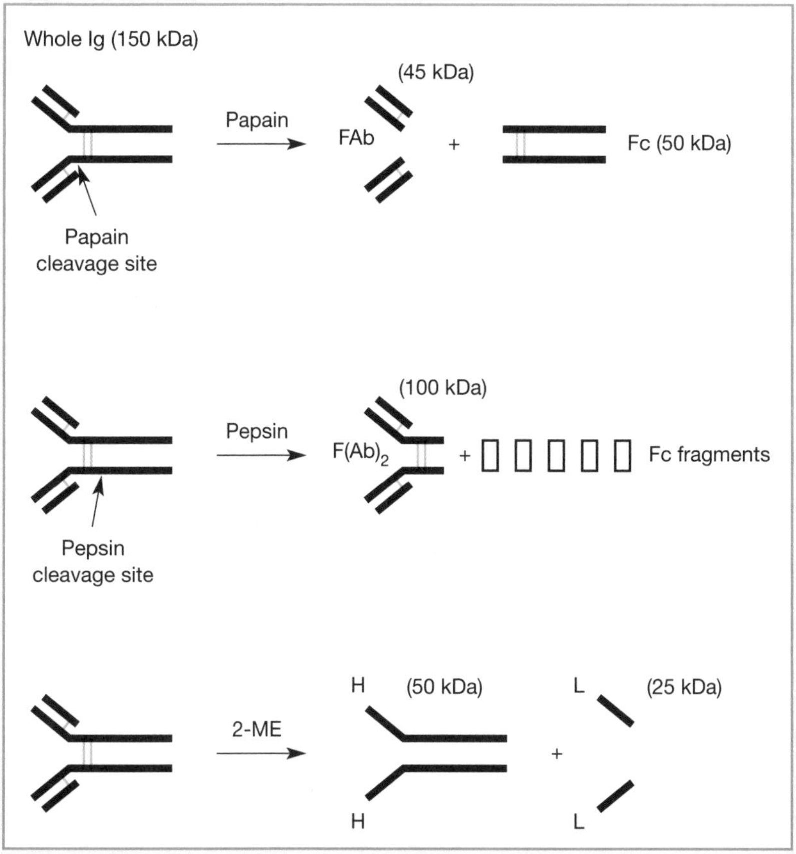

Whole Ig (150 kDa)
(45 kDa)
Papain
FAb
+
Fc (50 kDa)
Papain cleavage site
(100 kDa)
Pepsin
$F(Ab)_2$
+
Fc fragments
Pepsin cleavage site
H (50 kDa)
L (25 kDa)
2-ME
+
H
L

The approach used to address the problem was to see whether the antibodies could be broken down into smaller subunits. One approach used enzymes to break down antibodies. Brief incubation of Ig with the enzyme **papain** (found in the papaya latex) was found to produce two fragments, one of M_r 45 000, which was called **FAb** (for 'fragment antigen binding') because this fragment could still bind antigen, and one of M_r 50 000 called **Fc**, which crystallised upon storage at 4°C (Fc stood for 'fragment crystallisable'). Another enzyme, pepsin, generated a single fragment of M_r 100 000, which became called **F(Ab)$_2$**. Another reagent used was 2-mercapto-ethanol (2-ME), which breaks interchain disulphide bonds. Treatment of Ig with 2-ME produced two fragments of M_r 50 000 and M_r 25 000, which were the heavy and light chains respectively.

Eventually the structure of IgG was deduced and the basis for the different patterns of cleavage explained as shown in the figure. This work led to the award of the Nobel Prize to Rodney Porter in the UK and Gerald Edelman in the USA.

Hyper-variable regions and complementarity-determining regions

The terms 'hyper-variable region' and 'complementarity-determining region' mean the same thing but it is not always easy to understand why. For an antibody to bind to an antigenic epitope, the shape of the epitope must fit that of the antibody. Another way to say this is that the antigen-binding site of the antibody is **complementary** in shape to the antigenic epitope (Figure 3.4). It is the hyper-variable regions of an antibody that

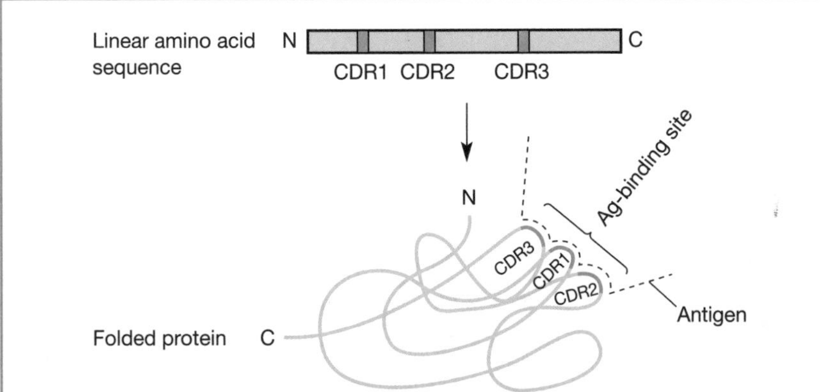

Figure 3.4 Hyper-variable regions of antibody chains. These fold to form the antigen-binding site. The three hyper-variable regions are spaced apart in the linear amino acid sequence of the variable region of the heavy and light chain proteins. However, when protein folding occurs the three hyper-variable regions come together to form a single antigen-binding site. This occurs for both H- and L-chains so that the H + L combination forms a single antigen-binding site composed of six hyper-variable regions – three from the H-chain and three from the L-chain.

determine whether it is complementary to an antigenic epitope, and so the hyper-variable regions of an antibody are also called the **complementarity-determining regions**, or CDRs for short.

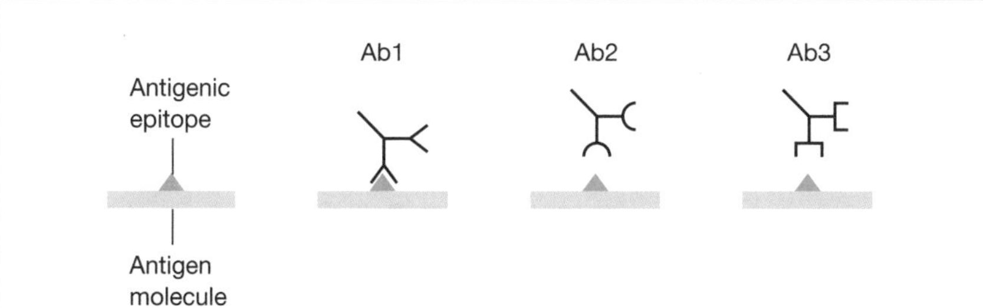

Figure 3.5 Different antibodies bind different antigenic epitopes. An antigen is shown schematically with an antigenic epitope depicted as a triangle. Ab1, Ab2 and Ab3 have different-shaped antigen-binding regions depicted by a ∧, ∩ or ⊓ shape on the ends of the arms of the antibody. Only Ab1 will bind the antigenic epitope.

3.3.2 The nature of antibody–antigen binding – specific recognition

The chemical interactions between antibody and antigen are non-covalent. Four types of non-covalent interaction are involved: hydrogen bonds, electrostatic forces, van der Waals forces and hydrophobic forces (Figure 3.6). These non-covalent interactions are weak unless the two molecules forming the bond are very close together in molecular terms; thus a lot of these interactions are required for strong binding of antibody to antigen. However, as the molecules come close together, electrons of atoms in the antigen and antibody repel each other because they are negatively charged. The strength of binding between an antibody binding site and an antigenic epitope is therefore determined by the net balance between the attractive and repulsive forces. This is why there must be a good fit between the antibody-binding site and the antigenic epitope for strong binding to occur (Plate 6 and Figure 3.7). Weaker binding may occur if the fit between the antibody-binding site and the antigenic epitope is not so good. The strength with which the antigen-binding site of an antibody binds to an antigenic epitope is known as the **affinity** of the antibody for the antigen.

In many situations one amino acid in the antigenic epitope can determine whether it will bind antibody. The antigen to which the antibody is binding in Plate 6 is called hen-egg lysozyme; this is a common antigen used for experimental purposes in studying immune responses and antigen–antibody interaction. One amino acid, a glutamine, in the antigenic epitope of hen-egg lysozyme forms strong hydrogen bonds with the binding site of the antibody. Turkey-egg lysozyme is almost identical to hen-egg lysozyme except that it has substituted another amino acid for the glutamine

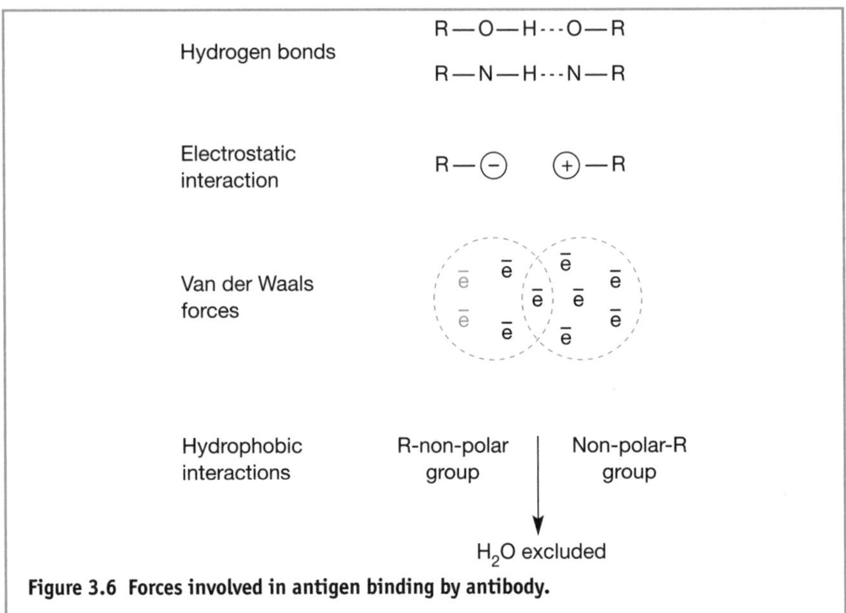

Figure 3.6 Forces involved in antigen binding by antibody.

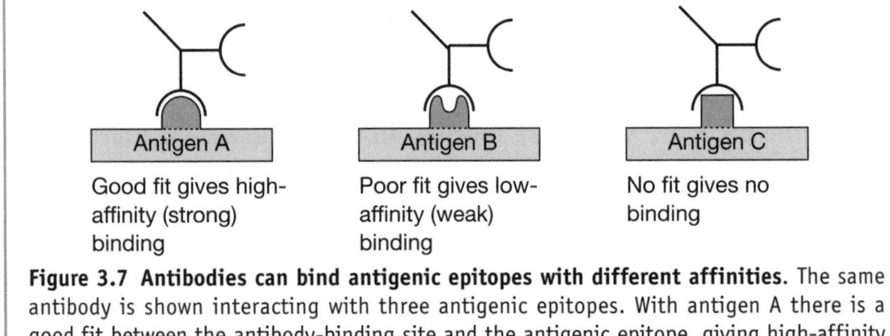

Figure 3.7 Antibodies can bind antigenic epitopes with different affinities. The same antibody is shown interacting with three antigenic epitopes. With antigen A there is a good fit between the antibody-binding site and the antigenic epitope, giving high-affinity (strong) binding. There is a poor fit between the antibody and the epitope on antigen B, resulting in some binding of low affinity (weak binding). The antibody-binding site does not fit the epitope on antigen C, so there is no binding.

and as a result the antibody will not bind to turkey-egg lysozyme. Therefore antibodies are capable of binding to one protein but not to another protein that differs in amino acid sequence by only one amino acid. This means that even if a protein on the surface of a pathogen differed from an individual's own protein by one amino acid, the antibody would still be able to bind to the pathogen but not the host cells. This provides an incredibly powerful way of distinguishing between foreign and self molecules.

3.3.3 The strength of antibody binding to antigen is measured by affinity and avidity

These two terms often cause confusion. Antibodies have at least two antigen-binding sites, each made up of a heavy-chain/light-chain pair of polypeptides. As mentioned above, the strength with which an individual binding site of an antibody binds its epitope is called the **affinity** of binding. However, if an antibody is using both binding sites to bind to two epitopes on the same particle the overall strength of the binding is increased and the total strength of this binding is called the **avidity** of the antibody for the antigen (Figure 3.8). Avidity is more than a simple sum of the affinities of each antibody-binding site for its antigenic epitope. To break the interaction between multiple binding sites it is necessary to break the binding at every binding site at the same time and this requires much more energy than breaking the binding at a single binding site.

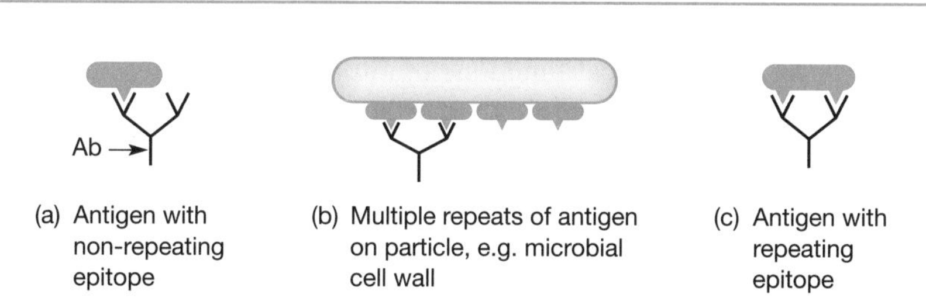

Figure 3.8 Affinity and avidity. (a) With most soluble protein antigens, such as secreted bacterial toxins, each epitope occurs only once on each protein molecule, although there may be a number of different epitopes on each protein molecule – these are called non-repeating epitopes and the strength of binding of the antibody is determined solely by the affinity of the antibody-binding site for the antigenic epitope. (b) If several copies of the protein are present on the surface of a particle, such as a bacterium, each antibody-binding site can bind to an epitope, and therefore the overall strength of binding of the antibody to the **particle**, or avidity of binding, is determined by the number of binding sites and the affinity of each binding site. (c) Some antigen molecules, especially carbohydrates, have more than one copy of an epitope and therefore both antibody-binding sites can bind to the antigen, resulting in increased avidity as in (b).

A way that is often used to illustrate the difference between affinity and avidity is to imagine that you are hanging on to a high wire by one or both hands. If you were hanging by one hand, the strength with which you could hang on to the wire would be determined solely by the strength of your grip. The strength of grip can be likened to the affinity of a single antibody-binding site for its antigenic epitope. If sufficient force was applied to loosen your grip on the wire you would immediately fall to whatever fate awaited you. Now imagine that you are hanging on to the wire by both hands. The same force that could loosen the grip of only one hand would not be strong enough to loosen the grip of both hands. Therefore the overall strength by which you are hanging on to the wire is much greater; this represents the avidity of antibody-binding to an antigen using two or more binding sites.

3.4 There are different antibody classes with different biological functions

So far we have concentrated on how antibodies can bind to antigens, which may be on the surface of a bacterium, virus or other microbe. However this binding is only useful if the antibody can then carry out a biological function that helps protect against the bacterium or virus. In fact antibodies have a number of different biological functions which are described in Chapter 8. The biological functions of antibody are carried out by the Fc part of the antibody so antibody can be considered to have antigen-binding properties at the FAb end and functional properties at the Fc end (see Figure 3.1c). These different functions require different properties of the Fc region and it is not possible for one Fc region to carry out all the different biological functions of antibodies. To overcome this problem, different antibodies exist with different Fc regions and different functions. These different types of antibody are called **classes** of antibody. There are five different classes of antibody in humans, called IgM, IgG, IgA, IgE and IgD (Figure 3.9). Slightly different variations of IgG and IgA exist, which are called **subclasses** of antibody. There are four subclasses of IgG and two of IgA. Each class or subclass of antibody has a unique set of functions determined by the Fc part of the molecules (except for the two subclasses of IgA, which appear to have identical function). However, the specificity of antibody for antigen is related not to its class but to the shape of the antigen-binding site. Different classes of antibody can have the same antigen specificity.

3.4.1 Immunoglobulin G (IgG)

This is the most abundant antibody in serum (see Table 3.1) and exists as the basic 2H + 2L chain antibody molecule (Figure 3.9). The four subclasses of IgG in the human are IgG1, IgG2, IgG3 and IgG4, all having the same 2 heavy + 2 light chain structure (Table 3.1). The different IgG subclasses have slightly different structures and functions, although they are closely related to each other. IgG1 is the most common of the IgG subclasses, comprising about 70% of total IgG.

Table 3.1 Antibody classes

Class	*Heavy chain*	*Molecular mass (kDa)*	*Serum concentration (mg/ml)*	*Serum half-life (days)*
IgM	μ	900	1.5	10
IgG1	γ1	150	9	21
IgG2	γ2	150	3	20
IgG3	γ3	165	1	7
IgG4	γ4	150	0.5	21
IgA1	α1	160	3	6
IgA2	α2	160	0.5	6
IgE	ε	190	5×10^{-5}	2
IgD	δ	185	0.03	3

3.4.2 Immunoglobulin M (IgM)

IgM is the earliest antibody to be produced after first contact with a new antigen. Structurally it consists of a pentamer of five Ig molecules joined together by disulphide bonds and an extra protein called the joining or J-chain (Figure 3.9). The H-chains of IgM differ from those of IgG in having four constant domains instead of the three in IgG. The J-chain of IgM has a molecular weight of 15 000 and is coded by a gene on a separate chromosome from the genes coding immunoglobulin heavy and light chains. It binds through disulphide linkages to the heavy chain of two of the five Ig molecules that make up IgM (Figure 3.9). The J-chain is required for the proper polymerisation of IgM into pentamers and in the absence of a J-chain pentamers are not formed. IgM potentially has ten antigen-binding sites although steric hindrance means that in practice it can usually use only up to six of them at once.

3.4.3 Immunoglobulin A (IgA)

IgA is present in the serum and is also the main class of antibody found in various secretions such as mucus in the intestinal and respiratory tracts, saliva, sweat, breast milk and colostrum. The two subclasses of IgA in the human, IgA1 and IgA2, appear to have the same functions. IgA has different structures depending on whether it is in serum or secretions. In serum, IgA adopts the basic 2 heavy + 2 light chain Ig structure (Figure 3.9). IgA in secretions has a different structure and consists of two Ig molecules joined together by a J-chain (as in IgM) and an additional protein called the secretory piece. The secretory piece is not actually made by the antibody-producing cell but is added to the IgA in a special way (see Section 7.4). The secretory piece helps the transport of the IgA into secretions and also helps to protect the IgA from breakdown by proteolytic enzymes that are found in secretions.

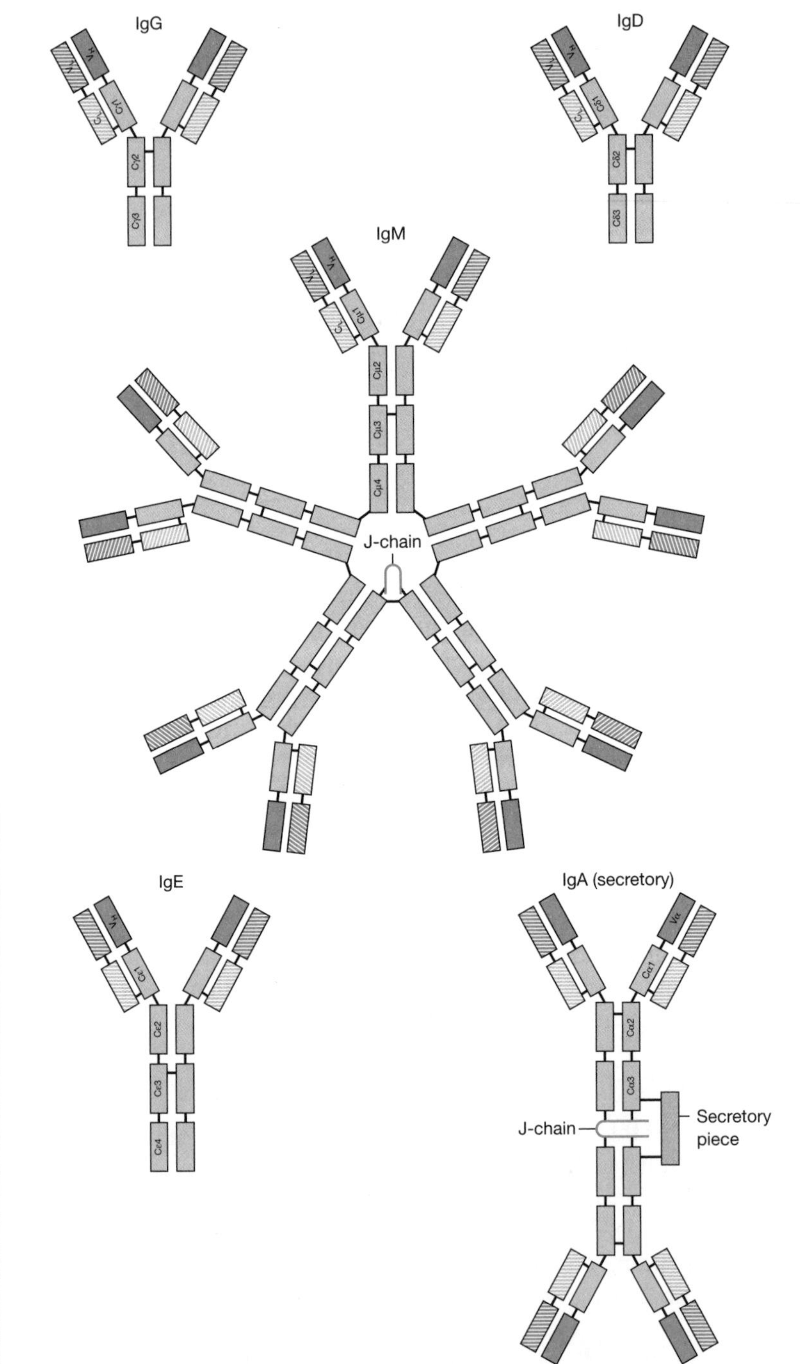

Figure 3.9 Antibody classes. The heavy chains of IgM and IgE have four C_H-domains compared with the three of other Ig classes. Secretory IgA is shown. IgA also exists as a monomer in serum, without the J-chain and secretory piece.

3.4.4 Immunoglobulin E (IgE)

IgE is present at the lowest concentration of all antibody classes in serum. It exists as a monomer consisting of the basic two heavy and two light chain Ig structure. Like IgM, its heavy chain also has four C-domains (Figure 3.9). IgE has very special functions and is involved in asthma and allergy (see Chapter 13).

3.4.5 Immunoglobulin D (IgD)

IgD generally has a low serum concentration and is unstable in serum, being quickly degraded by serum plasmin. It exists as a monomer of the basic Ig structure (Figure 3.9) but the functional significance of serum IgD is not clear.

3.5 Antibody can be secreted or expressed on the cell surface of B lymphocytes

Antibodies are soluble proteins that are secreted and circulate in the blood or are found on mucosal surfaces where they provide important protection. The cells that make antibody are called plasma cells (Plate 4). These are derived from B lymphocytes (see Chapter 6). Although the nature of antibody specificity was known to some extent in the late 1800s it was not until the 1950s that it was realised that B cells used immunoglobulin on their cell surface to recognise antigen.

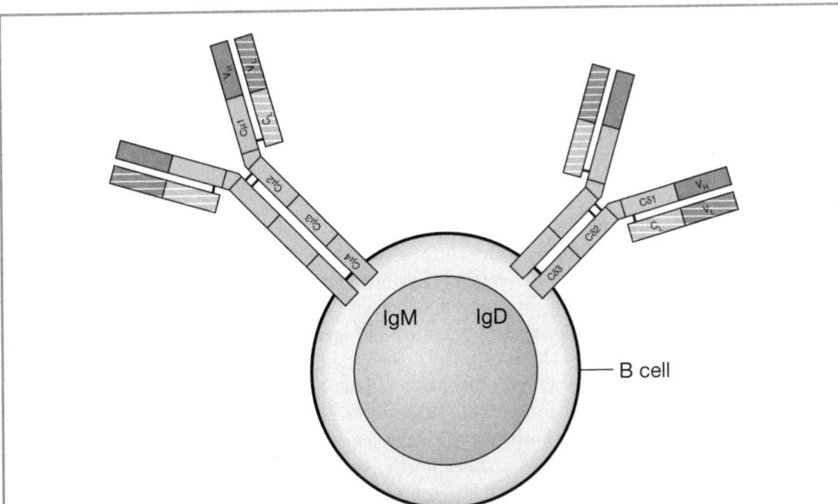

Figure 3.10 B cells express both IgM and IgD as cell-surface membrane proteins. The V_H region is the same on both IgM and IgD and both the IgM and IgD molecules contain identical L-chains.

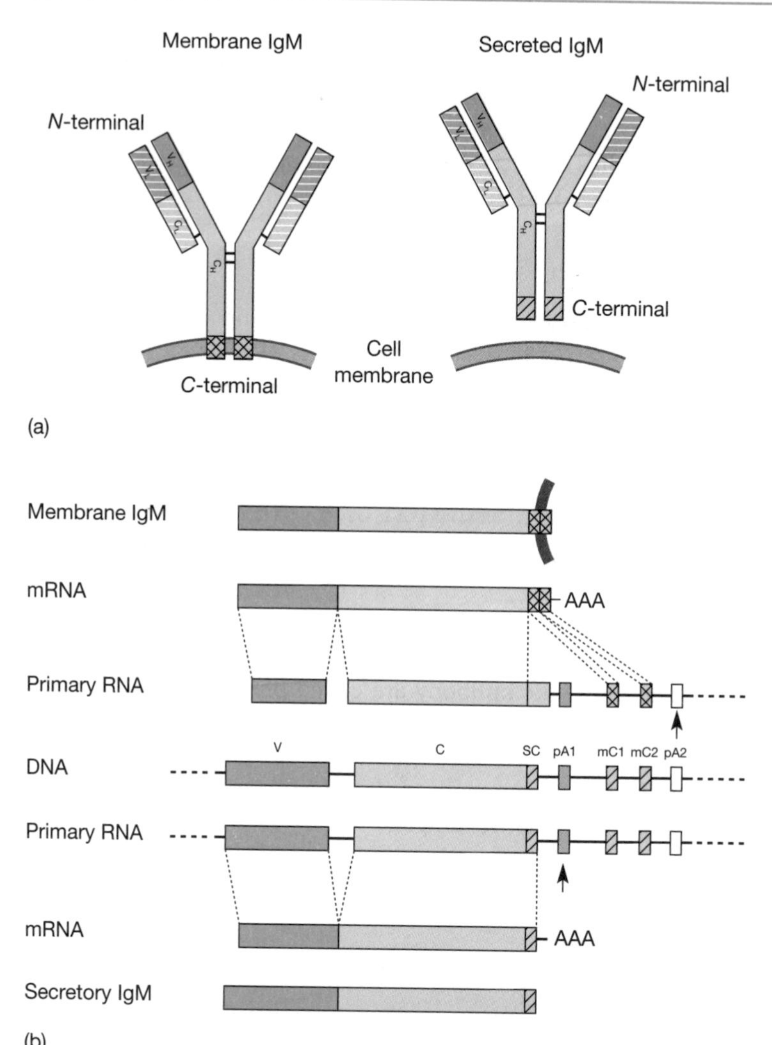

Figure 3.11 Production of membrane and secreted IgM. (a) A B cell can both express IgM on its cell surface and secrete IgM. The surface and secreted IgM are identical except for the *C*-terminal ends of their H-chains (shown by different shading). The *C*-terminal end of membrane IgM contains many hydrophobic amino acids that anchor it in the cell membrane; secreted IgM does not have these. Both secreted and membrane Ig H-chain associate with the same light chain made by the B cell. (b) A simplified diagram of the genes coding for IgM show alternative polyadenylation sites, pA1 and pA2, and genes coding for the secretory component (SC) or membrane component (MC1 and MC2) of the IgM H-chain. The primary RNA transcript is the same for both secretory and membrane IgM. Polyadenylation at pA1 and RNA splicing leads to mRNA that codes for the secreted IgM, while polyadenylation at pA2 and different RNA splicing lead to mRNA coding for membrane IgM.

B lymphocytes express the immunoglobulin molecules IgM and IgD on their cell surface as integral membrane proteins (Figure 3.10). Putting Ig into the cell membrane rather than secreting it is achieved by differential processing of the heavy chain RNA (Figure 3.11). All the IgM and IgD molecules on the surface of any one B cell have the same heavy chain variable regions and are associated with the same light chain; only the heavy chain constant regions differ between the IgM and IgD. Each B cell therefore has one set of heavy and light chain variable regions and one antigen specificity. Different B cells will have different sets of heavy and light chain variable regions and therefore **each B cell has a different antigen specificity** (Figure 3.12).

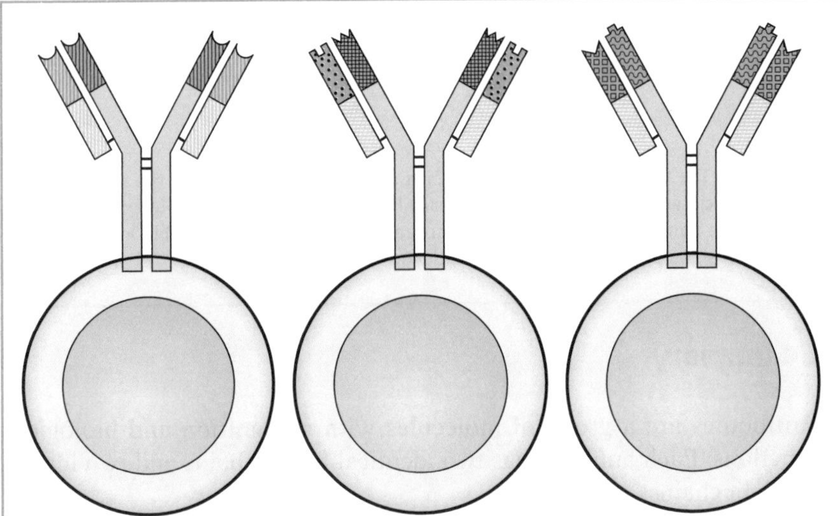

Figure 3.12 Different B cells express membrane Ig molecules with different antigen specificity. All the Ig molecules expressed by a single B cell have the same H- and L-chains. However, different B cells will have different V_H and V_L regions (indicated by different shading and shapes at the end of the V regions) and therefore different B cells will have different antigen specificity.

The membrane Ig on the B lymphocyte surface forms a signalling complex with two other proteins called Igα and Igβ. This signalling complex actually consists of eight chains: two identical Ig heavy chains (μ or δ), two identical Ig light chains (κ or λ) forming the Ig molecule, and two dimers each consisting of an Igα and an Igβ chain (Figure 3.13).

Therefore antibody can exist in two forms: a soluble form that has a number of biological activities depending on the class of antibody (see Chapter 8) and an integral cell-membrane protein on the surface of B lymphocytes that enables B lymphocytes to recognise a specific antigen and respond to it.

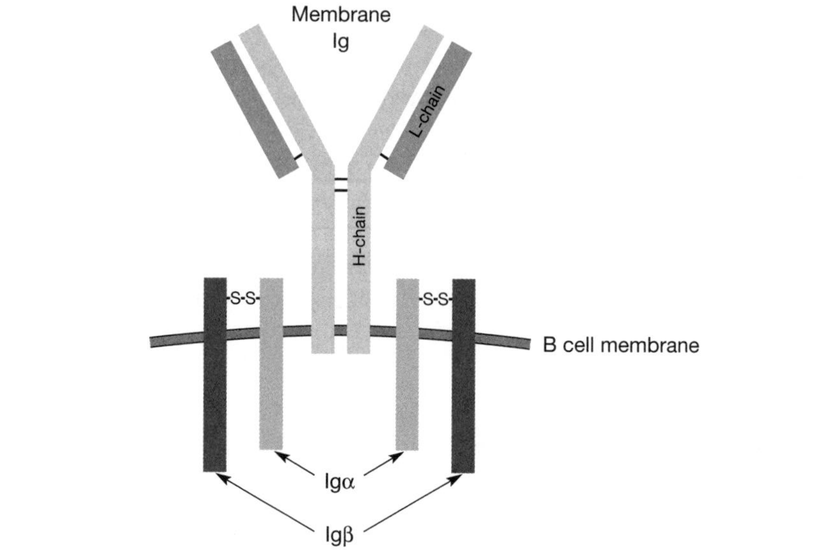

Figure 3.13 The B cell receptor complex. Each membrane Ig molecule on the surface of a B cell is associated with two heterodimers; each heterodimer consists of an Igα chain linked by disulphide bonds to an Igβ chain. The Igα and Igβ chains are responsible for initiating intracellular signalling when the Ig binds antigen.

3.6 Summary

- Antibodies are a group of molecules with recognition and biological functions. Each antibody has two identical heavy chains and two identical light chains.
- The *n*-terminal parts of the heavy and light chains are called variable regions. These fold together to form the antigen-binding site, which binds to antigens on pathogens or other foreign objects. Antigens are most often proteins but can be carbohydrates or, less commonly, lipids. Antibodies bind to small parts of antigens called antigenic epitopes.
- Recognition of antigens by antibodies differs from recognition by the innate immune system in that antibody-binding is specific. This means that one antibody molecule will bind to some, but not all, antigenic epitopes. Different antibodies will bind to different epitopes, i.e. they will have different specificities.
- Because each antibody will bind to only a very small proportion of the total number of antigenic epitopes, lots of different antibody specificities are required to recognise all the different antigens.
- The *c*-terminal part of the antibody molecule is called the constant region. Different classes of antibody exist with different constant

regions. The different classes of antibodies are called IgM, IgG, IgA, IgE and IgD and they have different functions.

- Immunoglobulin can be secreted as a soluble protein or can be put into the cell membrane of B lymphocytes, where it can be part of a cell signalling complex.

3.7 Questions

1) What are the three major types of lymphocytes?

2) Draw and label a diagram of an antibody molecule.

3) Which of the following is not an antigen?

   A) Protein
   B) Carbohydrate
   C) Magnesium
   D) DNA
   E) Lipid

4) What is the difference between an antigen and an antigenic epitope?

5) Which of the following forces are not involved in the binding of antibody to antigen?

   A) Hydrogen
   B) Disulphide bridges
   C) Electrostatic
   D) Hydrophobic
   E) Van der Waals

6) What is the difference between affinity and avidity of antibody binding to antigen?

The answers to these questions can be found on page 334.

3.8 Further reading

1) Roux K. (1999) Immunoglobulin structure and function as revealed by electron microscopy. *International Archive of Allergy and Immunology*. 120:85–99.

2) **www.path.cam.ac.uk/~mrc7/mikeimages.html.** Mike Clarke Immunoglobulin Structure/Function Home Page.

CHAPTER FOUR

T lymphocytes and MHC-associated recognition of antigen

Learning objectives

To know the structure of the MHC and the roles of MHC molecules in antigen presentation to T cells. To know how T cells recognise antigen. To learn about antigen processing. To appreciate the unique polymorphism of class I and class II MHC molecules and the advantage of this.

Key topics

- T lymphocyte subsets
- Major histocompatibility complex (MHC)
  - Genetic organisation of the MHC
  - MHC gene products – class I and class II MHC proteins
  - Polymorphism of MHC genes
  - Expression of MHC proteins
- Recognition of antigen by T cells – structure of the T cell receptor for antigen
- Antigen processing and presentation by class I and class II MHC

4.1 There are different types of T lymphocytes

The previous chapter described one of the important recognition molecules of the specific immune system – the antibody molecule – which can be secreted in soluble form or be present on the surface of B lymphocytes. Through antibody on their cell surface, B lymphocytes can recognise antigen and this triggers a cellular response in the B lymphocyte. In this chapter we will describe a different type of lymphocyte that can also recognise antigen but in a different way from B lymphocytes. These cells are the **T lymphocytes.**

The T in T lymphocyte stands for **thymus**-derived. The thymus is a bi-lobed organ situated in the mediastinum above the heart (see Chapter 5). How the importance of the thymus in T cell development was discovered is described in Box 4.1. T lymphocyte precursors, like those of B lymphocytes, originate in the bone marrow. Whereas B cells complete their maturation in the bone marrow, T lymphocyte precursors migrate to the thymus where they develop into mature T lymphocytes (T cell development is described in Chapter 5). The mature T lymphocytes then leave the thymus and circulate through the bloodstream and lymphoid tissue (see Chapter 6).

In addition to helper T cells there is another major type of T cell, which is also produced in the thymus and has different functions from helper T cells. Both of these cells are called T cells or T lymphocytes. The two types of T cell are distinguishable phenotypically by the expression of mutually exclusive molecules on their cell surface. Helper T cells express a molecule called CD4 on their cell surface and are therefore called **CD4 T cells.** Cells of the other T lymphocyte subset express a different molecule called CD8 on their cell surface and are called **CD8 T cells.** T lymphocytes in the periphery express either CD4 or CD8 but not both.

BOX 4.1: DISCOVERY OF THE IMPORTANCE OF THE THYMUS

Prior to the work of Jacques Miller in the late 1950s and early 1960s, the role of the thymus was obscure because its removal from adult mice or humans had no apparent adverse effects. Miller was working for the Chester Beatty Cancer Research Institute investigating the role of the thymus in the development of leukaemia caused by the injection of a virus, called Gross leukaemia virus, into mice. (The virus was called Gross not because it was particularly disgusting but because it had recently been isolated by Ludwik Gross in the US.) The virus caused leukaemia when injected into neonatal (newborn) mice. However, if the mice were thymectomised (i.e. the thymus was removed) when they were weaned (at four weeks of age), the mice did not develop leukaemia. Therefore the thymus was essential for the development of leukaemia in this model.

Miller wished to test the effect of thymectomy *before* the neonatal mice were inoculated with virus. He therefore neonatally thymectomised many mice of different strains for these experiments. What he observed was that the neonatally thymectomised mice grew well until they were weaned when, whether they had been infected with virus or not, they suffered a wasting disease and died early.

Histological examination showed that the mice had reduced lymphocytes in the blood and lymphoid tissue, and liver lesions suggestive of viral infection. Further immunological studies revealed that the neonatally thymectomised mice could not reject skin grafts from other strains of mice, or even other species, showing that the mice were profoundly immunodeficient. Further evidence showed that if the mice were kept in germ-free conditions they did not suffer the wasting disease and survived for much longer, indicating that early death was due to infection. These studies, on the completely different subject of the development of leukaemia, were among the first to establish that the thymus did have an important role in immune function.

4.2 T cells recognise antigen through their T cell receptor (TCR)

Although there are similarities, T cells recognise antigen in a fundamentally different way from B cells. Both CD4 and CD8 T cells use a receptor called the T cell receptor (TCR) for recognising antigen. The TCR is related to, but is different from, antibody, and the genes coding for the TCR are on different chromosomes from those coding for antibody. The TCR consists of two glycoprotein chains called α and β (Figure 4.1). The α-chain has a molecular size of 40–50 kDa and the β-chain 35–47 kDa. Both chains show a typical Ig-like domain structure showing that TCR and Ig evolved from a common gene. Similar to Ig, each chain of the TCR has a constant

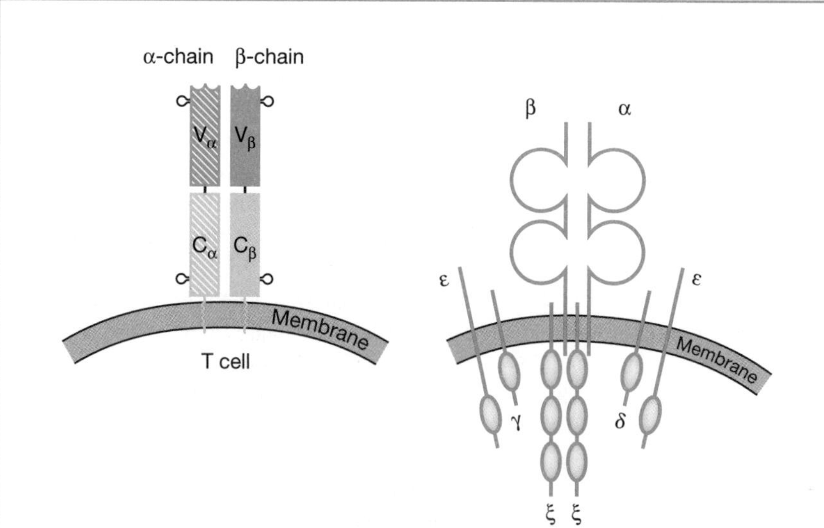

Figure 4.1 Structure of T cell receptor for antigen (TCR). The TCR consists of two glycoprotein chains, an α-chain of M_r 40–50 kDa and a β-chain of M_r 35–47 kDa. The Cα, Cβ, Vα and Vβ domains are structurally similar to the Ig domains. The α– and β–chains are associated with the CD3 complex of γ, δ, ε and ξ chains (left).

domain near the cell membrane and variable domain farthest from the cell membrane (Figure 4.1). The variable domains contain hyper-variable regions and, again like antibody, the TCR adopts a globular structure and the hyper-variable domains of the α- and β-chains fold to provide an antigen-binding site. The TCR is never secreted like antibody but exists only as a receptor on the surface of T cells.

Just as antibody on the B lymphocyte cell surface is associated with intracellular signalling molecules, the TCR is associated with a group of proteins with signalling function called CD3. There are five CD3 proteins called gamma (γ), delta (δ), epsilon (ε), zeta (ζ) and eta (η). The ε-chain can form heterodimers with γ or δ and the ζ-chain can form a homodimer with another ζ-chain or a heterodimer with a η-chain (see Figure 4.1). All CD3 complexes with the TCR contain γε and δε dimers, about 90% contain ζζ dimers and 10% contain ζη dimers. The CD3 complex is not involved in antigen recognition but initiates intracellular signalling pathways when TCRs recognise antigen.

T cells follow some of the same rules as B cells; different T cells will recognise different antigens because they have TCRs with different variable regions on their cell surface, but each T cell will express only one TCR specificity. Both CD4 and CD8 T cells use the same genes to produce their TCR. However, T cell recognition of antigen differs in one fundamental aspect from the way in which antibody recognises antigen. T cells do not recognise free antigen in the way antibody can. They recognise antigen that is associated with molecules on the surface of cells called **major histocompatibility complex (MHC) molecules**. Before describing how T cells recognise antigen, it is necessary to know about the major histocompatibility complex.

4.3 The major histocompatibility complex

The terminology surrounding the major histocompatibility complex tends to make the whole subject sound more complicated than it actually is. Therefore some of the terminology will be explained before describing the MHC in more detail.

4.3.1 Terminology of the MHC

Major histocompatibility complex (MHC)

The term major histocompatibility complex actually refers to a region of DNA spanning some 4 Mbp (base pair) and containing over 200 genes. It is located on chromosome 6 in the human and chromosome 17 in the mouse. The organisation of the genes in humans is shown in Figure 4.2. All vertebrate species have an MHC, and therefore the term MHC can be used in reference to any species. Each species also has a unique name for its MHC; in humans the MHC is also called HLA (**h**uman **l**eukocyte **a**ntigen) and in mouse it is called H-2. Therefore, in the human, the terms MHC and HLA mean the same thing. The origin of the terminology surrounding the MHC is described in Box 4.2.

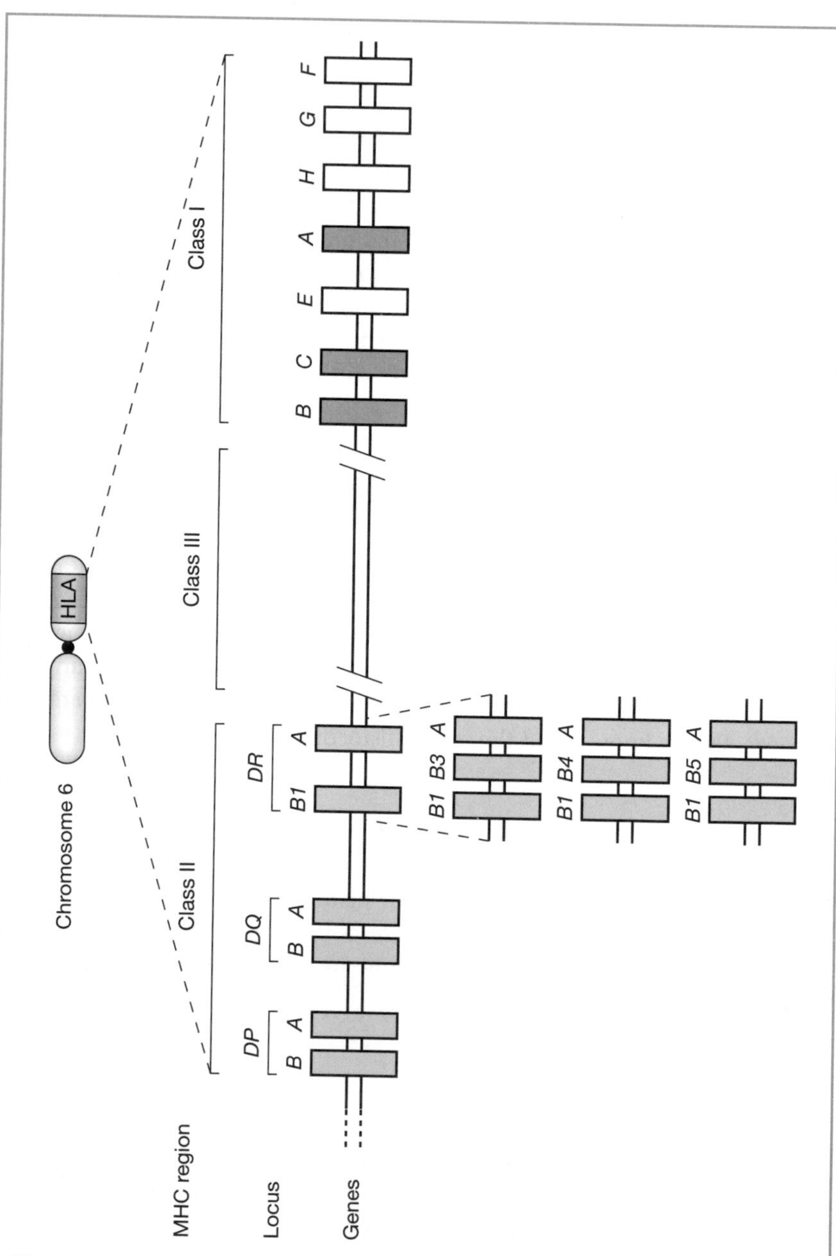

Figure 4.2 Organisation of HLA genes. The class III region contains numerous diverse genes. Each of the *HLA-DRB* genes is polymorphic. The organisation of the *DRB* genes is such that, depending on the form of gene inherited, an individual possesses an *HLA-DRB1* allele alone or a *HLA-DRB1* allele plus one of *DRB3*, *DRB4* or *DRB5*. Each of the *DRB1*, *DRB3*, *DRB4* and *DRB5* loci are polymorphic.

BOX 4.2: TERMINOLOGY OF MHC IN DIFFERENT SPECIES

The major histocompatibility complex was first discovered in mice, where it was called H-2. The H stood for histocompatibility. The word histocompatibility is derived from the Greek words *'Histos'* meaning tissue and *'Compatibilitos'* meaning compatible. Early in the twentieth century tumour biologists were transplanting tumours between mice. They observed that the tumour grafts suffered one of three fates: they grew and killed the mouse, they were rejected slowly (>50 days) or they were rejected rapidly (<20 days). When inbred strains of mice were used it was found that the three fates of the tumour grafts were observed between the different strains. Two strains of mice that could exchange tumour grafts without destruction of the graft were said to be histocompatible, i.e. their tissues could co-exist with each other. By contrast two strains of mice that rejected each other's tumours were said to be histo-incompatible, i.e. their tissues could not co-exist with each other. If the rejection was rapid the strains were said to show major incompatibility and if the rejection was slow the strains were said to show minor incompatibility. Using classical genetic breeding studies it became clear that rapid rejection segregated as a single locus and this was originally called the major histocompatibility locus. Later it was realised that the locus contained many genes and the word 'locus' was replaced with 'complex' – hence the term major histocompatibility complex and the abbreviation MHC. Because it was the second histocompatibility locus to be identified in the mouse it was called H-2 (standing for histocompatibility-2).

It was natural to look for a similar system in humans, especially with the possibility of human transplantation. Indeed an MHC was discovered in humans and called HLA, standing for **h**uman **l**eukocyte **a**ntigen because the MHC **proteins** in humans were first identified as transplantation antigens on white blood cells.

Therefore, although there was a logic to the nomenclature in both mouse and human, not everyone spoke the same language. When the rat MHC was called RT1 things threatened to get out of hand. Fortunately there is now consensus about the nomenclature of the MHC in other species and the human-style terminology has been adopted. Thus dog MHC is called DLA, pig MHC is called SLA (swine LA), chimpanzee MHC is called ChLA, etc. However, mouse MHC is still called H-2 and rat RT1.

MHC gene products

The proteins coded for by the genes of the MHC are called MHC proteins, MHC molecules or MHC products; these terms mean the same thing and are used interchangeably. Because they were initially identified as antigens involved in transplant rejection, MHC proteins are also often referred to as MHC antigens.

MHC locus

Initially it was thought that the MHC was one gene, or locus, but it quite soon became apparent that there were many genes. The term 'MHC locus' is now used to describe the region of DNA coding for an individual MHC product; each locus may contain one or more genes.

There are different classes of MHC genes or products

The MHC contains over 100 genes. However it is clear that some of the genes are closely related to each other and their products perform a similar function. On this basis the genes of the MHC, and the proteins coded by them, have been divided into three classes called class I, class II and class III (Figure 4.2). The class I and class II MHC molecules are related to each other but fall into two distinct families of proteins. They are nearly all cell surface proteins and are involved in T cell recognition of antigen (see below). Class III MHC molecules are a diverse collection of proteins that have many immune-related functions but are not related to class I or class II MHC or necessarily to each other. Table 4.1 shows some of the major types of MHC class III products.

Table 4.1 MHC class III gene products

Gene product	*Function*	
Steroid 21 hydroxylase	Enzyme	
C2	Complement component	see Chapter 8
C4	Complement component	see Chapter 8
B	Complement component	see Chapter 8
Hsp 70 (heat shock protein)	Intracellular trafficking of proteins	
TNFα	Inflammatory cytokine	
TNFβ	Inflammatory cytokine	

4.3.2 Class I MHC – a ubiquitously expressed set of cell surface proteins

Structure of class I MHC molecules

Class I MHC molecules are glycoproteins that are expressed on the cell surface of most nucleated cells (Figure 4.3). They consist of two protein chains. The longer chain, called the α-chain, is coded for by the MHC class I genes and has a molecular size of 45 kDa. The α-chain is non-covalently linked to a 12 kDa chain called β2-microglobulin (also referred to as β_2-m), which is coded for by a gene on a different chromosome (Chr 15 in humans) from that for the MHC.

The α-chain has three external domains, called α1, α2 and α3, that are homologous to immunoglobulin domains. The α1 domain is located at the *N*-terminal end of the molecule while the α3 domain is situated closest to the cell membrane. Like most cell surface molecules the α-chain has a short hydrophobic region spanning the membrane and a short cytosolic region. There are no known signalling functions associated with class I MHC. β2-microglobulin has one Ig-like domain and is located exclusively on the outside of the membrane; it has no transmembrane or intracellular regions. Diagrammatically class I MHC molecules can be depicted as two sets of domains (the α1 and α2) sitting on top of the other two (α3 domain and β2-microglobulin) (Figure 4.3).

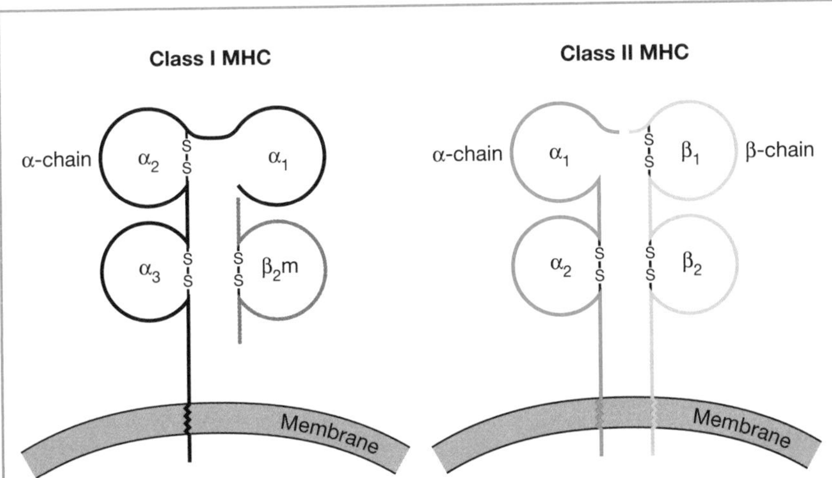

Figure 4.3 Diagram of MHC class I and class II molecules. Class I MHC consists of a 45 kDa α-chain non-covalently attached to β2-microglobulin. The α-chain of class I MHC has three domains called α1, α2 and α3. Class II MHC consists of a 30–34 kDa α-chain non-covalently linked to a 26–30 kDa β-chain. Both the α- and β-chains of class II contain domains (α1 and α2 and β1 and β2 respectively). The domains of the class I α-chain, the class II α- and β-chains and β2-microglobulin are all structurally related to the domains of the Ig and TCR molecules. Note that the α1 and α2 domains of the class I MHC α-chain are not the same as the α1 and α2 domains of the class II MHC α-chain.

The detailed molecular structure of class I MHC was determined by X-ray crystallography in 1987 and shows that the α1 and α2 domains of the α-chain do indeed sit on top of the rest of the molecule (Plate 7). This structure has been likened to sausages in a pan. The α1 and α2 domains each contribute four strands of an eight-stranded β-pleated sheet upon which sit two α-helices formed by other regions of the α1 and α2 domains. In between the two helices is a groove that is typical of MHC structure.

There are three class I loci in the human, called *HLA-A*, *HLA-B* and *HLA-C*. Each locus has one gene coding for an α-chain, which pairs with β2-microglobulin to give a complete class I protein.

There is a second family of class I MHC genes

When most people refer to the class I MHC they are talking about the three genes or proteins called *HLA-A*, *HLA-B* and *HLA-C*. However, there are other class I-like genes and products in the MHC. Some of these other MHC molecules and their properties are shown in Box 4.3. For the purpose of this book, class I MHC refers to *HLA-A*, *-B* or *-C* or the equivalent in other species.

4.3.3 Class II MHC genes are less widely expressed than class I

Class II MHC molecules also consist of two glycoproteins, an α-chain of molecular size 30–35 kDa and a β-chain of 26–30 kDa. However, unlike class I MHC, they are much more restricted in their expression, being found primarily on cells of the immune system such as monocytes/macrophages and B cells. Both the α- and β-chains are coded for by genes in the MHC. Unlike class I MHC, both the α- and β-chains of the class II MHC molecule have transmembrane regions and cytoplasmic tails (Figure 4.3). Class II MHC molecules are also capable of delivering intracellular signals, another property distinguishing them from class I MHC molecules.

BOX 4.3: CLASS IB MHC

Class IB MHC genes and function

When the MHC was examined at the genetic level it was found that there are about 30 genes coding for class I MHC-like molecules, although many of them are pseudo-genes and are not expressed. The main class I-like genes in the human are called *HLA-E, HLA-F* and *HLA-G*. Some of the others, such as *HLA-H* and *HLA-J*, are pseudogenes and are not expressed. Because they were discovered after *HLA-A, -B* and *-C* they have been referred to as class IB or non-classical class I MHC products. *HLA-A, -B* and *-C* are consequently sometimes referred to as class IA or classical class I MHC molecules.

The cellular expression of the class IB MHC proteins is generally more restricted than HLA-A,-B or-C. Although HLA-E is expressed on many cell types, HLA-F is only expressed in the bladder, liver, placenta and on lymphoblasts (dividing lymphocytes) and HLA-G is expressed in the placenta, thymus and on dendritic cells. Like HLA-A,-B, and-C, the HLA-E,-F and-G proteins can bind peptide fragments of 8–10 amino acids and HLA-E and-G can present peptides to TCRs which may play a role in protection against infection. HLA-E,-F and-G can also bind to different receptors on natural killer cells and another suggested function of these class 1B MHC molecules is protection of the body's own cells from killing by natural killer cells. The placental expression of HLA-G appears to play an important role in pregnancy.

Both the α- and β-chains of class II MHC have two external domains. The domains furthest away from the cell membrane are the α_1 and β_1 domains of the α- and β-chains respectively and those nearest the cell membrane are called α_2 and β_2. The chains of the class II MHC molecule fold to form a structure that is remarkably similar to class I MHC (Plate 7). The α_1 and β_1 domains form the β-pleated sheet with the two *a*-helices on top forming the typical 'sausages in the pan' structure seen with class I MHC.

There are three class II MHC loci in the human MHC, called *DP*, *DQ* and *DR* (Figure 4.2). The *DP* and *DQ* loci each contain a functional α-gene, called *DPA* and *DQA* respectively, which codes for the α-chain, and a functional β-gene, called *DPB* and *DQB*, which codes for the β-chain. The *DR* locus is more complicated; it contains one functional α-gene, called *DRA*, but different versions (or alleles) of the locus contain one or two functional *DRB* genes, each coding for a separate β-chain (see Figure 4.2).

4.3.4 Class I and class II MHC genes are unique in the extent of their polymorphism

One of the extraordinary features of class I and class II MHC genes is the degree of polymorphism shown at the different loci. A particular gene can exist in slightly different forms in different individuals; these different forms are called alleles. Alleles arise when the gene for a protein mutates to give a different DNA sequence, which may result in a different amino acid sequence in the protein. Many of these mutations are lost by chance or because the mutated protein does not function as well as the original. However, in some cases the different forms of the gene become established in the population as alleles. Where a gene expresses different alleles it is said to be polymorphic.

Many genes are not polymorphic and only one version of the protein exists; these genes are said to be monomorphic. Where genes are polymorphic they usually have only a few alleles (fewer than five) and the different forms of the protein differ by one or a few amino acids. Class I and II MHC genes are unique both in the number of alleles at the different loci and also by the degree to which the amino acid sequence of the alleles can vary. Some MHC loci in the human have over 1000 alleles and the amino acid sequence between alleles can vary by more than 20 amino acids (see Table 4.2). Therefore the variety of different MHC molecules that can be expressed in a population vastly exceeds the variety of any other protein. The possible reason for this extraordinary polymorphism will be explained when the function of MHC molecules has been described.

4.3.5 Nomenclature of MHC genes, proteins and alleles

The extraordinary polymorphism of the class I and class II MHC loci led to a predicted confusion in terminology. There is now a standard nomenclature for the alleles of the different MHC genes, which ensures consistency but is not that easy to understand unless a little is known about its development.

Table 4.2 Number of alleles of class I and class II MHC genes

Gene	*Number of alleles*
HLA-A	965
HLA-B	1540
HLA-C	626
HLA-DPA	27
HLA-DPB	138
HLA-DQA	35
HLA-DQB	107
HLA-DRA	3
HLA-DRB	847[a]

These are the number of DNA sequences at each locus and do not necessarily represent different protein structures.

[a]Total number or alleles for all DRB loci.

Initially the different alleles of the MHC were recognised using antibodies. Individuals who are exposed to foreign MHC molecules can make antibodies against them just as they would against any foreign antigen. Therefore women who have had babies (especially more than one), people who have had blood transfusions and some people who have had transplants have all been exposed to foreign MHC molecules and their sera may contain antibodies that are specific for the foreign MHC molecule. In the case of pregnancy mothers are exposed to the MHC molecules that the baby has inherited from the father and make antibodies against them. Using these sera it is possible to identify specific MHC molecules on the surface of leukocytes so that the particular alleles of MHC an individual possesses can be established. This process is known as **tissue typing** (Figure 4.4).

These sera showed that there were many alleles at *HLA-A*, *-B*, *-C* and *-DR* (they were not so useful for *DP* and *DQ*) and many different forms of the proteins. The different alleles of a particular MHC molecule that were identified by antibodies were given different numbers. Therefore different alleles of *HLA-A* were called *HLA-A1*, *HLA-A2*, *HLA-A3*, etc. Similarly alleles of *HLA-B* were called *HLA-B1*, *HLA-B2*, etc. and the same was adopted for *HLA-C* and *DR*. These were the **serologically determined** alleles.

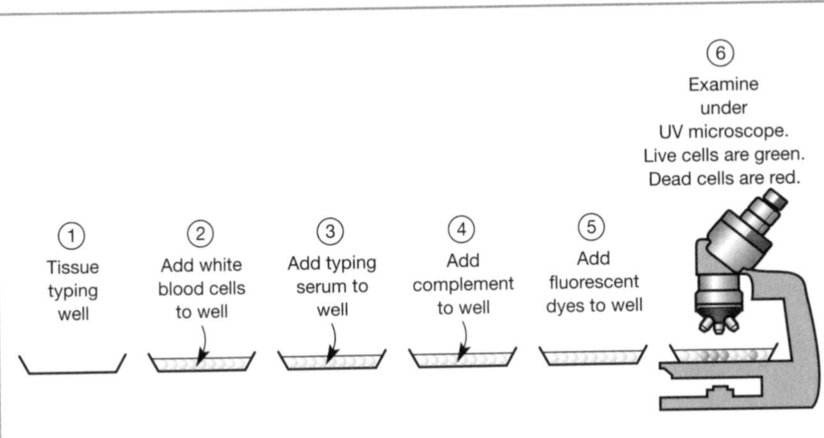

Serum containing antibodies to HLA	*HLA-A* of cells		
	Person ① A1 A1	Person ② A1 A2	Person ③ A2 A2
Serum (a) Anti-A1 + A2	✓	✓	✓
Serum (b) Anti-A2 + A3	✗	✓	✓
Serum (c) Anti-A1 + A3	✓	✓	✗
Reactivity pattern of cells with serum a, b, c	✓ ✗ ✓	✓ ✓ ✓	✓ ✓ ✗

Figure 4.4 HLA-typing. The table shows a simplified reaction pattern of three typing sera with cells of three different *HLA-A* types. A tick indicates that the serum recognises the HLA type of the cells and kills them. A cross indicates that the serum does not recognise the HLA on the cells and does not kill them.

With the advent of DNA sequencing and other molecular genetic techniques for examining HLA polymorphism it became apparent that there were many more HLA alleles than could be identified by antibodies. For example *HLA-A2* was one serologically determined allele of *HLA-A*. However, when people who were serologically *HLA-A2* were typed genetically (including sequencing the DNA of the *HLA-A2* gene in different individuals) it was found that there were 13 different genetic variations of the serologically determined *HLA-A2*. Because these genetic alleles are similar to each other, the antibodies recognising *HLA-A2* could not distinguish between them and serologically they all appeared the same. The nomenclature of HLA was therefore amended to take into account both the serological and genetic bases of the polymorphism.

As described in Section 4.3.2, the genes for class I MHC are called *HLA-A*, *HLA-B* and *HLA-C*. The genes for class II MHC α-chains are *DPA*, *DQA* and *DRA*. The genes for the class II MHC β-chains are *DPB*, *DPQ* and *DRB1*, *DRB3*, *DRB4* and *DRB5*; there are four different genes coding for the DR β-chain and an individual will have either *DRB1* alone or *DRB1* and *DRB3*, *DRB4* or *DRB5*. The alleles for each gene are given a four-number designation. The first two numbers refer to the serologically determined allele and the last two numbers refer to the genetically determined allele. Continuing with *HLA-A2* as an example, the serological allele is now called *HLA*A02*. The 13 different genetic alleles have been assigned the numbers 01 to 13. The full description of the alleles is now *HLA-A*0201*, *HLA-A*0202*, etc. to *HLA-A*0213*. The same nomenclature was adopted for the other MHC genes and Table 4.2 shows the number of different alleles of the class I and class II MHC genes.

4.3.6 How many different class I and class II MHC proteins does an individual express?

Like any other gene, individuals inherit one copy of an MHC gene from their mother and one from their father. Given the extreme polymorphism of the MHC genes it is likely that an individual will inherit two different alleles of a gene. Such an individual is said to be heterozygous for that gene. Individuals who inherit two copies of the same allele are homozygous for that gene. MHC genes are expressed co-dominantly; that is, if you inherit two different alleles of a gene you will make both forms of the protein. Therefore individuals can express many different MHC molecules but may vary in exactly how many. It is possible to calculate the number of different MHC proteins any one person may express.

Class I MHC

The situation with class I is relatively simple. An individual can express one (if they are homozygous) or two proteins of each of HLA-A, -B and -C. Therefore an individual will express between three and six class I MHC proteins.

Class II MHC

This is more complicated because both the α- and β-chains are polymorphic. When the two class II MHC chains associate to form class II molecules, the α-chain from one locus can associate only with a β-chain from the same locus. Therefore DPα can associate with DPβ but not with DQβ and DQα can associate only with DQβ. The DRα can associate with any of the DRβ chains expressed by an individual.

However, each α-chain for a locus can associate with more than one β-chain to generate more than one class II MHC product for each locus. To take DQ as an example, you will inherit a maternal *DQA* gene and a paternal *DQA* gene as well as a maternal and a paternal *DQB* gene. The α-chain coded by the maternal *DQA* gene can associate with the β-chain coded for by the maternal *DQB* gene or the paternal *DQB* gene (Figure 4.5). Similarly the paternal α-chain can associate with the paternal or the maternal β-chain. If you are heterozygous for both the α and β genes, you can make four different DQ molecules. If you are homozygous at both α and β, you can make only one DQ molecule. The situation is exactly the same with DP: you can make up to four different DP molecules if you are heterozygous at *DPA* and *DPB*.

The situation with DR is simpler in some ways but more complicated in others. The *HLA-DRA* gene is not functionally polymorphic and therefore there is only one form of the DRα-chain. However, an individual may make between one and four different DRβ-chains depending on which alleles they inherit (see Figure 4.2) and whether they are heterozygous. Therefore it is possible to express between one and four HLA-DR molecules.

Overall then the fewest different class II MHC molecules you could make would be three (one each of DP, DQ and DR) and the most would be 12 (4DP + 4DQ + 4DR). Most individuals express a number between these two extremes.

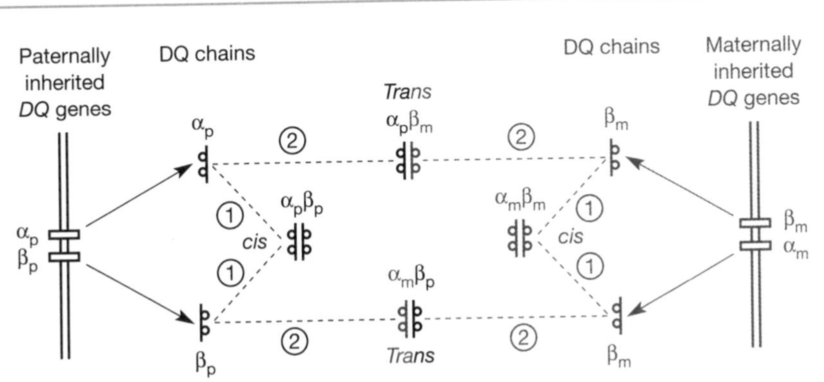

Figure 4.5 *Cis*- and *trans*-association of class II MHC α- and β-chains. The example uses *HLA-DQ* and assumes the individual is heterozygous for DQα and DQβ. An individual inherits one set of *DQ* genes from their father and one set from their mother. Paternally inherited genes and their products are shown in dark blue and maternally inherited in light blue. Because MHC genes are expressed codominantly, two different DQα chains, α_p and α_m, and two different DQβ chains, β_p and β_m, are made. Each α-chain can associate with either β-chain. The association of α_p with β_p and the association of α_m with β_m are called *cis*-association because the genes coding the α-chain and the β-chain are on the same chromosome. By contrast, association of α_p with β_m or α_m with β_p is called *trans*-association because the genes coding the α- and β-chains are on different chromosomes.

4.4 Recognition of antigen by T cells

The similarity in structure between class I MHC and class II MHC molecules and the unique level of polymorphism in the class I and class II MHC genes suggest that class I and class II MHC perform a similar function. In fact they do, and this function is to help T cells recognise antigen. Both class I and class II MHC molecules have a groove on top of the molecule (Figure 4.3 and Plate 7). Small peptides derived from larger protein antigens bind to the groove of the MHC molecules so that MHC molecules on the surface of a cell display these antigenic peptides (Figures 4.3 and 4.6 and Plate 7). The complexes of peptide/MHC are sticking out from the cell surface so that they are easily accessible to the TCR on a T cell. The TCR can bind to the complex of the antigenic peptide and the MHC molecule; some regions of the TCR antigen-binding site bind to parts of the antigenic peptide and some bind to parts of the α-helices of the MHC molecule (Figure 4.6 and Plate 8). There is one very important difference in the way CD4 and CD8 T cells recognise antigen. **CD4** T cells recognise antigen presented by **class II** MHC molecules and **CD8** T cells recognise antigen presented by **class I** MHC molecules.

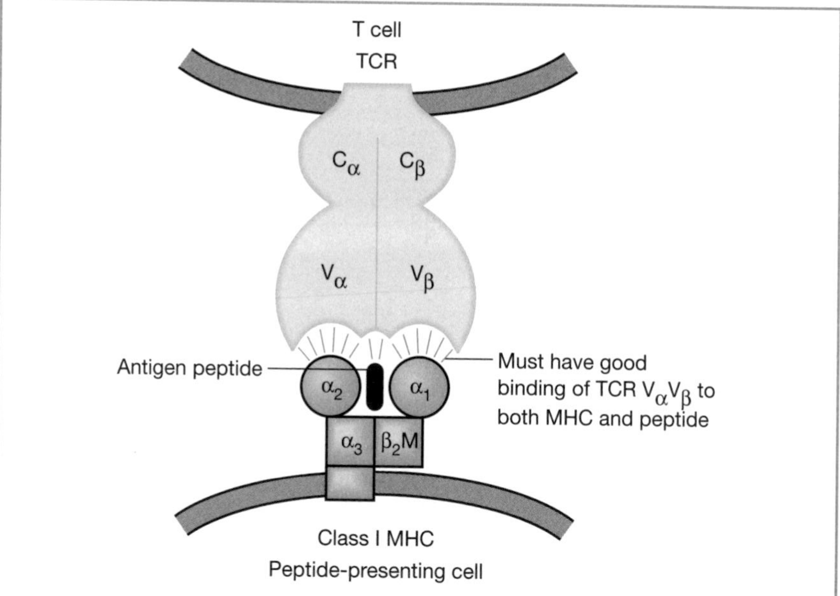

Figure 4.6 Interaction of the TCR with antigen/class I MHC. The antigen-binding site of the TCR is made up of parts of the Vα and Vβ chains. Parts of the binding site interact with the antigenic peptide bound in the groove of the MHC molecule and parts of the binding site interact with the MHC molecule.

The binding of the TCR to the antigen peptide/MHC complex is specific for both antigen and MHC. The variable regions of the α- and β-chains of the TCR fold to form the antigen/MHC binding site. Both CD4 and CD8 T cells use the same TCR genes to generate their TCRs but, just like antibody, each T cell will have different amino acid sequences in the hyper-variable regions of the α- and β-chains of the TCR and therefore will have a different shape in the antigen/MHC binding site of the TCR. The binding of the TCR to the antigen/MHC complex involves the same types of non-covalent bonds as antibody binding to antigen (see Section 3.3.2). Only if the

TCR fits closely to ***both*** the antigenic peptide and the MHC molecule will binding occur. Therefore different T cells will have different specificities for antigen/MHC.

An individual T cell will recognise antigen only in association with one MHC molecule. Therefore the same antigen being presented by different MHC molecules will be recognised by different T cells (Figure 4.7). For instance a CD8 T cell that can recognise an antigenic peptide presented by an *HLA-A* molecule will not be able to recognise the same peptide

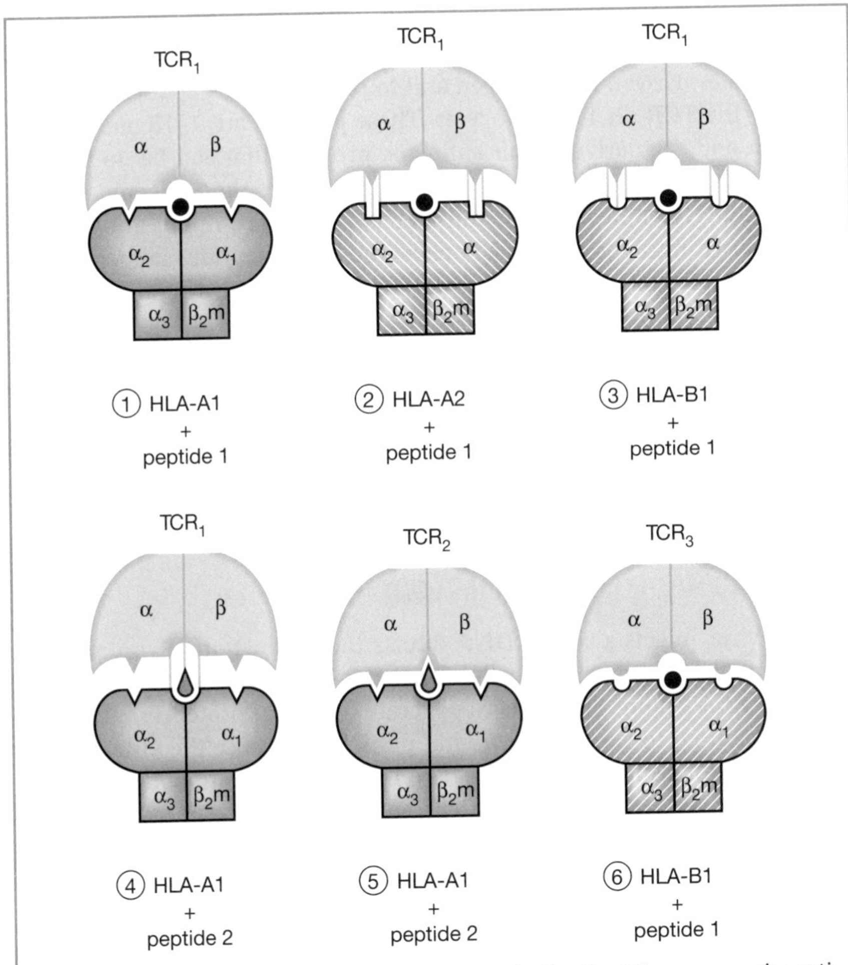

Figure 4.7 MHC-restricted recognition of antigen by T cells. TCR_1 can recognise antigenic peptide 1 being presented by HLA-A1 ①. TCR_1 cannot recognise antigenic peptide 1 presented by HLA-A2 ② or HLA-B1 ③; nor can it recognise antigenic peptide 2 being presented by HLA-A1 ④. TCR_1 is therefore said to be specific for antigenic peptide 1 restricted to HLA-A1. TCR_2 can recognise a different antigenic peptide (peptide 2) being presented by the same HLA-A1 ⑤ as TCR_1 and is therefore specific for antigenic peptide 2 restricted to HLA-A1. Finally TCR_3 can recognise peptide 1 in association with HLA-B1. Antigenic peptides are shown in black or grey.

presented by an *HLA-B* molecule. In fact it will not even be able to recognise a peptide presented by a different allele of an HLA-A, e.g. one TCR will recognise antigen presented by *HLA-A1* but not the same antigen presented by *HLA-A2*. This is what is meant by **MHC-restricted recognition of antigen** that is seen with T cells but not B cells.

4.5 Antigens must be processed before they can be presented by MHC molecules

We have seen how the role of class I and class II MHC molecules is to bind peptides derived from antigen and to present these peptides for recognition by the TCR on T lymphocytes. These peptides are 7–18 amino acids in length and are derived from antigenic proteins that can be up to 1000 amino acids in size. The question arises: how does the antigenic peptide get to the MHC molecule? The answer is that large protein antigens are broken down into peptides inside cells and these peptides associate with MHC molecules intracellularly. The MHC molecules bearing the antigenic peptides are then transported and expressed on the cell surface. However, the way in which this occurs is quite different for class I and class II MHC molecules and is also influenced by where the original antigen is located. Endogenous antigens are produced within the cell (e.g. viral proteins) and are processed and presented by class I MHC. Exogenous antigens derive from outside the cell (e.g. from an extracellularly living bacterium or parasitic worm) and are processed and presented by class II MHC.

4.5.1 Class I MHC is involved in the presentation of intracellular antigen

When a virus infects a cell its DNA directs the production of viral proteins within the cytoplasm of the cell (Figure 4.8). The object of this exercise from the virus's point of view is to generate new viral proteins that eventually will be used to generate new viral particles. Because the viral proteins are in the cytoplasm of the cell they are inaccessible to antibody. However, peptides derived from these viral proteins can be presented by class I MHC. This occurs in the following way.

Proteasomes degrade cytoplasmic antigens

Within the cytoplasm of cells are structures called proteasomes (Figure 4.8). Proteasomes are made up of 12–15 protein subunits, which assemble to form a tube-like structure. Two of the major proteins are called βi1 and βi5. As their name implies, proteasomes have proteolytic activity and are able to degrade cytosolic proteins into peptide fragments of 8 or more amino acids. The optimal size of peptide fragments for binding to class I MHC is 8–10 amino acids.

Transporters carry antigenic peptides into the rough endoplasmic reticulum

The next problem is how to get the peptides generated by proteasomes to bind to class I MHC molecules. Like any other cell surface protein, class I MHC molecules are synthesised in the rough endoplasmic reticulum (RER), modified in the Golgi apparatus and transported to the cell surface in transport vesicles. The peptides and MHC molecules are in separate compartments of the cell. The situation is resolved by the presence of special peptide transporter molecules that transport the peptide fragments from the cytoplasm to the RER. These transporters are made up of two proteins called TAP-1 and TAP-2 (Figure 4.8). TAP stands for **t**ransporter associated with **a**ntigen **p**rocessing. The transporter composed of TAP-1 and TAP-2 is a member of a family of transporter molecules called ABC transporters. These have ATP-binding cassettes (hence the term ABC) and use the energy from the hydrolysis of ATP to move ions or small proteins across membranes. The TAP transporter is most efficient at transporting peptides of 9–16 amino acids in length but can transport peptides of other sizes.

Antigenic peptides are required for the correct assembly of class I MHC molecules

Once the antigen-derived peptides have been transported to the RER by the TAP transporters they need to bind in the groove of the class I MHC molecules. This process is more complex than simple binding of the peptide

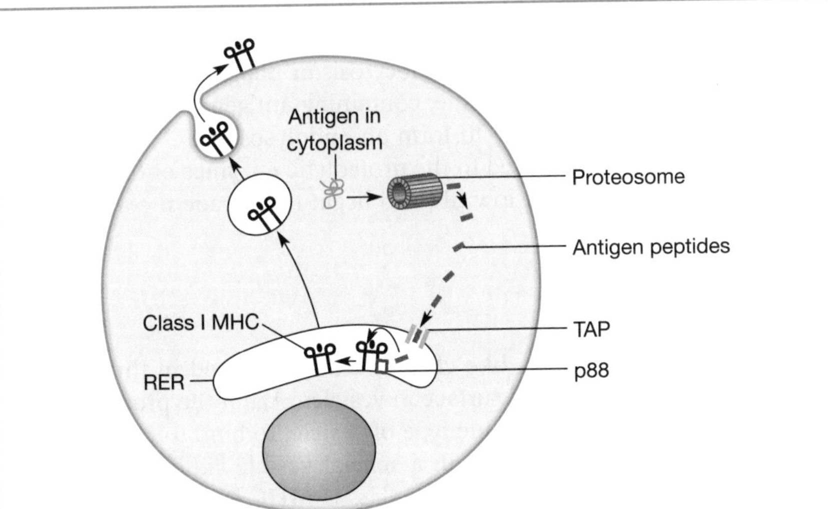

Figure 4.8 The class I MHC pathway of antigen processing. Endogenous antigen (produced within the cell) is degraded in proteasomes and antigenic peptides are transported by TAP to the RER where they associate with the class I molecule being assembled. The class I molecule bearing the antigenic peptide is then transported to the cell surface and presents the antigenic peptide. A protein, p88, is initially associated with class I in the RER but dissociates when antigen peptide is bound.

to the groove of a completed class I MHC molecule. The proper association of β2-microglobulin with the class I MHC α-chain to form a stable class I MHC molecule requires the presence of an antigenic peptide. Thus the whole complex of class I MHC α-chain, β2-microglobulin and antigenic peptide is assembled. The class I MHC with the antigenic peptide in its groove then travels to the Golgi and subsequently is transported to the cell surface in a transport vehicle and is expressed on the cell surface where it can present the antigenic peptide to CD8 T cells. In the absence of antigenic peptide the class I MHC molecules formed in the RER are unstable and do not travel to the cell surface. This process ensures that only class I MHC molecules bearing antigenic peptide, and hence able to present the peptide to T cells, are expressed on the cell surface and empty class I molecules, which serve no useful purpose, do not clutter up the cell surface.

4.5.2 Class II MHC is involved in the presentation of extra cellular antigen

Class II MHC molecules present antigenic peptides that are derived from antigens that are outside the cell. The fundamental problems are the same as with class I MHC – how do you generate antigenic fragments from larger proteins and get them to bind to class II MHC? The pathway for this is quite different from that for class I MHC.

Extracellular proteins are endocytosed and degraded in lysosomes

Extracellular proteins, which could be antigens derived from an extracellular dwelling pathogen, are taken up by cells by endocytosis. This could be pinocytosis, receptor-mediated endocytosis or phagocytosis (Figure 4.9). The endocytic vesicle, or endosome, containing antigen proteins lowers its pH and fuses with a lysosome to form an endolysosome. The contents of the endosome are now exposed to the proteolytic enzymes of the lysosome. These enzymes are active at low pH and begin to degrade the antigen into smaller fragments.

Antigenic peptides associate with class II MHC in specialised compartments

Class II MHC molecules, like class I, are synthesised in the RER and travel via the Golgi to the cell surface in vesicles. Again the problem is how to get the peptide, which is in one type of vesicle, to bind to class II MHC, which is in another. This occurs in a special vesicle called the compartment for peptide loading (CPL). When class II MHC is initially synthesised in the RER, the α- and β-chains associate with a third protein chain called the **invariant** chain (Figure 4.9). The invariant chain is thought to have two important functions. It binds to the groove of the class II MHC molecule and therefore stops self-peptides which are being made in the RER from binding to the MHC. The invariant chain is also thought to direct the class II MHC/invariant chain complex from the Golgi to the CPL. Because

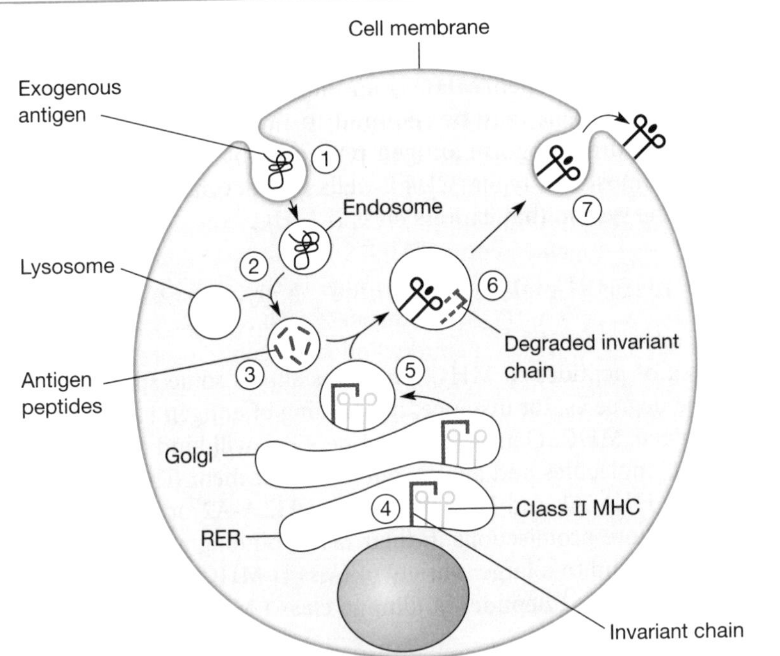

Figure 4.9 The class II MHC pathway of antigen processing. Exogenous antigen (produced outside the cell) is taken up by endocytosis ①. The endosome fuses with a lysosome ② and the antigen is degraded ③. Class II MHC is initially synthesised in the RER with an associated protein, the invariant chain ④. The endocytic vesicle delivers the antigenic peptides to a compartment for peptide loading ⑤ where the antigenic fragment displaces invariant chain from class II MHC and becomes bound to the MHC molecules ⑥. The MHC molecules are then transported to the cell surface ⑦ where they present antigenic peptide.

class I molecules are not associated with invariant chain they do not go to the CPL. In this way class I and class II MHC molecules, which are both present in the same Golgi network, are directed to separate vesicles – class I to a standard transport vesicle and class II to the CPL.

The CPL now fuses with the endolysosome containing the degraded antigenic peptides and the invariant chain is degraded by the enzymes that were in the endolysosme. This allows antigenic peptides of the right size (5–13 amino acids) to displace the invariant chain and become bound in the groove of the MHC molecule. The class II MHC molecule bearing the antigenic peptide is then transported to the cell surface where it is expressed and can present the antigenic peptide to CD4 T cells.

4.5.3 Why is it important to have two antigen-processing pathways?

The two pathways of antigen processing ensure that exogenous antigen is targeted to class II MHC for recognition by CD4 T cells and endogenous antigen is targeted to class I MHC for recognition by CD8 T cells. This is very important in relation to the different biological function of CD4 and CD8 T cells (see Chapters 6 and 9). Both cell types interact with other cells and recognition of antigen/MHC is an important part of this interaction. However, CD8 T cells may be required to interact with almost any cell type and therefore recognise antigen presented by class I MHC, which is expressed on most cell types. CD4 T cells interact mostly with other cells of the immune system that express class II MHC.

4.5.4 Different MHC molecules can bind many, but not all, different peptides

The binding of peptides to MHC molecules shows some specificity but not to the same degree as, for instance, the binding of antigen to antibody or the TCR to antigen/MHC. Generally class I peptides will bind to many different class I MHC molecules and alleles, but not all of them. Therefore a peptide may bind to HLA-A1 and HLA-B1 but not HLA-A2 or HLA-B2. Class II peptides are more promiscuous in their MHC binding and generally a class II peptide will bind to a larger variety of class II MHC molecules and alleles than is seen for class I peptides binding to class I MHC.

4.5.5 Why is there so much polymorphism in the MHC?

Two questions raised by the arrangement of the MHC complex are: why is the MHC so polymorphic and why do you have three different class I loci doing the same thing and three class II loci doing the same thing? The answers to these questions are related and can best be explained if you imagine a situation where you had only one class I MHC locus and one class II MHC locus and there was no polymorphism. One survival strategy for pathogens is to evolve antigens that produce peptides that cannot bind to MHC molecules in the host; obviously in this situation there is no possibility of T cells recognising the antigen. With only one MHC locus and no polymorphism a pathogen would need only to evolve antigens that did not bind to this one MHC molecule and the whole population would be at risk from infection and possible elimination (Figure 4.10). If an individual has three loci for each type of MHC molecule and therefore expresses up to six different MHC molecules, it is much more difficult for a pathogen to evolve antigens that do not bind to any of the six different MHC molecules. This makes it more likely that an individual will continue to respond against the pathogen. If you now add extreme polymorphism it means that the population as a whole expresses hundreds of different variants of MHC molecules. This makes it almost impossible for a pathogen to evolve antigens that would not bind to at least some of the MHC variants. Therefore the population as a whole is protected, even though some individuals may not respond to pathogen antigens. Evidence is beginning to emerge that MHC polymorphism can be driven by selection of resistance to endemic infectious agents. In parts of Africa where malaria is common, the frequency of certain HLA alleles is higher than in other populations where malaria does not occur. These alleles have been shown to be associated with increased resistance to the parasite.

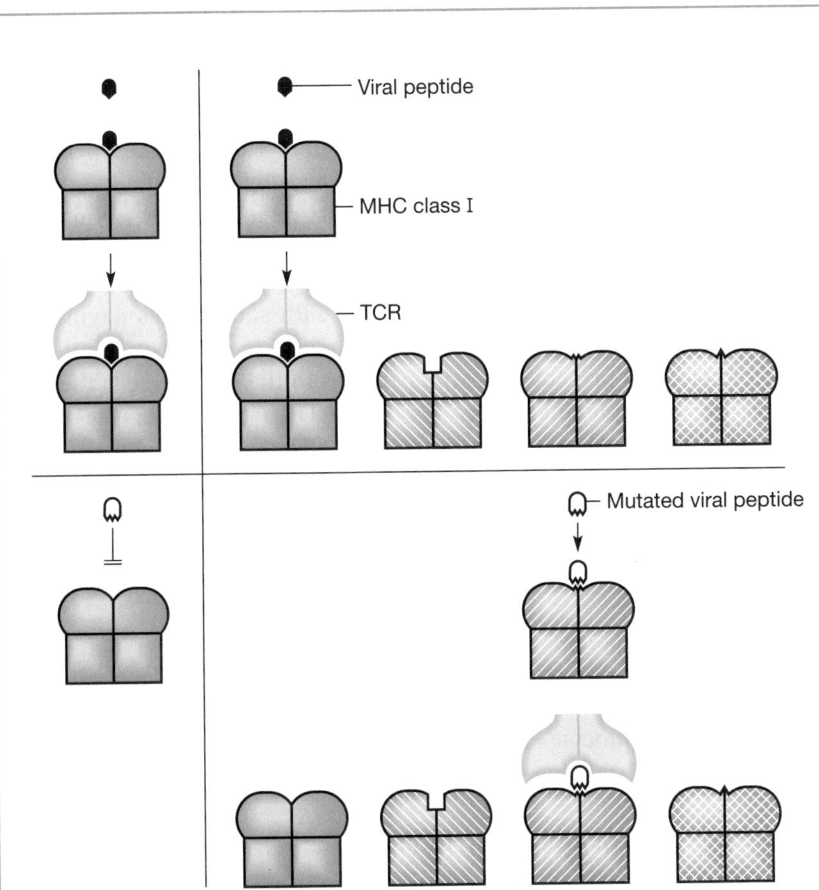

Figure 4.10 The advantage of multiple MHC loci and MHC polymorphism. The left-hand panel illustrates the situation if we possessed only one class I HLA locus that was not polymorphic. Although a viral peptide may bind to the HLA molecule (top panel), if the gene for the peptide mutated the antigenic peptide may no longer bind to the HLA and therefore no recognition of the viral peptide would be possible and the individual would be susceptible to disease caused by the virus. In the right-hand panel it can be seen that if an individual has multiple HLA class molecules, each with a different capacity to bind antigenic peptide, a similar mutation in the pathogen may result in an antigenic peptide that does not bind to the original HLA molecule but the peptide will be bound and presented by a different HLA class I molecule, thereby maintaining the ability to respond to the virus.

4.6 Summary

- There are two types of T lymphocytes called CD4 T cells and CD8 T cells. Both are produced in the thymus from precursors that migrate from the bone marrow.
- T cells recognise antigen using a different receptor from B cells. The T cell receptor for antigen (TCR) has two chains, an α-chain and a β-chain. It is found in the membrane of T cells but is not secreted like antibody. Each chain of the TCR has a variable region, which contains hyper-variable regions that differ from T cell to T cell; these hyper-variable regions give the different T cells different antigen specificities.
- The major histocompatibility complex is a region of DNA containing over 100 genes. There are three classes of MHC genes and products called class I, class II and class III MHC genes. Class I and class II MHC genes are extremely polymorphic, with some genes having over 100 alleles. Class I and class II MHC proteins are expressed on the surface of cells and present antigenic peptides to the TCR. T cells therefore recognise a complex of antigen + MHC. CD4 T cells recognise antigen presented by class II MHC molecules and CD8 T cells recognise antigen + class I MHC.
- Special mechanisms of antigen processing ensure that endogenous antigens are presented by class I MHC molecules and exogenous antigens are presented by class II MHC molecules.

4.7 Questions

1) Why are T lymphocytes so called?
2) What are the two main types of T cells?
3) Draw a diagram of a class I MHC molecule.
4) The diagram opposite shows a T cell recognising antigen. What labels would you give to the arrows A, B and C? Is this a CD4 or a CD8 T cell?
5) (i) What is the unique feature about polymorphism of class I and II MHC genes? (ii) What is the advantage of this unique feature?
6) How are exogenous and endogenous antigens processed?

The answers to these questions can be found on page 335.

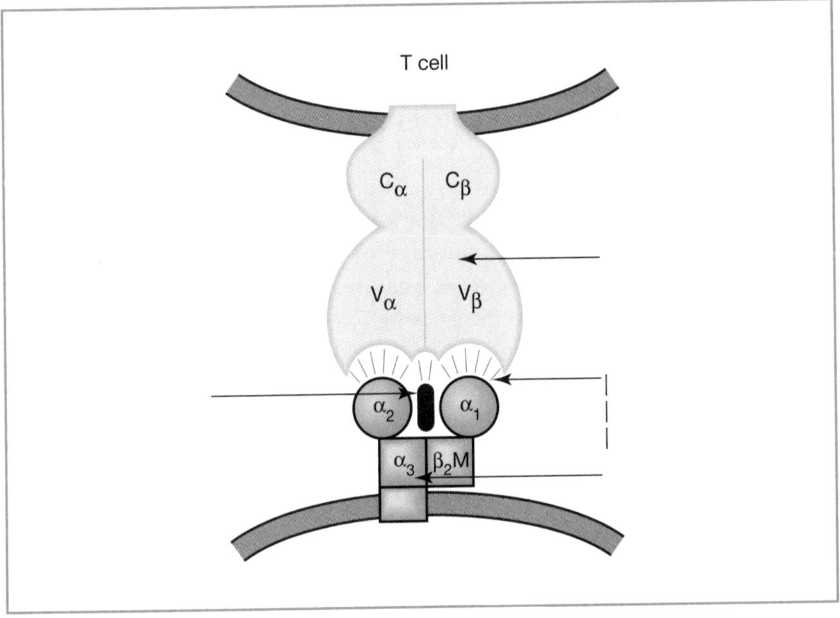

4.8 Further reading

1) Horton R, Wilming L, Rand V *et al*. (2004) Gene map of extended human MHC. *Nature Reviews Ogenetics* 5:889–899.

2) Garcia KC, Teyton GL Wilson IA. (1999) Structural basis of T cell recognition. *Annual Review of Immunology* 17:369–398.

3) Krogsgaard M, Davis MM. (2005) How T cells 'see' antigen. *Nature Immunology* 6:239–245.

CHAPTER FIVE

Lymphocyte development and the generation of antigen receptors

Learning objectives

To learn how B and T cells develop from bone marrow precursors. To know about the thymus and its role in T cell development. To understand the molecular mechanisms that are used to generate a vast diversity of Ig and TCR with specificity for different antigens.

Key topics

- Production of B cells
- Production of T cells
- Structure of genes coding for Ig and TCR
- Ig and TCR gene rearrangements
    - Imprecise joining
- Number of Ig and TCRs
- Control of lymphopoeisis

5.1 The production of lymphocytes: lymphopoiesis

Lymphocytes are white blood cells whose job is to recirculate through blood and tissues looking for their antigen. Like all other blood cells lymphocytes are bone marrow-derived cells and are produced originally from a pluripotential stem cell (Figure 5.1). This is a cell in the bone marrow that is ultimately responsible for maintaining numbers of all blood cells. It is capable of division, thereby maintaining the number of stem cells, but can differentiate into any of the different haematopoietic cells.

Pluripotential stem cells produce lymphocytes through processes of cell division and differentiation. Differentiation is the process by which cells change function through changes in gene expression. As the stem cells differentiate they become more and more specialised. In the production of lymphocytes some of the pluripotential stem cells become lymphopoietic stem cells, which are also called common lymphocyte progenitors. These cells are now committed to becoming B lymphocytes or T lymphocytes. At this point there is a big divergence in the development of B and T cells. B cells complete their development in the bone marrow and eventually leave the bone marrow and enter the bloodstream. T cells, on the other hand, finish their development in the thymus (Figure 5.1). During the development of lymphocytes the pluripotent stem cells differentiate and gradually acquire the functional ability of a B or T cell.

Another crucial event during lymphocyte development is that the cell expresses a receptor for antigen, whether it be Ig for B cells or the TCR for T cells. This poses major problems for the immune system. Each lymphocyte will recognise only one, or a few, antigen epitopes. Therefore you need to generate populations of B and T cells with enough different receptor specificities to recognise the large numbers of different antigenic epitopes that may be present on all the microbes and other pathogens that we are exposed to. So how many epitopes are there out there? It is impossible to know the exact number but if you consider all the bacteria, viruses,

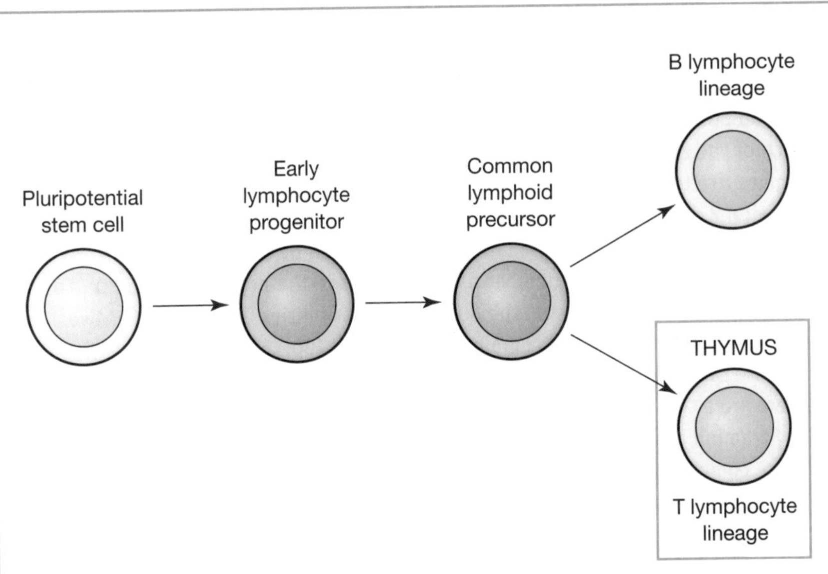

Figure 5.1 Lymphopoeisis. Both B and T lymphocytes begin differentiate from pluripotential stem cells. At the common lymphoid precursor stage some cells commit to the B lymphocyte lineage and complete their diffferentiation in the bone marrow while other cells move to the thymus and differentiate into T lymphocytes.

yeasts and parasites that exist, each with tens to thousand of antigens, you can imagine the number is very large. In fact you can calculate the range of possible numbers of epitopes and a conservative estimate is that there are in excess of 10^{11} epitopes (see Box 5.1). Therefore the immune system must have a way of producing lymphocytes which as a population have a similar number of different receptors, i.e. 10^{11}. Given that the human genome contains about 30 000 genes it is clear that you cannot have a separate gene for each receptor specificity. But, as you will see below, the immune system has evolved amazing mechanisms to generate trillions of different receptor specificities from only a few hundred genes. These mechanisms come into operation during lymphocyte development.

BOX 5.1: HOW MANY ANTIGENIC EPITOPES ARE THERE?

It is impossible to know how many epitopes there are. It is possible to calculate an estimate of the number of *potential* epitopes. X-ray crystallography of antibody–antigen interactions has revealed that an antigenic epitope comprises 8–22 amino acids depending on the nature of the antigen. There are 20 different amino acids. Therefore, in a peptide, each amino acid position has 20 alternatives. With a peptide of one amino acid there are 20 possible alternatives (the number of amino acids) = 20^1. With a two-amino-acid peptide (dipeptide) there are many more different forms possible. If the first amino acid is, say, alanine, it is possible to combine this with any of the 20 amino acids at position 2 giving rise to 20 alternative dipeptides. Changing the first amino acid to arginine enables this to be combined with any of the 20 amino acids at the second position, giving another 20 dipeptides. This process can be repeated for each of the 20 amino acids at position 1; each of the 20 amino acids at position 1 gives rise to 20 dipeptides by being combined with a different amino acid at position 2. Therefore the number of combinations is 20×20, or 20^2. You can see that the number of combinations is determined by the number of different amino acids, 20, and the number of amino acids in the peptide, n, i.e. 20^n. For one amino acid, where $n = 1$, the number of possibilities is $20^1 = 20$. For two amino acids the number of combinations is $20^2 = 400$.

If we take the lower value of eight amino acids for an antigenic epitope, there are a total of $20^8 = 2.56 \times 10^{10}$ combinations. The number of combinations of 22 amino acids is $20^{22} = 2 \times 10^{28}$. It is probable that only a tiny fraction of these combinations exists but even if one in a million of these combinations existed, the total number of epitopes would be in excess of 10^{22}.

Although the exact number is not known it can be seen that the potential number of epitopes is very large indeed and a reasonable estimate is in excess of 10^{11}. Therefore the immune system needs to be able to generate a large number of different antigen specificities to be able to recognise this large number of epitopes.

The development of lymphocytes takes place in specialised environments in the bone marrow and, for T cells, in the thymus. Development can be followed in a number of ways and one of these is to look at changing gene expression. As they develop, lymphocytes begin to express different proteins, either on the cell surface or intracellularly. These new proteins are often indicators that the cells are becoming more and more 'B-like' or 'T-like'.

5.2 B lymphocytes are produced in the bone marrow

The differentiation of B lymphocytes in the bone marrow can be followed by looking at the expression of different molecules by the developing B cell. One of the first signs that a cell is committed to the B cell lineage is the expression of a surface protein called CD19 (Figure 5.2). The cell at this stage is called a pro-B cell. The cell then expresses the IgM heavy-chain (μ-chain) in its cytoplasm and is known as a pre-B cell. The next stage is called an immature B cell which expresses IgM on its cell surface. Finally the cell becomes a mature B cell expressing both IgM and IgD on its surface. This B cell leaves the bone marrow and enters the bloodstream. It is during this process that unique molecular mechanisms operate on the Ig genes to generate large numbers of different variations of the Ig which recognise different antigens.

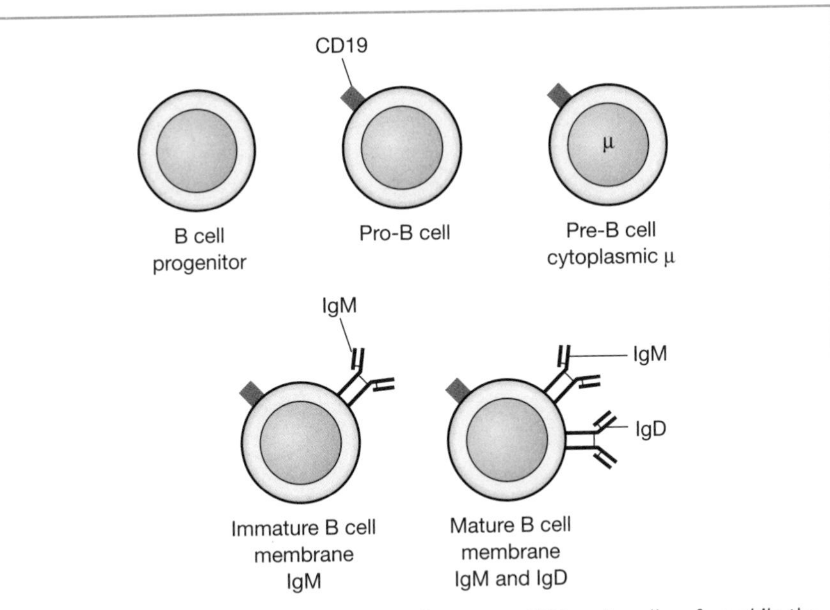

Figure 5.2 B cell development. The pro-B cell expresses CD19 on its cell surface while the pre-B cell expresses the Ig μ-chain in its cytoplasm. Immature B cells express IgM but not IgD on their cell surface while mature B cells express both IgM and IgD.

5.3 T lymphocytes finish their production in the thymus

Although the fundamental requirements for differentiation of B and T lymphocytes are similar, the situation with T cells is more complex than with B cells for two reasons. All B cells produced in the bone marrow are the same except for the specificity of their surface IgM and IgD. By contrast three types of T cell are produced in the thymus from the same type of precursor cell (Figure 5.3). Two of these cell types, the CD4 and CD8 T cells, express an α/β TCR, and the third type bears a different receptor for antigen, the γ/δ TCR. The rest of this section will cover the production of α/β TCR-bearing T cells; the production of γ/δ T cells is covered in Box 5.2. All of these cell types are produced in the thymus. The second reason why T cell development is more complicated is that T cells recognise antigen in association with MHC. T cells must be selected in each individual to recognise antigen in association with that individual's own MHC, a process known as thymic education. This is described in more detail in Section 11.3.1.

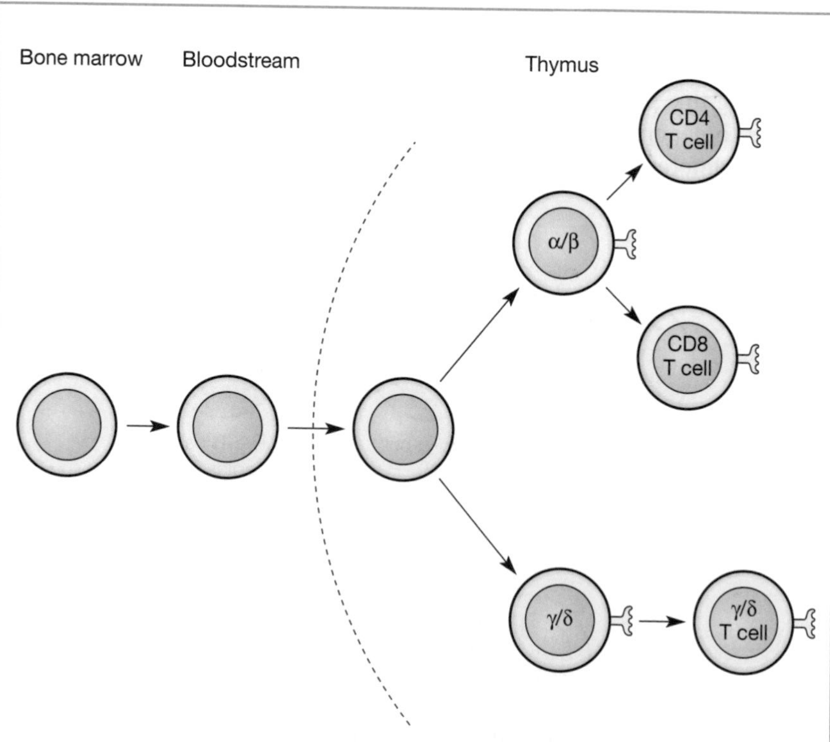

Figure 5.3 T cell production in the thymus. T cell progenitors enter the thymus from the blood and can develop into CD4 or CD8 T cells bearing the α/β TcR or T cells expressing the γ/δ TcR.

BOX 5.2: γ/δ T CELLS

A small population of T cells exists that express a different antigen receptor from the conventional α/β receptor expressed by CD4 and CD8 T cells. This receptor is composed of a γ-chain and δ-chain that pair to form a dimer that is expressed on the cell surface in association with CD3. The γ and δ genes are similarly arranged to the Ig and TCR genes, consisting of V, J and C gene segments plus a D segment in the case of the δ gene. These genes undergo rearrangement in the same way as the Ig and TCR genes and can generate a large diversity of antigen-specific receptors.

Less than 5% of T cells are γ/δ T cells and much about them is not clear. They are found at the highest numbers in epithelial sites such as the epidermis and epithelium of the gut, intestine and vagina. Most of the γ/δ T cells in one site express the same Vγ and Vδ gene segments and many have the same Jγ, Jδ and Dδ segments with little or no junctional diversity. This suggests that their recognition of antigen is limited. Different tissues will have γ/δ T cells using different V, D and J segments from those in another tissue. The way in which γ/δ T cells recognise antigen differs from α/β T cells in that it is not MHC-restricted. Some γ/δ T cells still need antigen to be somehow associated with MHC, but the allele of the MHC does not matter. Other γ/δ T cells recognise antigen without any apparent involvement of MHC.

The function of γ/δ T cells is also not clear. *In vitro* they are capable of killing tumour and other cells and secrete many cytokines. Cytotoxicity and cytokine secretion occur rapidly *in vitro* without any apparent need to stimulate the post-antigenic stimulation differentiation required by B and T cells (although it cannot be excluded that the cells are not continuously being stimulated by antigen *in vivo*). Because of their immediate response and location primarily in epithelial sites it has been suggested that they perform a sentinel role, waiting in the epithelium to provide an immediate response to infection.

The thymus is located in the mediastinum and lies above the heart. It has a two-lobed structure and each lobe is divided into many lobules separated by trabeculae of connective tissue (Figure 5.4 and Plate 9). Each lobule has two sections that can be easily distinguished histologically. The outer area stains more densely with haematoxylin and eosin (H & E) and is called the cortex. The less densely staining inner section is called the medulla. It should be mentioned that there is no apparent anatomical barrier between the cortex and medulla, although particulate material injected intravenously is able to access the medulla but not the cortex. The thymus reaches its peak size around one year of age, after which it shrinks or involutes. In adults it is composed mostly of fat cells and connective tissue, although it produces new T cells well into old age.

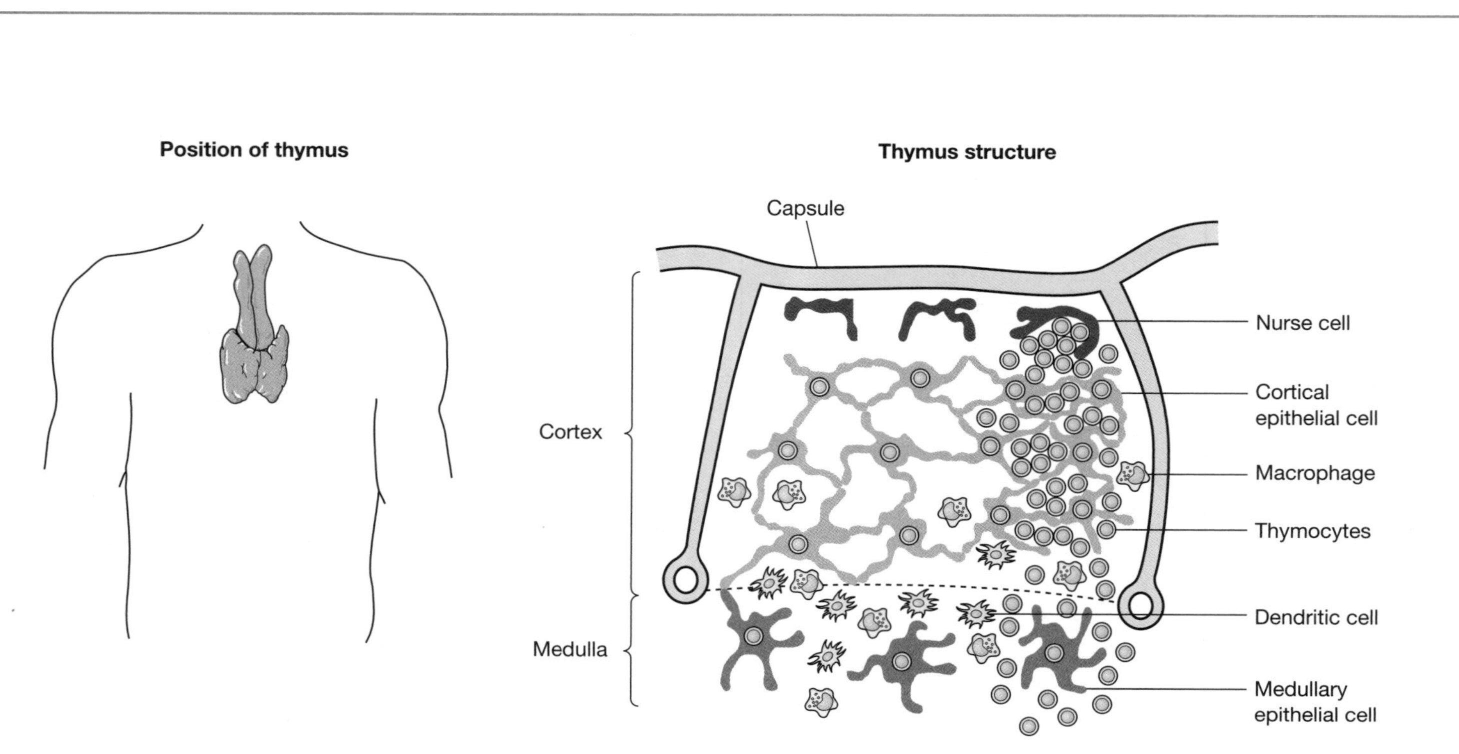

Figure 5.4 Position of thymus and thymus structure. The thymus is located above the heart.

Numerous different cell types can be found in the thymus which all play a role in T cell development:

- **Epithelial cells.** Three types of epithelial cell have been described in the thymus:
  - **Cortical epithelial cells**. These make up a mesh throughout the cortex. They provide structure and secrete factors that are essential for T cell development.
  - **Nurse cells.** These are another type of cortical epithelial cell and can be seen in close contact surrounding developing thymocytes. They express class I and class II MHC on their cell surface.
  - **Medullary epithelial cells.** These provide structure for the medulla and recent evidence suggests they play an important part in T cell tolerance (see Section 11.3).
- **Macrophages.** These are found in both the cortex and the medulla, although they are more numerous in the latter. They are class II MHC negative and play an important role in phagocytosing thymocytes that have died by apoptosis during development (see Section 11.3).
- **Dendritic cells.** These are another type of bone marrow-derived cell related to dendritic cells found in most tissues and lymphoid tissue. They express class II MHC as well as class I MHC.
- **Thymocytes.** These are the most abundant cells in the thymus and consist of T cells in various stages of development.

T cells develop from bone marrow precursors called prothymocytes. These precursors travel from the bone marrow through the bloodstream and enter the thymus through venules at the cortico-medullary junction (Figure 5.5). The most immature thymocytes are found in the subcapsular region of the cortex. These cells are rapidly dividing and give rise to large numbers of thymocytes. At this point they don't express either CD4 or CD8 which are expressed by mature T cells and they are called 'double-negative thymocytes'. As the developing thymocytes carry on with the maturation process they stop dividing and move deeper into the cortex. They then express both CD4 and CD8 and are called 'double-positive thymocytes'. At this stage the thymocytes begin to express the T cell receptor for antigen (TCR). Finally the double positive thymocytes either stop expressing CD4, and become CD8 T cells, or they stop expressing CD8 and become CD4 T cells. The T cells migrate to the medulla of the thymus and eventually leave the thymus either via venules at the cortico-medullary juction to enter the bloodstream, or via lymphatic vessels located at the cortico-medullary junction. Like B cells, the CD4 and CD8 T cells circulate through the blood and lymphatic systems looking for their antigen.

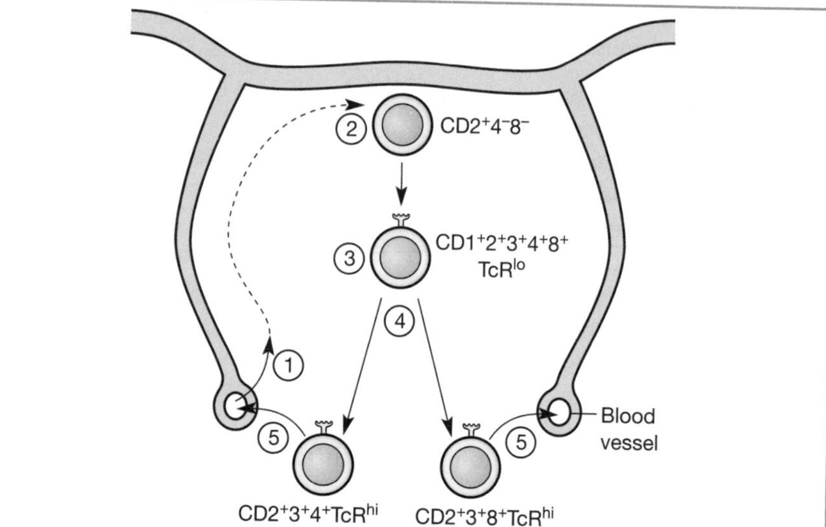

Figure 5.5 T cell development in the thymus. Cells enter the thymus from venules at the cortico-medullary junction ① and travel to the subcapsular area of the outer cortex ②. These cells, which do not express CD4 or CD8, divide rapidly for a while and then stop dividing and migrate into the cortex ③, where they express both CD4 and CD8. As the cells traverse from the cortex to the medulla they stop expressing CD4 or CD8 ④ and leave the medulla as naïve T cells through blood vessels at the cortico-medullary junction ⑤.

5.4 During their development lymphocytes must generate huge numbers of Ig and TCR receptors with different antigen specificities

One vital aspect of B and T cell differentiation is that the cells must generate antigen receptors, Ig or TCR, for huge numbers of different antigens. As described above, estimates are that in excess of 10^{11} different antigen specificities are required and these must be generated from a much smaller number of genes. So how is this achieved? Although the genes coding Ig H and L chains and TCR α and β chains are different, the B and T cells use the same molecular mechanisms to convert a limited number of genes into a huge variety of protein products. This is achieved through the specialised structure of the Ig and TCR genes and a unique molecular mechanism of Ig and TCR gene rearrangement.

5.4.1 Ig and TCR genes have special structures

Antibody molecules consist of two heavy chains and two light chains joined together by disulphide bonds (see Chapter 3). The TCR consists of an α-chain and a β-chain (see Chapter 4). The locations of the Ig and TCR genes are shown in Table 5.1. Each Ig or TCR protein chain has a variable (V) region and a constant (C) region. Like all other proteins, Ig and TCRs are coded for by genes that are transcribed and translated into the final protein. However there are unique features about the Ig and TCR genes that enable large numbers of different proteins to be generated from a limited number of genes. These features apply to all Ig and TCR genes but can best be explained by describing the human Ig κ-light chain gene, which has the simplest structure.

Table 5.1 Location of Ig and TCR genes

Gene[a]	Chromosome
IgH	14
κ	2
λ	22
TCRα	14
TCRβ	7

[a] The TCRδ genes are located between those coding for Vα and J.

Gene and protein structure of the human κ-light chain

At the protein level the κ-light chain consists of a variable region domain and a constant region domain as described in Chapter 3. The genes coding for the κ-light chain consist of exons, which are sequences of DNA coding for the protein, and introns, which are parts of the gene that do not code for the protein. In this respect the κ-light chain gene is similar to most other eukaryotic genes. Figure 5.6 shows the structure of a typical gene, that for α-globin, as well as the κ-light chain gene. Superficially they appear to be similar in that they both contain introns and exons. The α-globin gene has been chosen for its simplicity; it contains three exons and two introns. Other genes can contain dozens of introns and exons. The mRNA transcripts for α-globin are shown in Figure 5.6 and it can be seen that all of the exons are used to code for the final protein sequence. This is typical for most genes although sometimes the primary RNA can be processed in slightly different ways – so-called differential processing – to give different forms of the protein. However, this differential splicing results in only a few different forms of the protein.

If we now look at the κ-light chain gene, two important features are seen that distinguish this gene from typical genes. At the protein level the κ-chain consists of a variable and a constant region. A single exon codes for the constant region of the κ-chain; this is often referred to as the '*C*κ gene' although strictly speaking it is a gene segment. The variable region is coded by two exons. Most of the variable region is coded for by a single exon called the κ variable gene, or Vκ (Figure 5.6). Part of the CDR3 variable region is also coded for by the Vk gene but the part closest to the constant κ region is coded for by a separate gene segment called the J gene, or Jκ. The J stands for 'joining' because the J segment joins the V and C segments. Each of the 76 Vκ genes has a different nucleotide sequence, especially in the CDRs; similarly, the 5 Jκ genes have different nucleotide sequences.

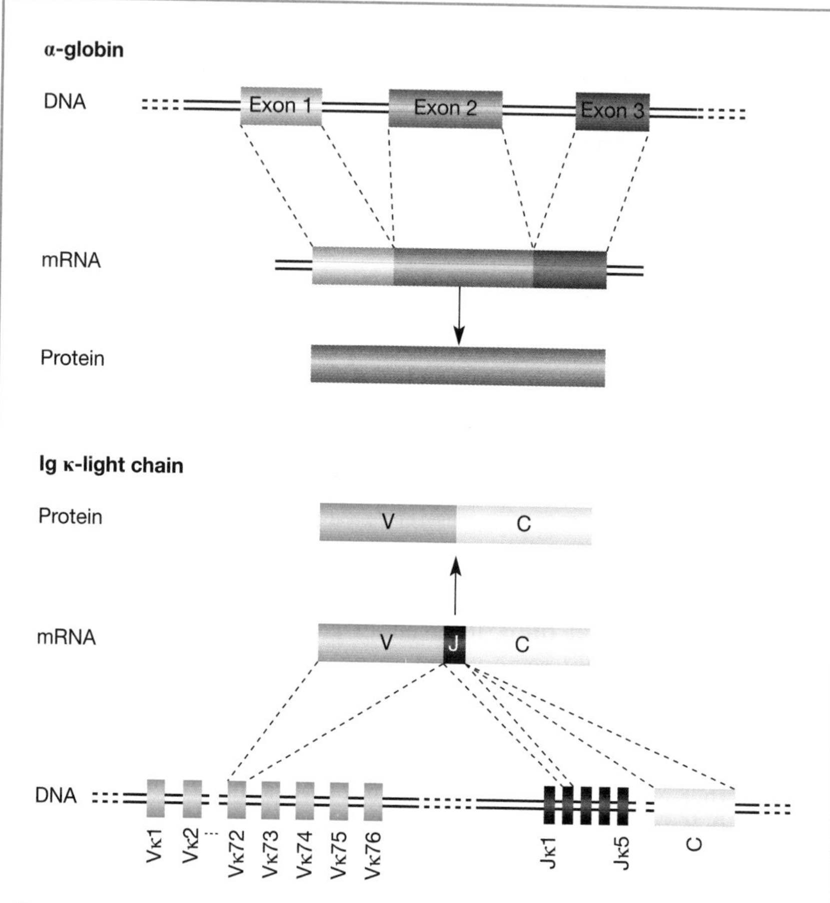

Figure 5.6 Structure of genes coding for α-globin and κ-light chain. All three exons of the α-globin gene are translated into RNA, which is processed and translated into protein. By contrast, only one of 76 Vκ genes and one of five Jκ genes are transcribed into RNA, along with the Cκ gene.

5.4.2 Each developing B cell chooses to use only one of its V and one of its J kappa genes

Although the developing B cell has a large number of V genes (or exons) and, to a lesser extent, J genes, **each B cell uses only one of the V genes and one of the J genes**. As shown in Figure 5.6, the DNA of the gene in a B cell has one of its 76 V genes joined to one of its five J genes and it is from this VJ DNA sequence that RNA will be transcribed. This 'choosing' to utilise only one out of the many exons occurs because of a specialised molecular process called **gene rearrangement** that occurs during the development of the B cell (Figure 5.7). This gene rearrangement involves the DNA encoding the κ-chain folding so that one of the V-region genes is positioned next to one of the J-region genes. The DNA is then cut so

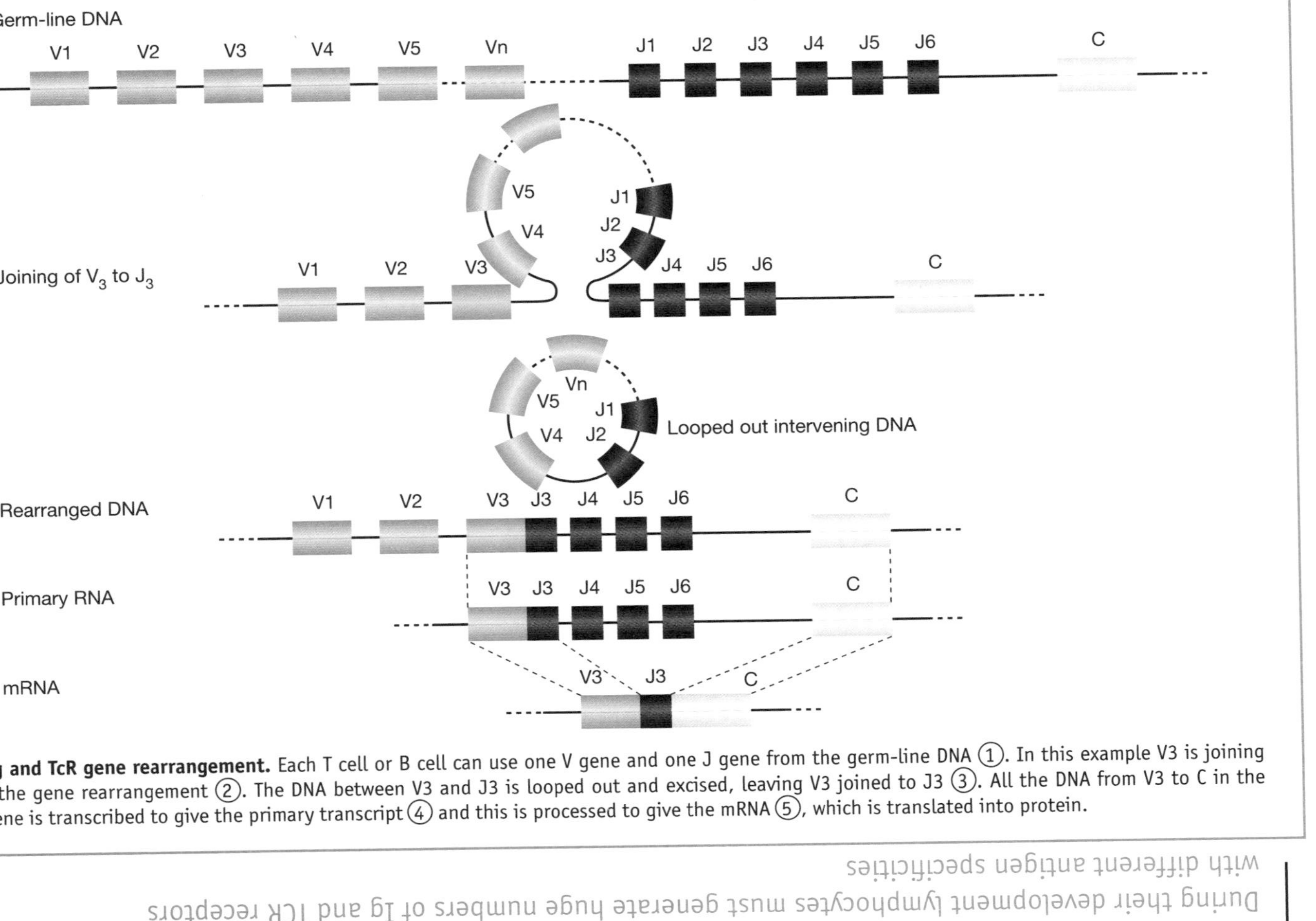

Figure 5.7 Ig and TcR gene rearrangement. Each T cell or B cell can use one V gene and one J gene from the germ-line DNA ①. In this example V3 is joining to J3 during the gene rearrangement ②. The DNA between V3 and J3 is looped out and excised, leaving V3 joined to J3 ③. All the DNA from V3 to C in the rearranged gene is transcribed to give the primary transcript ④ and this is processed to give the mRNA ⑤, which is translated into protein.

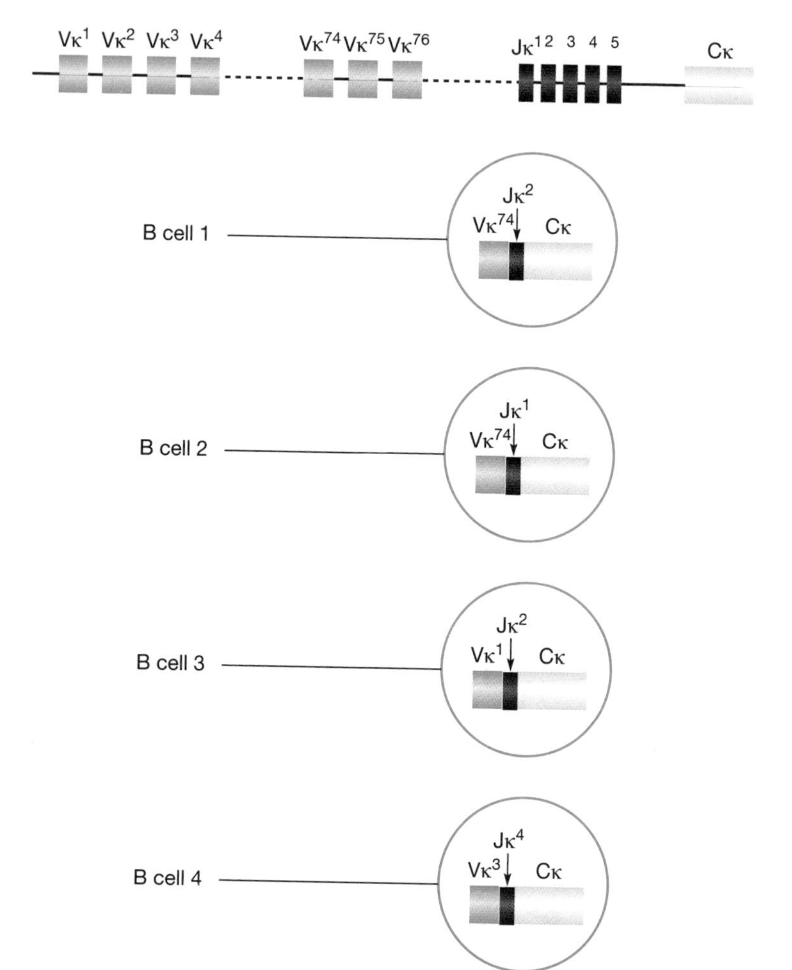

Figure 5.8 Different B cells utilise different Ig gene segments. Using the Ig κ-light chain as an example, there are 76 genes coding for Vκ and five genes coding for Jκ. Each B cell chooses at random one functional Vκ gene and one Jκ gene; the actual genes chosen will differ for each B cell. Therefore B cell 1 has chosen the Vκ74 and Jκ2. B cell 2 has chosen the same V, Vκ74, but a different J, Jκ1, and it can be seen that all four B cells have a different combination of V and J regions. Each V and J region has a unique nucleotide sequence and therefore each combination of V and J gives rise to a V-region protein with a slightly different amino acid sequence, and therefore different antigen specificity, from any other combination of V and J.

that the V gene can be joined to the J gene. The gene is now said to be rearranged and the rearranged DNA can be transcribed and translated into the κ-chain protein. Although each individual B cell uses only one V and one J segment, different B cells will 'choose' different V and J segments at random. Therefore the population of B cells in an individual will utilise all the different V and J segments (Figure 5.8).

The mechanism by which both the J gene and V genes code for the third complementarity-determining region (CDR3) of the variable region is important in increasing the variety of antigen specificities. We saw in Chapter 3 how Ig chains contain three regions that differ extensively in amino acid sequence in different antibodies. These regions contribute to the antigen-binding site. CDR1 and CDR2 are coded for exclusively by the V gene segment but will differ in different B cells because they will select different V gene segments. CDR3 is coded for partially by the V gene and partially by the J gene (Figure 5.9); therefore both the V gene and the J gene that a B cell selects will influence the antigen specificity of the Ig made by the B cell. This feature is very important in increasing the variety of amino acid sequences in CDR3.

Since there are 76 Vk genes and 5 Jk genes, the number of combinations V and J is 76 × 5 = 380. Therefore this mixing and matching gives 380 combinations (i.e. different DNA sequences have from 81 genes (76V + 5J). You can see already how the use of different gene segments increases variability.

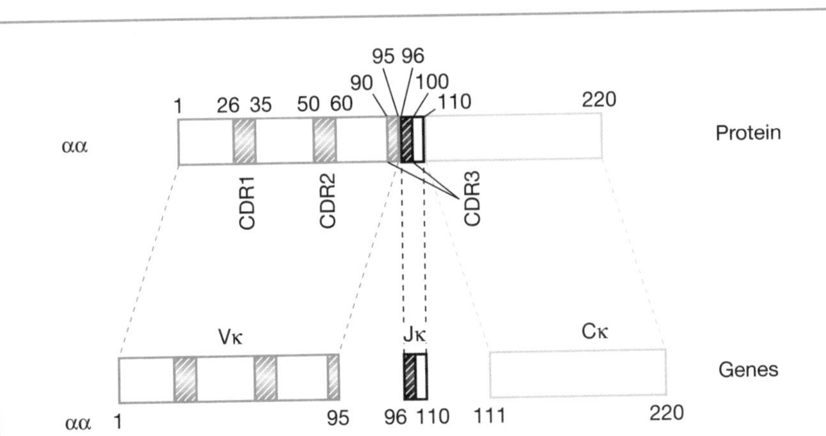

Figure 5.9 Relationship between Ig κ gene and protein. The Vκ-gene codes for the first 95 amino acids of V region of the protein. The Jκ gene codes for the final 15 amino acids of the V region of the protein. The CDR1 and 2 are coded for totally by the Vκ gene but the CDR3 is coded for partially by the Vκ gene and partially by the Jκ gene. Therefore different combinations of V and J will give different nucleotide sequences coding for CDR3 and therefore different amino acid sequences at CDR3. Because CDR1, 2 and 3 fold in the antibody molecule to form the antigen-binding site, the different amino acid sequences at CDR3, formed by different combinations of Vκ and Jκ genes, will have different antigen specificities.

The basic organisation of the genes and mechanisms of gene rearrangement are the same for the Igλ light chain and the Ig heavy chain although there are some differences in the details.

5.4.3 Structure of the λ-light chain gene

The human λ-light chain gene is slightly more complicated than the κ gene. There are 30 functional Vλ genes but instead of a series of J genes and a single C region gene as in the κ gene, there are four repeats of functional J and C genes (see Figure 5.10). The λ chains produced from each of the four Cλ genes are structurally the same.

5.4.4 Structure of Ig heavy chain gene

The basic organisation of the Ig H genes is similar to that of the κ-light chain genes but there are some important differences (Figure 5.10). Instead of the variable region being coded for by V and J genes there is an additional set of exons, called D, between the V and J genes. The D in this case stands for diversity. Each developing B cell 'chooses' one D gene as well as one V and one J gene to use in generating the H chain. Therefore two gene rearrangements of the heavy chain occur during B cell development, D to

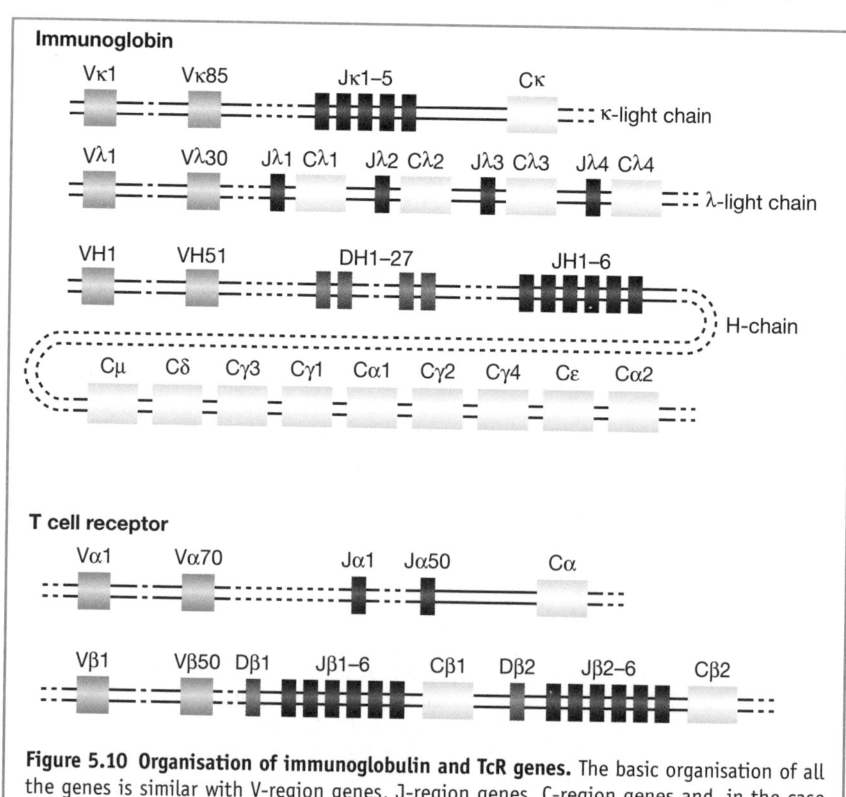

Figure 5.10 Organisation of immunoglobulin and TcR genes. The basic organisation of all the genes is similar with V-region genes, J-region genes, C-region genes and, in the case of IgH and TcRβ, D-region genes. Note that the C region for IgH actually consists of seven exons and the C region for TcRβ consists of four exons.

J and V to D. The D gene codes for part of the CDR3 of the heavy chain, so that three genes – part of V, and all of D and J – contribute to coding for the CDR3 of the Ig heavy chain.

Downstream (3') of the J genes are the constant region genes, which code for the constant region part of the molecule. There is one Ig heavy chain constant region gene for each class (and subclass) of antibody, giving a total of eight constant genes in humans. These are called Cμ, Cδ, Cγ1, Cγ2, etc. for the different Ig classes. Each constant region gene has seven exons (Figure 5.10).

5.4.5 Structure of the TCR genes

The TCR consists of an α-chain and a β-chain (see Chapter 4). The TCR genes are similar in arrangement to the Ig genes (Figure 5.10). The TCRα gene consists of one C region, which is coded for by four exons. There are about 50 Vα genes and a surprisingly high number of J genes – also 50 – in the α gene. Just as with the Ig light chain genes, each T cell 'chooses' one V gene and one J gene to use for its TCR α-chain. The CDR3 is coded for by part of the V and part of the J gene, while the CDR1 and CDR2 are coded for by the V gene.

The TCRβ gene is similar to the Ig heavy chain gene in that it utilises V, D and J segments to code for the variable part of the molecule. The TCRβ gene consists of about 50 V genes, followed by two repeats of a cluster of one D-region gene, seven J-region genes and a constant region gene coded for by four exons. There are a total of two D genes, 13 J genes (one J gene is a pseudo-gene and not expressed) and two constant region genes in the TCRβ gene.

Unlike the Ig heavy chains, whose constant regions give them different functional properties, the two constant regions in the β-chain do not have any different functions. Both CD4 and CD8 T cells use either constant region randomly and it would appear that the TCR is used just to recognise antigen. There is no secreted TCR equivalent of secreted Ig.

5.4.6 Additional mechanisms increase the diversity of Ig and TCR antigen specificities

We have seen how developing B and T cells choose V, D and J segments for their Ig and TCR genes during development. This 'mixing and matching' of V, D and J gives a large increase in combinations from a relatively small number of genes. The numbers of combinations of Ig and TCR genes are shown in Table 5.2. The number of Ig or TCRs that can be generated by this process is in the region of $2–5 \times 10^6$. This is quite an impressive number of different receptors to generate from fewer than 400 genes but is still well short of the estimated 10^{11} different receptors needed as calculated in Section 5.1. So how do we make up the difference? It turns out there are yet more unique mechanisms associated with gene rearrangement that increase the diversity of receptors through a process called **imprecise joining**. This can best be demonstrated by looking at the joining of V-D-J in the Ig heavy chain.

Table 5.2 The effect of having separate V, D and J genes on the number of different immunoglobulin and TcR variations possible

Receptor chain	Immunoglobulin							TcR				
	Heavy chain			κ-light chain		λ-light chain		TcRα-chain		TcRβ-chain		
Gene	V_H	D_H	J_H	Vκ	Jκ	Vλ	Jλ	Vα	Jα	Vβ	Dβ	Jβ
Number of genes	51	27	6	76	5	30	4	50	70	57	2	13
Without using D or J												
Number of chains	51 H				106 L			50 α			57 β	
Number of pairs	H + L		51 × 106 = **5406**					α + β			50 × 57 = **2850**	
Using D and J												
Genes used	VH	DH	JH	Vκ	Jκ	Vλ	Jλ	Vα	Jα	Vβ	Dβ	Jβ
Number of chains	51 × 27 × 6 = 8262			76 × 5 = 380		30 × 4 = 120		50 × 70 = 3500		57 × 2 × 13 = 1482		
				= 500 light chains in total								
Number of pairs		8262 × 500 = **4.13 × 10^6**						3500 × 1482 = **5.2 × 10^6**				

When the sequences of V, D and J in the IgH chain from different B cells were examined it was found that the nucleotide sequences differed from what would be predicted if there was precise joining of the 3' end of the V segment to the 5' end of the D segment and the 3' end of D to the 5' end of J (Figure 5.11). Some of the original nucleotides in the germ-line V, D and J segments had been lost and other nucleotides were present that were not in the original germ-line gene sequence. This lead to rearranged VDJ sequences that differed in the number of nucleotides and also had a greater variability in nucleotide sequence than in the original germ-line. The unique molecular mechanism that generated this diversity was called imprecise joining and the diversity was called **junctional diversity**. These mechanisms are described more fully in Box 5.3.

BOX 5.3: Ig/TCR REARRANGEMENT

The V, D and J gene segments of the Ig and TCR genes are flanked by specific nucleotide sequences, which are important recognition sequences in the process of DNA rearrangement. Flanking the 3' side of the V segment, both sides of the D segment and the 5' side of the J segment are special nucleotide sequences consisting of a heptamer (seven nucleotides) separated by 12 or 23 nucleotides from a nonamer (nine nucleotides) (a). The whole sequence of heptamer + 12 or 23 bps + nonamer is known as a recognition signal sequence (RSS). These RSSs are involved in the first stage of rearrangement, the stages of which are described below.

(1) Alignment of RSS nonamers

The nonamer from an RSS with a 12-nucleotide space is brought together with the nonamer from an RSS with a 23-nucleotide space ((a) 1). The rule that an RSS with a 12-nucleotide space can bind only to an RSS with a 23-nucleotide space ensures that when the IgH and TCRβ genes rearrange, V rearranges to D and not to J.

(2) Cleavage of DNA and generation of P-nucleotides

When the nonamers are aligned, the DNA strand within the heptamers is cleaved by endonucleases and the intervening DNA is excised, or 'looped out' ((a) 2). The ends of the V and D genes now contain parts of the heptamer sequences ((a) 3).

(3) Formation and cleavage of hairpins – generation of P-nucleotides

The two strands of DNA at the ends of the V and D genes join to form a hairpin ((b) 1). The hairpins are then cleaved at random ((b) 2) so that each DNA strand contains some nucleotides derived from the heptamer sequence ((b) 2). These nucleotides are called P-nucleotides and this process increases the variety of final DNA sequences that can be generated.

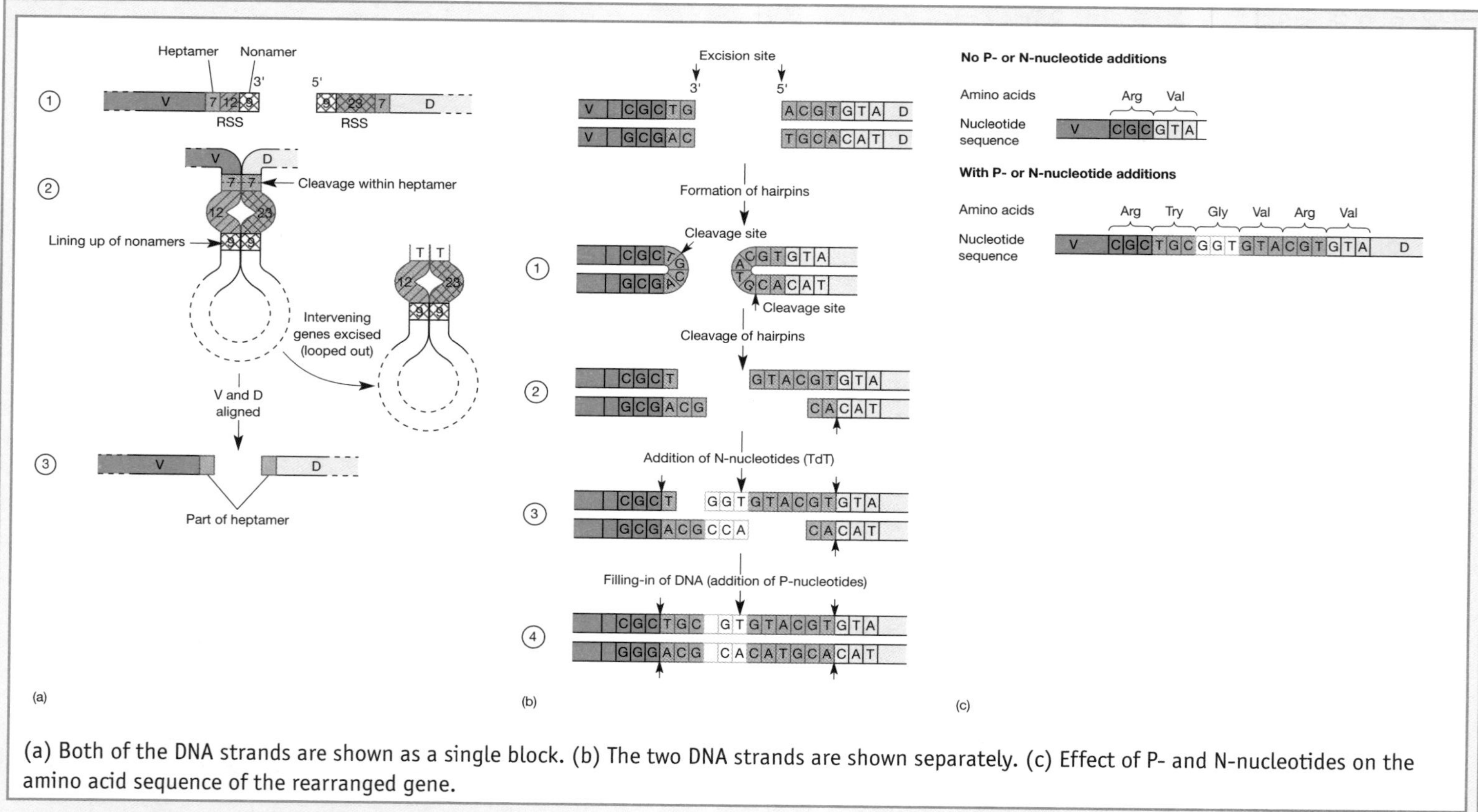

(a) Both of the DNA strands are shown as a single block. (b) The two DNA strands are shown separately. (c) Effect of P- and N-nucleotides on the amino acid sequence of the rearranged gene.

(4) Addition of N-nucleotides

The enzyme **terminal deoxynucleotide transferase** adds nucleotides at random to the ends of the DNA strands ((b) 3), further increasing variability of DNA sequences in different cells. These nucleotides are known as N-nucleotides.

(5) Filling-in of DNA

Exonucleases, DNA polymerase and DNA ligase repair the DNA and complete the formation of the joint with complete nucleotide sequences on both DNA strands ((b) 4).

The example shows that generation of P- and N-nucleotides has changed the amino acid sequence of the final protein from -Arg-Val- to -Arg-Tyr-Gly-Val-Arg-Val (c). Note that the actual nucleotides and amino acids shown are for illustrative purposes only and do not represent actual sequences.

5.4.7 So how many receptor specificities can be generated using the mechanisms of Ig and TCR gene rearrangement?

Junctional diversity greatly increased the number of different receptor specificities that can be generated. It is not known exactly by how much but estimates are that at each gene rearrangement junctional diversity increases the number of amino acid sequences by a factor of 100 to 10^6.

Therefore the variation in receptor amino acid sequences is generated in three ways:

1. The number of V, D and J genes for the different receptors determines the number of combinations. These are shown in Table 5.2.
2. The increase in the number of amino acid sequences because of junctional diversity.
3. The number of combinations of heavy and light chain for Ig and α- and β-chains for TCR.

We can now calculate an estimate of the number of different receptors as shown in Table 5.2; this uses the minimum estimate of a factor of 100 for junctional diversity at each gene rearrangement. With the addition of junctional diversity the number of receptor specificities is huge – in the region of 2.6×10^{12} for Ig and 5.2×10^{12} for TCR. So thanks to the unique features of gene rearrangement it is possible to generate more variations of antigen receptor than the minimum estimated number of 10^{11} antigenic epitopes (Table 5.2).

5.5 Developing lymphocytes rearrange their Ig or TCR genes in a carefully controlled order

The process of Ig and TCR gene rearrangement during lymphocyte development is carefully controlled. With regards to developing B cells, the first rearrangement of Ig genes is D to J in the Ig heavy chain gene at the

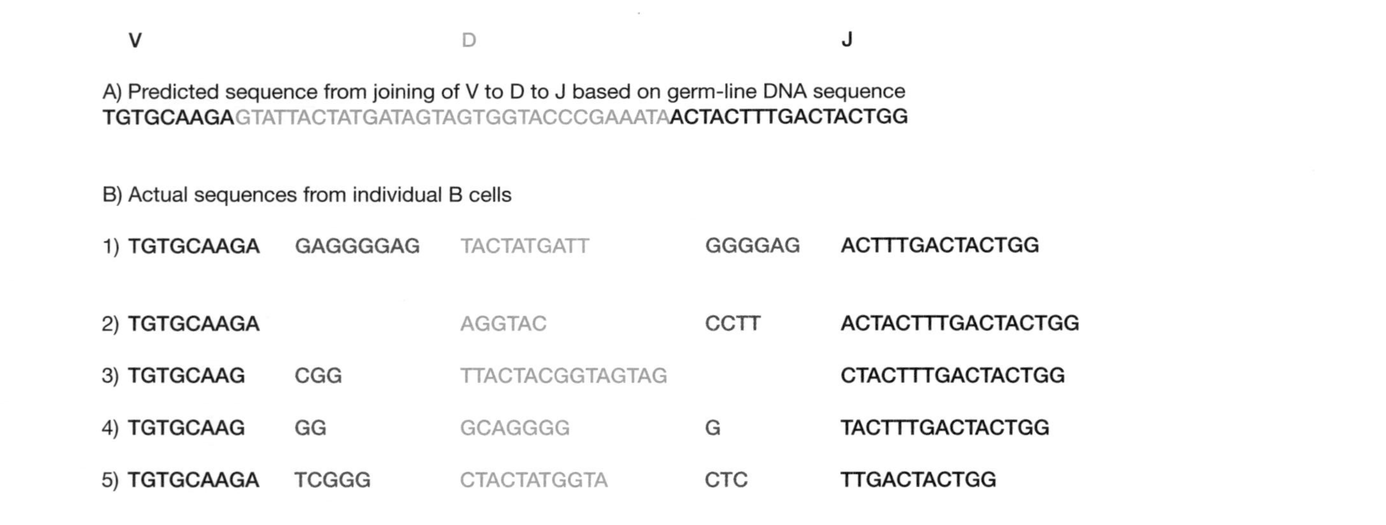

Figure 5.11 DNA sequences of CDR3 in B cells differ from those predicted by germ-line DNA sequence. (A) The top sequence shows the predicted sequence of rearranged DNA if the 3′ ends of V and D joined to the 5′ ends of D and J respectively. (B) The bottom five sequences are those obtained from individual B cells. Nucleotides from V and J are shown in black and D in blue. Nucleotides not present in the unrearranged germ-line genes are shown in grey.

pro-B cell stage. This is followed by rearrangement of an Ig heavy chain V segment to the rearranged DJ gene at the pre-B cell stage. These cells express the IgM heavy chain (μ-chain) protein in their cytoplasm. After rearrangement of the heavy chain genes is complete, the developing B cell rearranges V to J in one of the light chain genes. These cells can now make an Ig light chain protein and this associates with the Ig heavy chain protein and so the B lymphocyte can now make complete IgM and express it on its cell surface. This cell is now known as an immature B cell. Finally the B cell will also express IgD on its surface as well as IgM. At this point the B cell has completed the antigen-independent stage of maturation.

For developing T cells the first rearrangement is D to J of the β-chain. This is followed by rearrangement of Vβ to the rearranged DJ. The cell can now express the β-chain. The final rearrangement is V to J in the α-chain so the cell can now express the α-chain which associates with the β-chain to form the TCR on the cell.

Although the mechanisms of gene rearrangement have enormous power to increase the variety of receptor specificities, like many solutions they bring with them their own problems. At each gene rearrangement of Ig or TCR there is only a one-in-three chance that the rearrangement will be in frame. An out-of-frame rearrangement acts like a mutation and can create a stop codon or a nucleotide sequence that codes for a nonsense protein (Figure 5.12). So there is a good chance that gene rearrangements will not be successful. However things are not as gloomy as it would seem. Remember that lymphocytes have two copies of each receptor gene, one inherited from the mother and one inherited from the father. If the first receptor gene is not successfully rearranged the lymphocyte has another chance and can rearrange the second gene. The lymphocytes then have to go through a checkpoint where they are tested to see whether they have successfully rearranged the first gene (Box 5.4). Only those cells that have successfully rearranged their IgH or TCRβ gene are allowed to continue their development. This stops any more energy being wasted on producing non-functional lymphocytes.

In addition to checking that a lymphocyte has successfully rearranged one of each of its receptor genes it is important to ensure that lymphocytes do not rearrange both copies of a receptor gene. Therefore, if the first receptor gene is successfully rearranged, the lymphocyte must stop the second copy of the gene from rearranging. The process of stopping the second receptor gene rearranging is called **allelic exclusion** (Figure 5.13 and Box 5.5) and is very important. If allelic exclusion did not occur a lymphocyte could successfully rearrange both copies of a receptor gene and make two proteins. Each protein could form part of an antigen receptor but the two receptors would have different antigen specificity. Having lymphocytes with two receptors with different antigen specificities would make control of antigenic-specific responses chaotic.

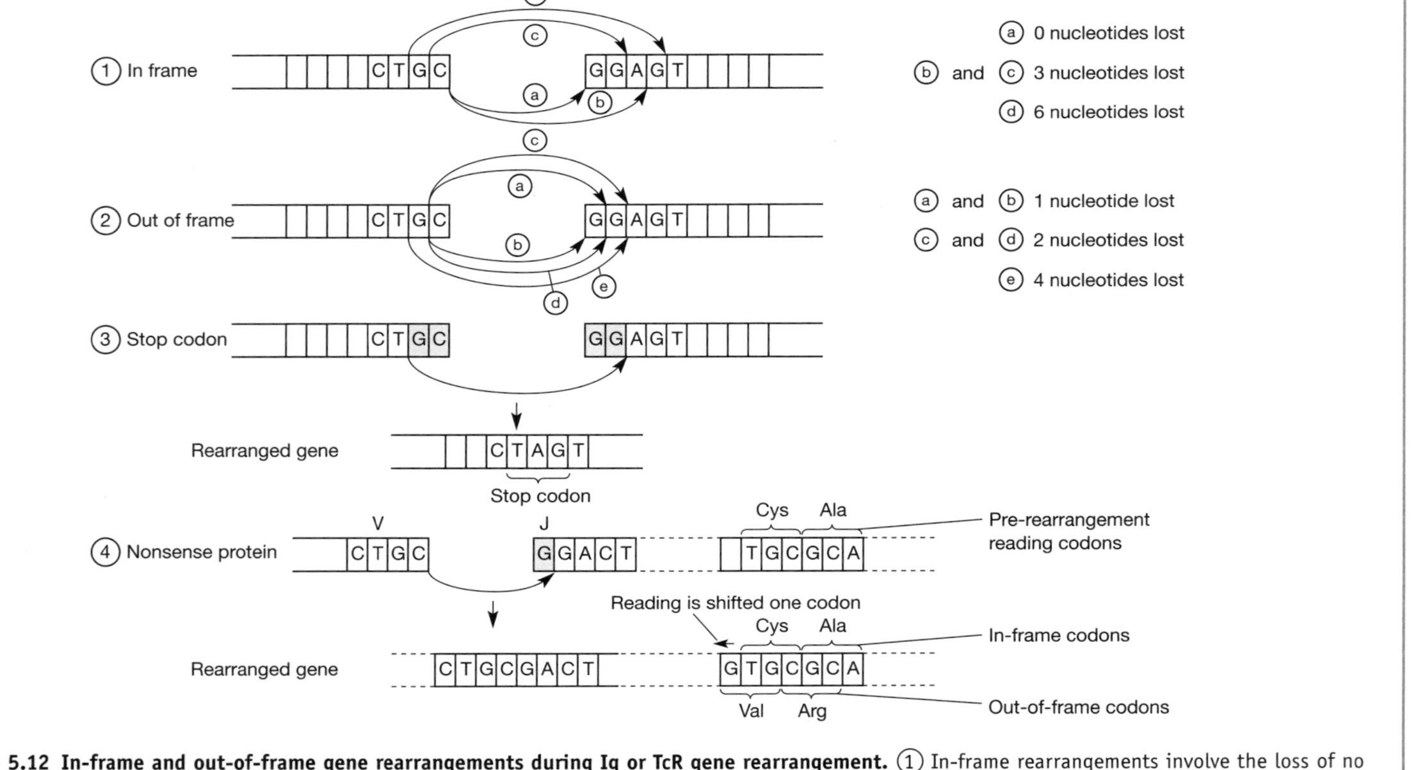

Figure 5.12 In-frame and out-of-frame gene rearrangements during Ig or TcR gene rearrangement. (1) In-frame rearrangements involve the loss of no nucleotides or multiples of three nucleotides (three, six, etc.), thereby maintaining the triplet codon sequence. (2) The number of nucleotides lost in an out-of-frame rearrangement is not a multiple of three (one, two, four, five, seven, etc.), resulting in loss of the original triplet codon sequence. (3) The out-of-frame rearrangement can result in the creation of a stop codon. (4) The out-of-frame rearrangement results in the loss of one nucleotide in the rearranged gene. This shifts the way the DNA is read in the constant region gene, creating new codons that result in a nonsense protein with no function.

BOX 5.4: DEVELOPING B AND T CELLS EXPRESS PRE-RECEPTORS TO CHECK FOR SUCCESSFUL GENE REARRANGEMENTS

If a B cell successfully rearranges one of its heavy chain genes it will begin to make heavy chain protein. The heavy chain protein is expressed on the cell surface in association with a non-rearranged Ig light chain made up of a VpreB variable segment and a light chain constant segment (see figure below). The expression of this receptor is essential for the future differentiation of the B cell and provides an important checkpoint. The exact way in which it works is not known but the receptor presumably engages some as yet unknown ligand, which results in the cell receiving a survival signal. Cells that have not successfully rearranged their H-chain genes will not be able to express the receptor and therefore will not continue the differentiation process and will die instead. This VpreBl5 light chain is expressed only at this stage of B cell development and has been called a 'surrogate' or 'pseudo' light chain because it is not involved in recognition of conventional antigen.

In a similar manner T cells that have successfully rearranged their TCR β-chain gene will express the β-chain on the cell surface in association with a pre-T cell α-gene product and CD3. Like B cells, the expression of this receptor is essential for the continued development of the thymocyte, although again the mechanism is not known. Thymocytes that do not successfully rearrange their β-chain genes cannot express the receptor and die by apoptosis.

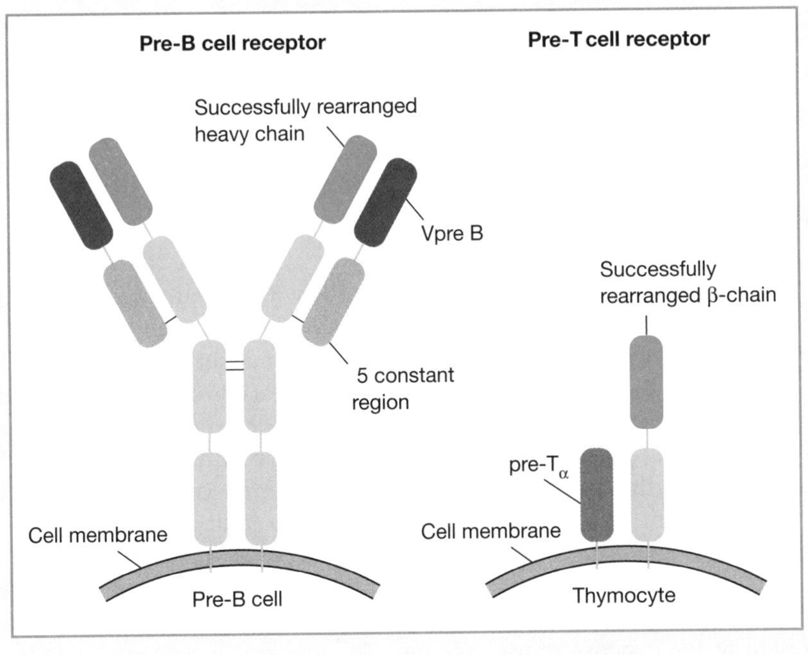

BOX 5.5: ALLELIC EXCLUSION

Allelic exclusion is a mechanism to stop lymphocytes expressing two different antigen receptors. The situation is relatively straightforward in B cells but not so well understood in T cells.

The model for B cells fits with the available data. This model proposes that developing B cells in the bone marrow begin to rearrange *one* copy of their heavy chain genes. If the rearrangements (D to J and V to D) are successful the heavy chain that is made stops rearrangement of the other copy of the heavy chain gene. If the first heavy chain gene rearrangement is unsuccessful the cell will rearrange the second copy of the heavy chain gene. Assuming one of the heavy chain genes has successfully rearranged, the cell will then begin to rearrange one copy of the κ-chain genes. If this is successful the κ-chain produced can associate with the heavy chain to form a complete antibody. This antibody stops rearrangement of the other κ gene and both λ genes. If the rearrangement is unsuccessful the cell will carry on rearranging the other κ gene and λ genes until it has tried to rearrange all four genes. At any stage, if a light chain gene rearrangement is successful the antibody produced as a consequence will stop rearrangement of any un-rearranged light chain genes.

A similar situation is thought to occur to some extent with T cells. A successful rearrangement of the first TCR β-chain gene stops rearrangement of the second heavy chain gene. However, the successful rearrangement of the first α-chain gene does not necessarily stop rearrangement of the second α-chain gene and some T cells can be produced that have two TCR receptors. Both the receptors have the same β-chain but they are associated with different α-chains.

5.6 Why is there continuous production of lymphocytes, most of which die?

The immune system generates lymphocytes with specificities for a huge range of antigens on any pathogen by randomly rearranging Ig and TCR genes during lymphocyte development. This means that following infection by a pathogen there will be some lymphocytes that are specific for antigens on the pathogen; these lymphocytes can be considered *useful* lymphocytes. However, many lymphocytes will have antigen receptors that are specific for an antigen that the immune system will never encounter, although the immune system cannot know this. Through no fault of their own these lymphocytes are *useless* to the individual.

Since we have room for only a fixed number of lymphocytes the immune system will operate most efficiently if many useful lymphocytes are present at the expense of useless ones. The way this is achieved is to have a continual process of producing lymphocytes with a finite lifespan. If, during its lifespan, a lymphocyte encounters antigen for which its antigen

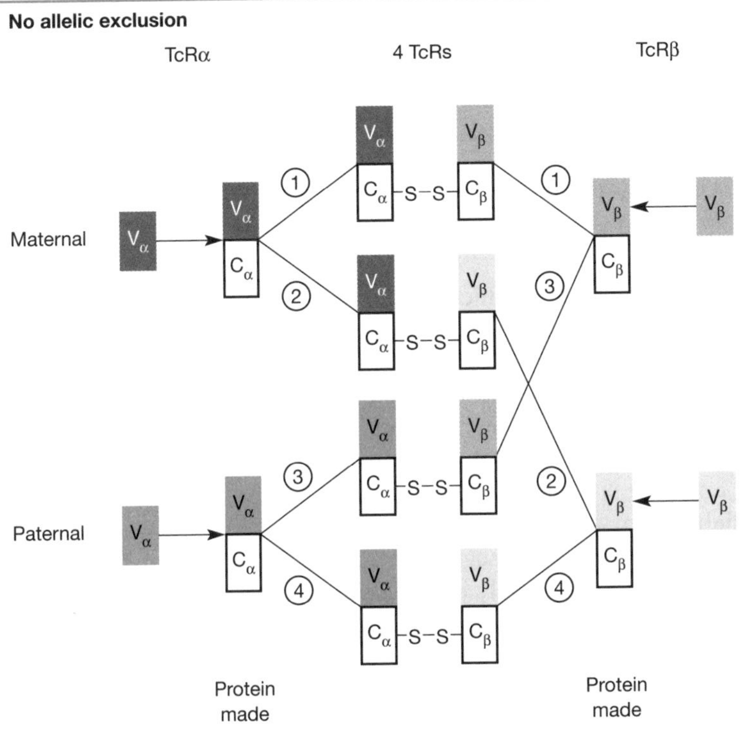

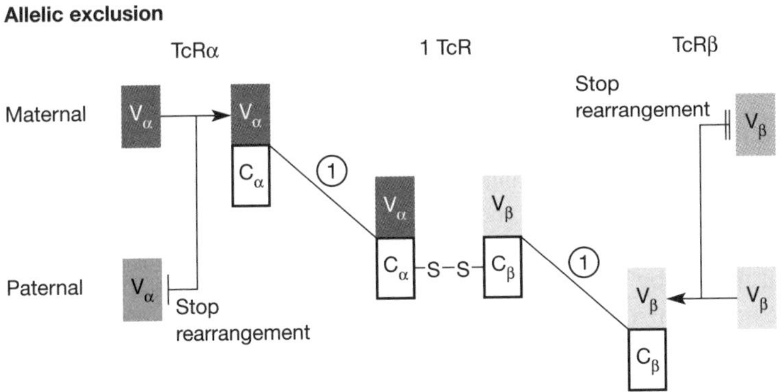

Figure 5.13 Allelic exclusion prevents lymphocytes expressing more than one receptor antigen specificity. The figure shows the situation how TCR expression would occur with and without allelic exclusion. Without allelic exclusion, both maternal and paternal Vα genes and Vβ genes may successfully rearrange, resulting in the production of two different α-chains and two different β-chains. The maternal Vα can associate with either the maternal or the paternal Vβ and the paternal Vα can associate with either the maternal or the paternal Vβ, giving four different TCRs with four different antigen specificities. The numbers in circles show pairings of Vα and Vβ. Allelic exclusion stops the second copy of the gene rearranging if the first copy rearranges successfully, therefore ensuring each T cell makes only one α-chain and one β-chain and therefore one TCR with one antigen specificity. For B cells the situation is even more complicated because a B cell could theoretically generate two different H-chains and four different L-chains (2κ and 2λ), giving a total of eight different receptor specificities.

receptors are specific, it is a useful lymphocyte and it will participate in an immune response. If the lymphocyte does not meet its antigen within a certain time period it is deemed useless and dies; it can then be replaced by a new lymphocyte that has the potential to be useful. In this way useful lymphocytes are retained and useless ones are lost.

5.7 Summary

- B lymphocytes are produced in the bone marrow where they differentiate from lymphopoietic stem cells.
- The genes coding for Ig chains are made up of segments. Some segments code for the C regions of the receptors and some for the V regions. The segments coding for the variable regions of the Ig heavy chains are called V_H, D_H and J_H and those coding for the Ig light chains are called V_κ and J_κ or V_λ and J_λ, depending on whether the B cell is using a κ or λ-light chain.
- There are many gene segments coding for V, D and J but each lymphocyte will express only one of each segment. Each B cell will use one V_H, one D_H and one J_H segment for the heavy chain and one V_κ and one J_κ segment or one V_λ and one J_λ segment for the light chain. These segments are chosen at random in each B cell during a process called gene rearrangement. The random selection of V, D and J segments generates combinatorial diversity.
- When the Ig genes rearrange the DNA does not join precisely, so that additional variation is introduced into the V-region DNA sequence at the site of the junction of V to D or J and D to J. This additional variation is called junctional diversity and greatly increases the number of different antigen receptor specificities.
- Initially developing B cells rearrange their Ig heavy chain genes. Those that do this successfully then rearrange their light chain genes. They then become immature B cells expressing IgM but not IgD. B cells finish their maturation by expressing both IgM and IgD on their surface. They leave the bone marrow and re-circulate through the blood and lymphoid tissue.
- T cells develop from immature precursors in the thymus, a bi-lobed organ in the mediastinum.
- The genes for the TCR are similarly organised as those for Ig. The TCR α-chain genes have multiple V and J gene segments and the β-chain genes have multiple V, D and J segments. Developing thymocytes rearrange their TCR β-chain gene first and then go on to rearrange their TCR α-chain.
- Lymphocytes are produced continually – those that are stimulated by antigen (e.g. on an infectious organism) participate in an immune response, those that do not see their specific antigen die in a determined time.

5.8 Questions

1) Which of the following is NOT ESSENTIAL for cellular differentiation?
   A) Change in gene expression
   B) Acquisition of new functions
   C) Movement of cells to new anatomical sites
   D) Intracellular signalling
   E) Protein synthesis.

2) What is missing from the following sentences?

During differentiation in the __________________ developing B cells rearrange their __________________ genes. The heavy chain genes rearrange _____ to _____ and ______ to ____ while the light chain genes rearrange ____ to ____. In a similar fashion, developing T cells in the ______________ rearrange their ______________ genes. The β-chain genes rearrange ____ to ____ and ____ to _____ while the α-chain genes rearrange ____ to _____. The purpose of this gene rearrangement is to ___________ ____ ____.

3) The table below gives the number of V, D and J genes for the TCR α- and β-chain genes. Assuming that junctional diversity increases the number of amino acid sequences by a factor of one hundred for each gene rearrangement, calculate the number of different TCR specificities that can be generated.

Number of genes	α-chain	β-chain
V	46	57
D	None	2
J	49	13

4) Why is allelic exclusion important?

5) Why do we continuously produce lymphocytes?

The answers to these questions can be found on page 335.

5.9 Further reading

1) Hardy RR, Hayakawa KB. (2001) B cell developmeny pathways. *Annual Review of Immunology* 19:595–621.

2) Zunicka-Pflucker JC. (2004) T-cell development made simple. *Nature Reviews Immunology* 4:67–72.

3) Goldrath AW, Bevan MJ. (1999) Selecting and maintaining a diverse T-cell repertoire. *Nature* 402:255–262.

4) Oettinger M. (1999) V(D)J recombination: on the cutting edge. *Current Opinion in Cell Biology* 11:325–329.

CHAPTER SIX

Anatomy of the immune system

Learning objectives

To understand the problems of generating immune responses *in vivo*. To learn the anatomy of the lymphoid system and how this promotes the generation of immune responses. To know how cells of the immune system move throughout the body and within tissues and the molecular basis controlling this movement.

Key topics

- Requirements of the immune system *in vivo* and the need to generate specific immune responses
- Anatomy of the lymphoid system
  - The lymphatic system
  - Lymph nodes
  - Spleen
  - Mucosal associated lymphoid tissue (MALT)
- Lymphocyte recirculation

6.1 Requirements of the immune system *in vivo*

The job description of the immune system *in vivo* is very simple: to protect every tissue and organ in the body against any pathogen. How to achieve this is not quite so simple. The immune system does not know what pathogens are waiting to infect an individual or where a particular pathogen will enter the body. It therefore has to be ready to respond to any situation.

The ultimate aim of an immune response is to eliminate or neutralise threats to the body posed by an infectious agent (pathogen). Sometimes this can be achieved by the innate immune system; for instance, phagocytes may engulf and destroy infectious bacteria. However, in many situations the innate immune system alone cannot cope and requires help from the specific immune system. The specific immune system consists essentially of the B lymphocytes and CD4 and CD8 T lymphocytes described in Chapters 3 and 4 and therefore a specific immune response involves one or more of these cell types.

6.2 Different pathogens require different types of immune responses

There are a number of ways in which a specific immune response can contribute to the elimination of pathogens:

- **The generation of antibody.** Antibody can interact with a large variety of components of the innate immune system to help neutralise or eliminate a pathogen. Antibody is produced by **plasma cells**, which are themselves derived from B lymphocytes. Therefore antibody production requires the differentiation of antigen-specific B lymphocytes into plasma cells that produce antibody against the antigen.
- **The production of CD8 cytotoxic T cells (Tcs).** Tcs can kill other cells that are expressing antigen on their class I MHC molecules. This response is particularly important in killing virally infected cells before the virus has a chance to replicate inside the cells.
- **Delayed type hypersensitivity (DTH) responses.** Most tissues contain some macrophages. These macrophages can respond to pathogens or their products and be activated to kill many infectious agents. Some pathogens, however, have evolved mechanisms to resist killing by these tissue macrophages; in many cases the pathogenic organism is actually able to survive and replicate within the tissue macrophages themselves. Additionally there may simply not be enough macrophages to deal with the number of infectious particles. This is where a DTH response is required. A DTH response involves two important factors: monocytes are recruited from the blood to the site of infection and both the recruited monocytes and the resident tissue macrophages are activated so that they are better able to kill the pathogen (see Section 9.4).
- **Inflammatory responses and the recruitment of neutrophils.** For some types of bacteria and yeast, specific immune responses are generated that involve the recruitment of neutrophils to the site of infection. This is done by specialised CD4 T cells as described in Chapter 9.

All of these specific immune responses are dependent on CD4 T cells (Figure 6.1). Without CD4 T cells antibody cannot be generated against most antigens, Tcs cannot be produced against many viruses, DTH reactions do not occur and specialised inflammatory responses are deficient. This explains to a large extent the devastating effect of the AIDS virus on the immune system. One of the main effects of the AIDS virus is to cause the loss of CD4 T cells, resulting in the reduced ability to generate any type of acquired response and thus profound immunosuppression.

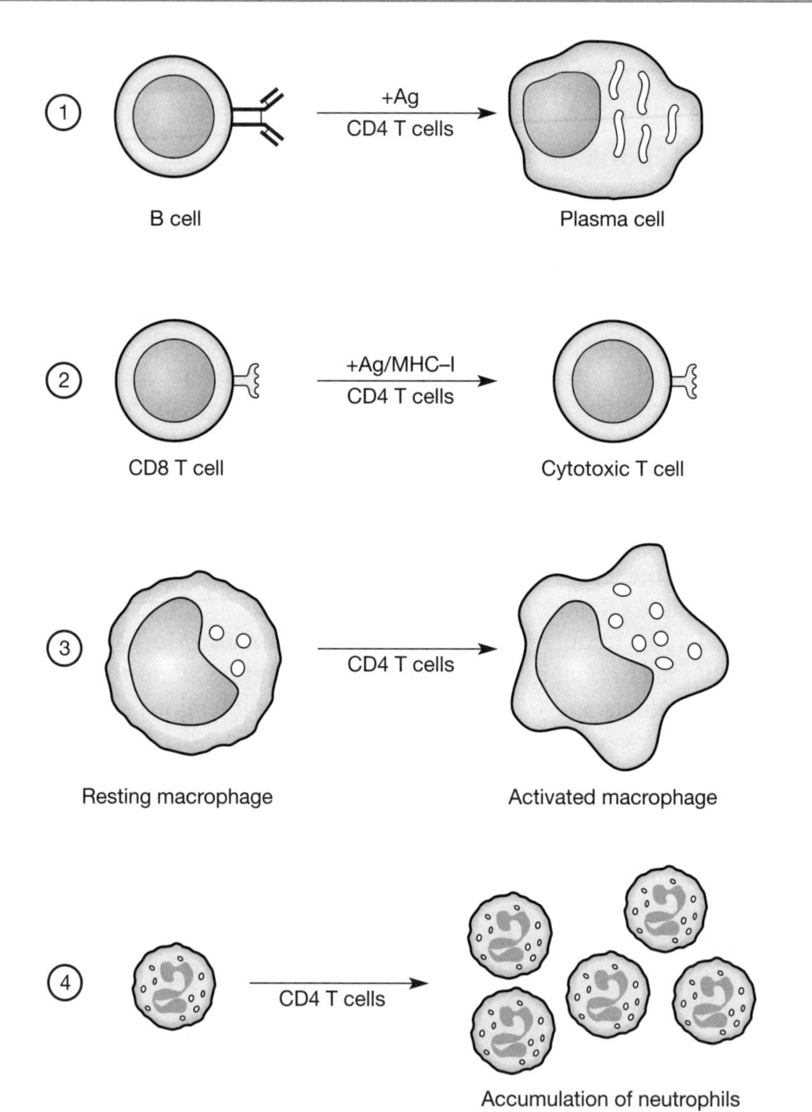

Figure 6.1 CD4 Th are required for all types of specific immune responses. Th are required for B cells to differentiate into plasma cells ①, for naïve CD8 T cells to become effector Tcs ②, for the activation of macrophages in a DTH response ③ and for accumulation of neutrophils in a an CD4 T cell-mediated inflammatory repsonse.

Therefore, during a specific immune response, CD4 T cells must interact with many other cell types including B cells, CD8 T cells, macrophages and neutrophils; the specific interactions depend on the nature of the immune response. Another important consideration is that if this is the first infection with a particular pathogen, B cells and T cells specific for the antigen will be quite rare. This poses two problems for specific immune responses. The first of these is that specific immune responses must be acquired.

6.2.1 Specific immune responses involving B and T cells must be generated after infection

The reason why specific immune responses must be acquired is best explained if we look at the situation with antibody. It was estimated in Chapter 5 that there will be in excess of 10^{11} different antigenic epitopes and in excess of 10^{11} different antibodies will be needed to recognise all these different epitopes. At first sight it might seem easiest just to make all of these different antibodies and then, when an infection occurs, antibodies with specificity for the antigens on the pathogen will bind to those antigens. This is not possible for the following reason. It has been calculated that a protective concentration of antibody, in terms of eliminating or neutralising an infectious agent, is 10 ng/ml. Therefore, to maintain this level of antibody for 10^{11} different antibody specificities in the blood of an adult human whose blood volume is 5 L would require:

5000 volume of blood (mls) ×
10^{11} number of different antibody molecules required ×
10 protective level of antibody (ng/ml)
$= 5 \times 10^{15}$ ng $= 5 \times 10^{6}$ g or 5000 kg of antibody

This is approximately 5 tonnes of antibody! It is clearly not possible to have this amount of antibody present all the time. An alternative strategy is therefore required to provide antibody against a particular pathogen when that antibody is required. Fortunately we are exposed to only a limited number of infections at one time and therefore, at any particular time, we only need to make antibody against a minute fraction of the total number of antigenic epitopes that exist. However, this poses a problem. Since the immune system does not know which antigens it will encounter, how does it select which antibodies to make? The only feasible way to do this is to make antibody specific for an antigen **in response to the presence of the antigen**. In other words the antibody response is **acquired** after the immune system is exposed to antigen.

Just as it is not possible to have pre-existing antibody against all possible antigens, it is not possible to have enough antibody-producing cells prior to exposure to antigen. The antibody-producing cells have to be produced after exposure to antigen. The situation is the same for T lymphocytes; although there will be some T cells with specificity for antigens derived from an infectious agent there will not be enough to deal with the infection and it will be necessary to generate new T cells specific for the antigen after exposure to antigen. Therefore specific immunity, whether it is antibody production or T lymphocyte responses, is acquired after exposure to antigen. The need to acquire specific immunity is another feature that distinguishes specific immunity from innate immunity. Innate immunity exists in the absence of infectious organisms, although the level of certain components can increase in the presence of an infectious agent.

6.2.2 Cellular interactions are required during specific immune responses

The requirement to generate specific immune responses after exposure to antigen poses another major challenge for the immune system *in vivo*. Simply put, how do you arrange things so that extremely rare antigen-specific B cells and T cells can interact both with antigen and with each other? If you consider a human being, it is quite a big place for small cells such as lymphocytes, and they need to be in the same place at the same time to encounter antigen. Normally we are infected with relatively small numbers of infectious particles and so the amount of antigen is quite low. Also, as described in Chapter 1, infectious agents, and the antigens associated with them, can arrive by various routes. Finally, antigen-presenting cells are required and essential for presenting antigen on MHC to T cells. So how does the immune system bring together rare antigen-specific T and B cells, antigen-presenting cells and antigen to a single anatomical site so that antigen stimulation and the cellular interactions involved in generating effector cells can occur?

Basically there are two strategies to achieve this. The first is to have collections of lymphocytes and the other cell types required to generate an immune response located in the sites of the body where infection is most likely to occur – these are the mucosa of the GI, respiratory and GU tracts. These collections of lymphoid cells are called the mucosal associated lymphoid tissue, or MALT, and are described in Section 6.3.4. The second strategy is to have lymphocytes circulate throughout the body and have specialised anatomical structures designed to capture antigen and enable lymphocytes specific for the antigen to meet the antigen and get together with the other cell types required to generate an immune response. These structures are the spleen and lymph nodes (Figure 6.2). By constantly recirculating through the spleen and lymph nodes, antigen-specific lymphocytes patrol the whole body. If antigen is present in a specific lymph node or the spleen, cells can stop recirculating and antigen-specific cells accumulate at the site of antigen.

6.3 The anatomy of the lymphoid system promotes the interaction of cells and antigen

The immune system consists of a series of organs and the vessels that connect them (Figure 6.2). Organs of the immune system have been divided into **primary** and **secondary** lymphoid tissue. The two primary lymphoid

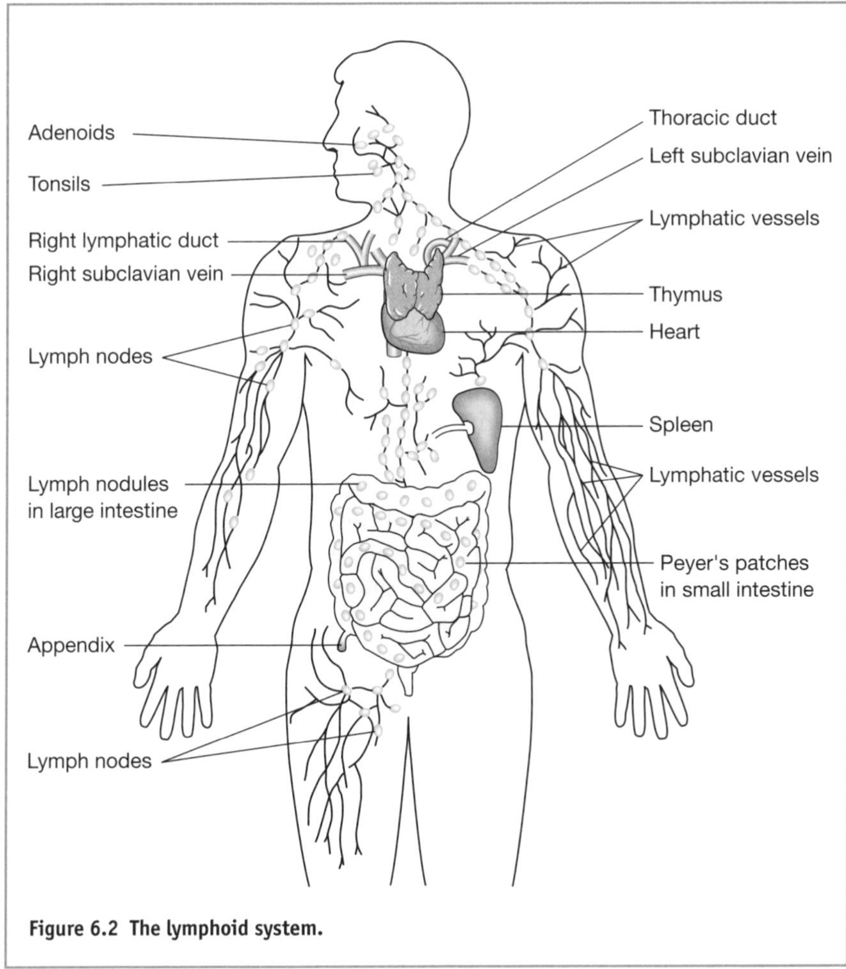

Figure 6.2 The lymphoid system.

organs in most mammals are the thymus and bone marrow. The term 'primary' is used because these are organs in which naïve lymphocytes develop from bone marrow precursors; B cells develop in the bone marrow and T cells in the thymus. This is an antigen-independent process and is described in Chapter 5. Once lymphocytes are formed they migrate to the secondary lymphoid organs and tissue. The secondary lymphoid tissue consists of the lymph nodes, spleen and MALT.

Other specialised structures of the lymphoid system are the **lymphatic vessels**. These are somewhat similar to blood vessels and connect most tissues in the body with lymph nodes and eventually with the bloodstream. Lymphocytes recirculate constantly through blood and lymphatic vessels. It should be noted that the spleen is an important lymphoid organ that does not have afferent lymphatic vessels; as will be seen later, it is devoted to dealing with antigens in the bloodstream.

6.3.1 Lymphatic vessels provide a transport system for cells and antigens

Lymphatic vessels are somewhat similar to blood vessels but there are important differences. One of these is that there is no heart, or equivalent to, pump the fluid in lymphatic vessels; instead the fluid is moved by the movement of muscles, including respiratory muscles, and valves in the lymphatic vessels stop the fluid from moving the wrong way. The smallest lymphatic vessels are called lymphatic capillaries. These are present in most tissues and internal organs (Figure 6.3). The lymphatic capillaries eventually join and become larger lymphatic vessels, which eventually re-enter the bloodstream via the largest lymphatic vessels, the thoracic duct and right lymphatic duct (Figure 6.2). The content of fluid and cells differs between blood and lymph. The fluid in lymphatic vessels is called lymph and is derived from the interstitial fluid surrounding cells in any tissue or organ. It is lower in protein content than blood plasma and the composition of cells is different from blood. Lymph contains lymphocytes and tissue-derived dendritic cells but does not normally contain red blood cells, monocytes or macrophages, or neutrophils, eosinophils or basophils. The lymphocytes and dendritic cells enter the lymphatic vessels from tissue (Figure 6.3).

6.3.2 Lymph nodes provide specialised environments for cells to meet each other

At regular intervals in the lymphatic system a number of lymphatic vessels meet at specialised structures called **lymph nodes** (Figure 6.3 and Plate 10). Lymph nodes are kidney-shaped structures; in humans they vary in size from a few millimetres to 2 cm in length (Figure 6.4). They are surrounded by a capsule consisting of connective tissue and are grey-white in colour. Immediately under the capsule is an area called the sub-capsular sinus. The lymphatic vessels that bring lymph fluid into the lymph nodes are called **afferent lymphatic vessels** and they pierce the capsule and empty their contents into the sub-capsular sinus. The lymph fluid filters through the lymph node, carrying with it lymphocytes and antigen. The lymph fluid eventually leaves the lymph node via the efferent lymphatic vessel, which exits the lymph node at a region called the hilum. The hilum is also the site of the lymph node where blood vessels enter and leave the lymph node.

Morphologically lymph nodes consist of a reticular framework made up of fibres and epithelial cells which provide structural support for the organ. Like many glands lymph nodes are divided into an outer cortex and an inner medulla. Within the cortex can be seen circular aggregates of lymphocytes which, under haematoxylin and eosin (H & E) staining, may or may not have a lighter staining centre. The aggregates with a darker staining area are called primary follicles and consist mostly of unstimulated B cells. The aggregates with a lighter staining centre are called germinal centres or secondary follicles and develop following antigen stimulation (see Section 7.4). Germinal centres are where plasma cells and memory B cells

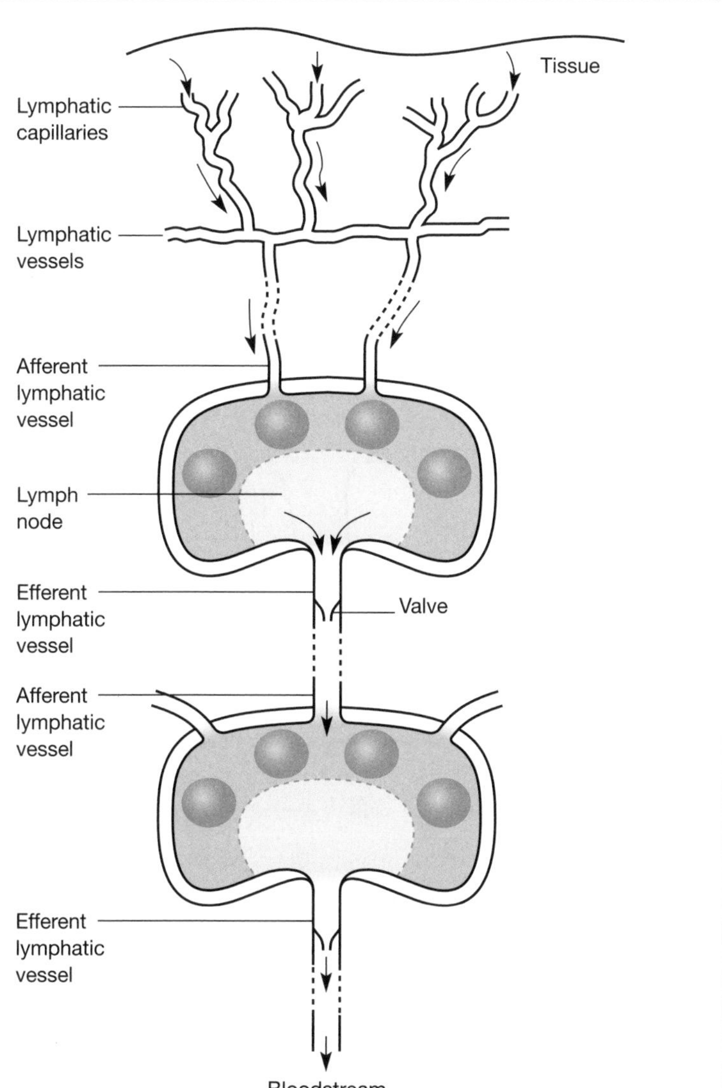

Figure 6.3 The lymphatic system. Lymphatic capillaries drain most tissues in the body. They join afferent lymphatic vessels, which eventually drain into lymph nodes. Lymphatic vessels leaving lymph nodes are called efferent lymphatic vessels, which may drain into other lymph nodes. Eventually efferent lymphatic vessels join to become larger and drain into the bloodstream via two major lymphatic ducts, the thoracic duct and the right lymphatic duct.

are generated. Between the follicles is an area called the paracortex, which contains densely packed T cells together with some dendritic cells and B cells. The medulla is less densely packed with cells and consists mainly of strands, or cords, of cells containing T cells, macrophages and large numbers of plasma cells.

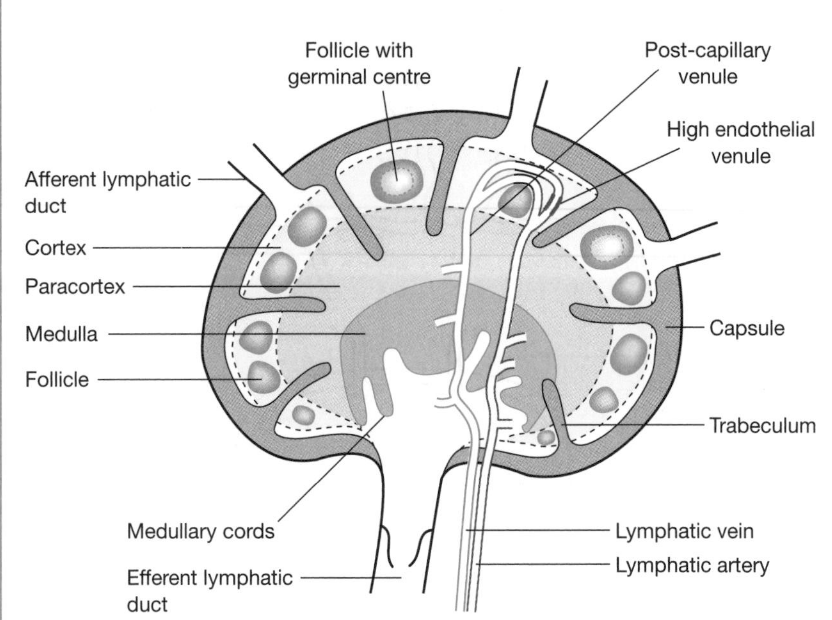

Figure 6.4 Lymph node. The innermost part of the lymph node is called the medulla and is where plasma cells secreting antibody are located. The outer part of the lymph node consists of the cortex, which is primarily a B cell area containing primary follicles and secondary follicles with germinal centres. The paracortex is located between the cortex and medulla and contains mainly T cells.

6.3.3 The spleen deals with antigen brought by the bloodstream

The spleen is located on the left side of the abdomen and in humans is oval with a length of about 12 cm and a diameter of 5 cm. It is surrounded by a connective tissue capsule. Examination of a cut spleen under the naked eye reveals a predominantly red surface with greyish white specks. These two areas are called the red pulp and the white pulp. The red and white pulp areas of the spleen can be clearly seen histologically (Figure 6.5 and Plate 10). The spleen is surrounded by a collagenous capsule and bundles of fibres called trabeculae extend from the capsule into the internal part of the spleen, partially separating it into compartments. The structure of the spleen is further maintained by a reticular framework. The function of the red pulp is to filter the blood and remove aged or damaged red blood cells and other debris; this function is described in Box 6.1.

The immune function of the spleen is performed primarily in the white pulp and the spleen is designed to filter antigen from the blood so that T and B cells can encounter the antigen and respond to it. Blood enters the spleen via the splenic artery, which divides into trabecular arteries; these in turn divide into many smaller arteries called the central arteries. The central arteries become thinner arterioles, which eventually enter the red pulp.

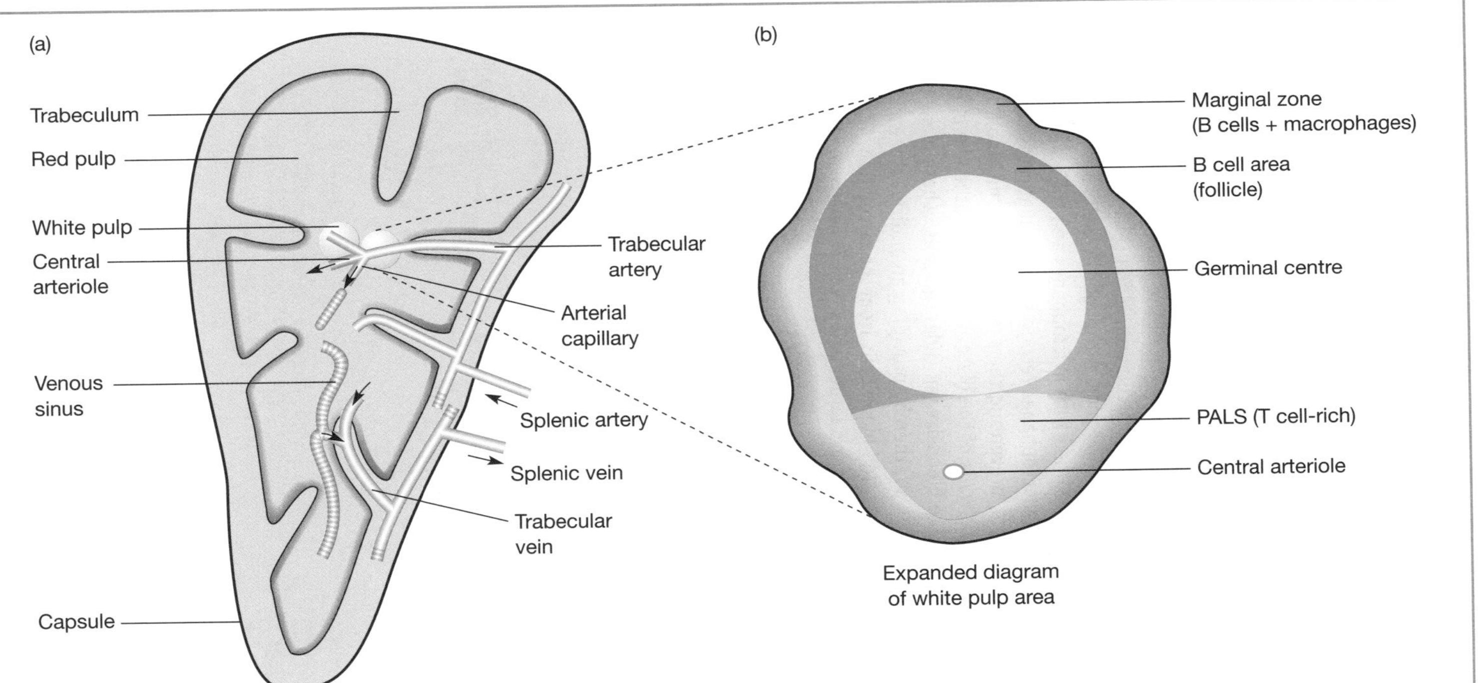

Figure 6.5 Spleen. (a) Transverse section of spleen. Arrows indicate direction of blood flow. Most of the spleen consists of red pulp, which is primarily involved in removing old erythrocytes and platelets. The lymphoid area of the spleen is the white pulp, which surrounds the central arterioles. (b) Expanded view of white pulp. The area immediately surrounding the central arteriole is the peri-arteriolar lymphoid sheath (PALS), which contains mostly T cells. Adjacent to the PALS is the B cell area containing primary follicles and secondary follicle with germinal centres. Between the white pulp and the red pulp is the area called the marginal zone, which contains B cells and macrophages.

BOX 6.1: THE SPLEEN AND RED BLOOD CELLS

The main function of the red pulp of the spleen is to filter the blood and remove old (effete) and damaged red blood cells. Not surprisingly the spleen has an extensive blood supply. Blood enters the spleen via the splenic artery, which divides into many arteries called central arteries. These arteries become thinner arterioles, which eventually enter the red pulp. The red pulp contains thin-walled blood vessels called venous sinusoids and in between these sinusoids are areas called splenic cords. Blood cells are emptied out of the arterioles directly into the splenic cords. To re-enter the blood circulation, blood cells must traverse the splenic cords and enter the venous sinusoids. Although the walls of the venous sinusoids are incomplete they are lined with macrophages and the splenic cords are full of macrophages. These macrophages recognise and phagocytose old or damaged red cells and platelets, preventing their re-entry into the blood. Blood cells must squeeze through the walls of the discontinuous endothelium of the venous sinusoids to re-enter the bloodstream. Older red cells, whose membranes are less elastic, are unable to do so, thereby providing an additional mechanism by which the spleen removes aged red blood cells from the circulation. The venous sinusoids eventually join to form larger veins, which eventually leave the spleen via the splenic vein.

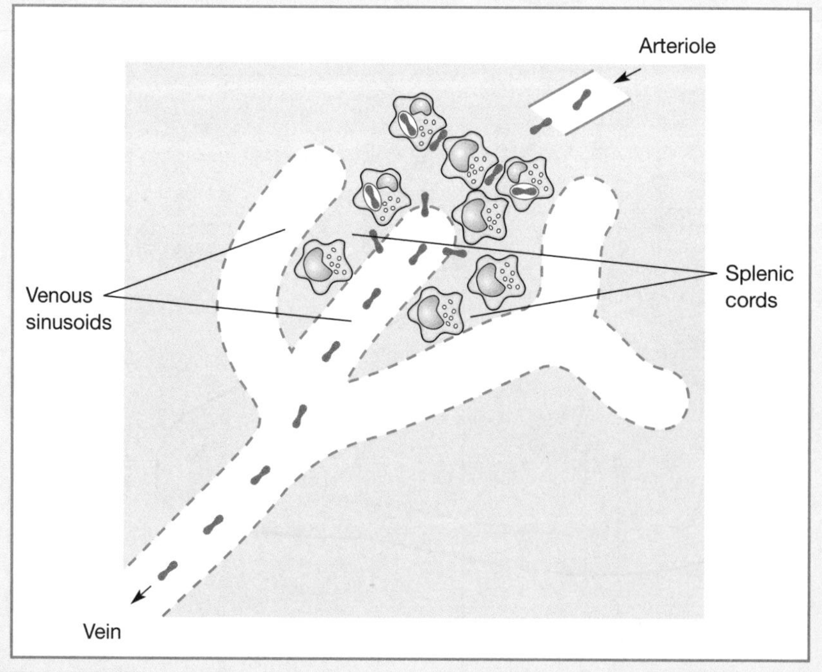

Surrounding the central arteries are sheaths of lymphoid cells consisting of T cells, B cells, macrophages and dendritic cells. T cells are usually located in the area immediately surrounding the arteries, which is called the peri-arteriolar lymphoid sheath, or PALS. The area immediately surrounding the PALS is called the marginal zone and contains B cells and macrophages. Structures called primary follicles and germinal centres are found at regular intervals along the sheaths; the germinal centres are the main sites of antibody production. As in the lymph node, the primary follicles represent collections of unstimulated recirculating naïve B cells and the germinal centres form after antigen stimulation.

6.3.4 Mucosa have special lymphoid structures to deal with antigen

All mucosa, including those of the gastro-intestinal, respiratory and genito-urinary tracts have specialised lymphoid structures associated with them. This has been termed mucosal associated lymphoid tissue, or MALT. MALT consists of more or less structured lymphoid tissue distributed throughout the mucosa of the GI, respiratory and GU tracts. The more structured elements of MALT include the tonsils, adenoids, appendix and Peyer's patches, which line the ileum (Figure 6.2). Less structured lymphoid follicles are also found in the lamina propria of the intestine and the mucosa of the respiratory and GU tracts. MALT contains lymphoid follicles like those seen in the spleen and lymph nodes (Plate 11) and is described in more detail in Chapter 7.

MALT actually comprises the largest component of the immune system in terms of cell numbers and of these the biggest is the lymphoid system associated with the gut. This is not surprising because the gut, having a surface area of 200 m^2, is the largest area that is exposed to potentially infectious pathogens. However the situation in the gut is more complicated because of the presence of huge numbers of commensal bacteria. These are bacteria that are not normally pathogenic and can be beneficial such as helping in the breakdown and digestion of complex polysaccharides. Therefore the immune system in the gut must be able to deal with pathogenic bacteria and eliminate them but at the same time not eliminate the useful commensal bacteria.

6.4 Lymphocytes continually recirculate through blood, tissues and lymphatic vessels

As mentioned above, the second way in which rare antigen-specific cells can meet with each other and interact with antigen and APCs is through lymphocyte recirculation. Lymphocytes differ from other leukocytes in that they constantly migrate between blood, the lymphatic vessels and the lymphoid organs. This process is called **lymphocyte recirculation** and the extent of this cell movement is quite remarkable. Lymphocytes circulate in the bloodstream for an average of 30 minutes. Approximately 45% of the cells enter the spleen where they have a transit time of about

5 hours before returning to the bloodstream via the splenic vein. However, lymphocytes are unique in having the ability to migrate directly from the bloodstream into lymph nodes. They do this by leaving the blood in **high endothelial venules** (HEVs) of the lymph nodes. High endothelial venules are specialised blood vessels that have a cuboidal (high) endothelium (Plate 10). About 40% of the lymphocytes in the bloodstream enter lymph nodes via this route and reside in the lymph nodes for 12 hours or so before leaving via the efferent lymphatics, which eventually transport the lymphocytes to the thoracic duct and back into the bloodstream. Lymphocytes also enter lymphatic vessels in tissues and can travel along the lymphatic vessels to lymph nodes. About 15% of lymphocytes entering a lymph node enter via the lymphatic vessels, and 85% enter across HEVs from the blood. Because of this extensive recirculation, lymphocytes migrate from blood to spleen and lymph nodes one or two times/day.

The migration of lymphocytes from blood into lymph nodes involves the same basic processes as seen with leukocytes entering sites of inflammation (see Chapter 2). The blood flow in the lymph node slows because of vasodilation. However, this vasodilation is not caused by inflammation but occurs because the high endothelial venules that are the site of migration from blood to lymph nodes are located where small-diameter capillaries become larger-diameter venules. This change in diameter slows the blood in the same way as the change in diameter caused by inflammation. This slowing of blood flow allows the lymphocytes to roll along the vessel wall. The subsequent stages of weak binding of adhesion molecules on the lymphocytes to those on endothelial cells, activation of adhesion molecules, firm attachment and crossing of the endothelium are the same processes as seen for leukocytes entering sites of inflammation, although the adhesion molecules and chemokines involved are different (see Table 6.1). The HEVs are located in the paracortex of the lymph node so that the lymphocytes enter the lymph node in a T cell area. Since B cells also leave the bloodstream and enter the lymph node they must then migrate to the B cell area, the cortex and follicles. The movement of the B cells to the follicles is promoted by the chemokine B lymphocyte chemo-attractant (BLC), which is produced by stromal cells of the B cell areas and follicular dendritic cells and binds to the chemokine receptor CXCR5, which is expressed on the B cells.

Because of this recirculation rare antigen-specific cells continually patrol the body looking for antigen. It is within lymph nodes, spleen and MALT that lymphocytes can meet antigen and also interact with each other and, as described in Chapter 7, these structures are designed to maximise the interactions of cells with antigen and each other.

Table 6.1 Adhesion molecules involved in migration of lymphocytes to lymphoid tissue

Adhesion molecule	*Lymphocyte distribution*	*Endothelial cell ligand*	*Type of migration affected*
L-selectin	B + T	GlyCAM-1	Migration to peripheral lymph node
		MadCAM-1	Migration to mucosal lymphoid tissue
$\alpha_4\beta_7$ integrin	T	MadCAM-1	Migration across mucosal HEV
$\alpha_L\beta_2$ integrin	T	ICAM-1,2,3	Migration across HEV

6.5 Summary

- There are a number of types of specific immune response: the production of antibody, the generation of CD8 cytotoxic T cells, delayed hypersensitivity responses and responses involving the recruitment of neutrophils.
- Immune responses must be generated in response to infection and this requires the interaction of rare antigen-specific cells with each other and with antigen.
- The immune system has specialised anatomical structures that promote the association of antigen with antigen-presenting cells and antigen-specific B and T cells.
- These structures include the lymphatic vessels, which connect together lymph nodes and eventually join with the blood system.
- The major lymphoid organs are the lymph nodes and spleen, together with a vast collection of lymphoid tissue associated with the mucosa, which is called mucosal associated lymphoid tissue or MALT.
- MALT consists of aggregates of structured lymphoid tissue such as the tonsils, appendix and Peyer's patches and loose aggregates of lymphoid cells called nodules.
- Lymphocytes recirculate extensively between the bloodstream and lymph nodes and spleen. This recirculation is controlled by adhesion molecules and chemokines and increases the chances of lymphocytes meeting their specific antigen.

6.6 Questions

1) What do the letters refer to in the diagram overleaf?

2) How do lymphatic vessels differ from blood vessels?

3) What is lymphocyte recirculation and why is it important?

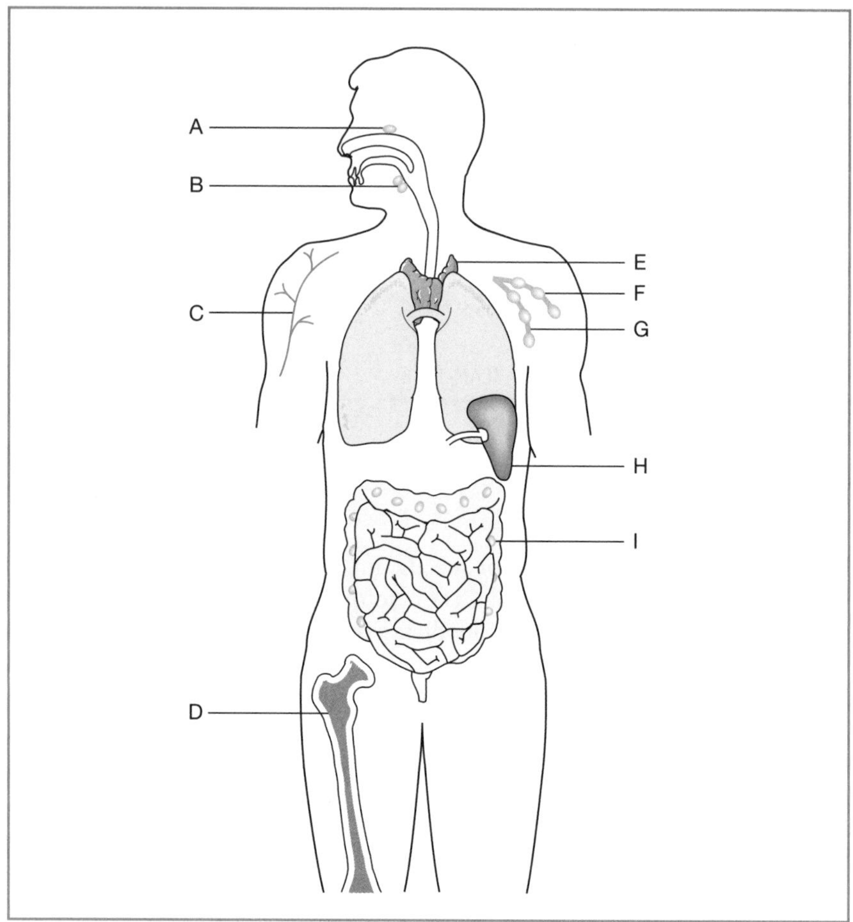

4) Where would you expect immune responses to the following antigens to be generated?

   (i) A blood-borne parasite
   (ii) A pathogenic gut bacterium
   (iii) A bacterium that has penetrated the skin
   (iv) A respiratory virus

5) What is the percentage of lymphocytes that enter a lymph node via the afferent lymphatic vessels compared to from the blood?

   A) 15:85
   B) 30:70
   C) 50:50
   D) 70:30
   E) 85:15

The answers to these questions can be found on page 336.

6.7 Further reading

1) Von Andrian UH, Mempel TR. (2003) Homing and cellular traffic in lymph nodes. *Nature Reviews Immunology* 3:867–878.

2) Mebius RE, Kraal G. (2005) Structure and function of the spleen. *Nature Reviews Immunology* 5:606–616.

3) Nagler-Anderson C. (2001) Man the barrier! Strategic defences in the intestinal mucosa. *Nature Reviews Immunology* 1:59–67.

CHAPTER SEVEN

Anatomical and cellular aspects of antibody production

Learning objectives

To learn how antibody is produced *in vivo* in response to antigen. To know about the role of CD4 T cells and cytokines in antibody production. To understand how B cells switch the class of antibody they make and increase the affinity of their antibody during a response to antigen and why this makes the response much more efficient. To know that B cells can become either plasma cells or memory B cells.

Key topics

- Requirements for antibody production
- CD4 T cells and cytokines in antibody production
- B lymphocytes and antibody production
  - Clonal selection
  - Class switch
  - Affinity maturation
  - Differentiation into plasma or memory cells
- The advantages of class switch and affinity maturation

7.1 Overview of antibody production

One of the most important immune responses to infection is the production of antibody. As described in Chapter 2, antibody can bind to antigens on pathogens and has a number of functions that can protect against the pathogen bearing the antigen. But, as explained in Chapter 6, antibody has to be made after you are infected with the pathogen. Antibody is actually produced by plasma cells that differentiate from antigen-specific B cells.

B lymphocytes that have never encountered antigen before have IgM and IgD on their surface (see Figure 5.2). They are able to recognise and bind antigen through IgM and IgD on their surface but at this stage cannot secrete antibody. For an individual B cell the IgM and IgD have identical variable Ig Vh regions and identical light chains and therefore the same antigen specificity. However, different B cells will have specificity for different antigens. To become a plasma cell, a B cell must undergo a process of **cell differentiation**. Many cell types, not just those of the immune system, undergo differentiation, usually during development or as a result of external signals such as hormones. Cell differentiation at its most basic involves changes in gene expression that alter the function of a cell. B cells must undergo a number of changes in their differentiation into plasma cells. This includes changing from a cell with Ig on its surface to a cell that can secrete enormous amounts of antibody. Plasma cells are basically cellular antibody-producing factories. B lymphocytes also usually go through two specialised differentiation processes during their differentiation into plasma cells. These processes are designed to increase the quality and range of antibodies produced and are:

- **Affinity maturation of antibody.** This is a specialised process that results in antibody with a high affinity for antigen and that is more effective at dealing with pathogens.
- **Antibody class switch.** As described in Chapter 3, there are many different classes of antibody, IgG, IgA and IgE, in addition to IgM and IgD. These classes of antibody have different biological functions. During differentiation into plasma cells, B cells can change the antibody on their cell surface from IgM and IgD to one of the other classes of antibody, which means that when they differentiate into plasma cells they secrete different classes of antibody with different biological function. It should be noted that an individual plasma cell will secrete only one class of antibody.

It can be seen that the differentiation of B lymphocytes into plasma cells is a complex process that must be carefully regulated. An important cell in the regulation of B cell differentiation is the CD4 T cell. The CD4 T cells that are involved in antibody production are called helper T cells, or Th for short. They were originally named helper T cells because although they didn't make antibody themselves, they were required for B cells to make antibody, i.e. helped in antibody production. Most antibody production is dependent on Th although some antigens can stimulate antibody production without the need for Th (Box 7.1). However, just as B cells that have never seen antigen before cannot secrete antibody and have to differentiate into plasma cells to do so, CD4 T cells that have never seen antigen before have no helper function and must undergo their process of differentiation into Th following stimulation by antigen peptide/class II MHC.

Another important feature of specific immune responses is the need for antigen-specific cells to proliferate after appropriate stimulation by antigen. If you are infected with a microbe for the first time you will have very few lymphocytes that are specific for the microbial antigens. Therefore it is important to increase the number of antigen-specific cells as quickly as possible and this is achieved by rapid cell proliferation. When you are infected, lymphocytes that are specific for antigens on the microbes are stimulated to proliferate. Because only lymphocytes that are specific for the

BOX 7.1: T-DEPENDENT AND T-INDEPENDENT ANTIGENS

B cells, like T cells, require two stimuli for activation by most antigens. The first stimulus is recognition of antigen by membrane Ig (mIg) and the second stimulus comes from the CD4 T cell. There are certain antigens, however, that can deliver a strong enough signal through the mIg complex to activate B cells without any help from CD4 T cells. These are called T-independent antigens because they can stimulate antibody production without T cell help. Some of these antigens can stimulate TLRs (see Chapter 2) and this seems to get round the need for Th. Others are polysaccharides with highly repetitive epitopes causing extensive cross-linking of the mIg and strong signalling (see figure). Responses to these antigens in the absence of T cells usually involve only IgM production although some other classes, especially IgG3, can be produced. Most antigens cannot cause such strong signalling in B cells and the cells require additional signals or co-stimuli and T cell help. The role of Th is to provide additional signals for the initial activation of B cells and to control the subsequent proliferation and differentiation of B cells both by secreting cytokines and by delivering signals to B cells by cell-to-cell contact between the B cell and CD4 T cell.

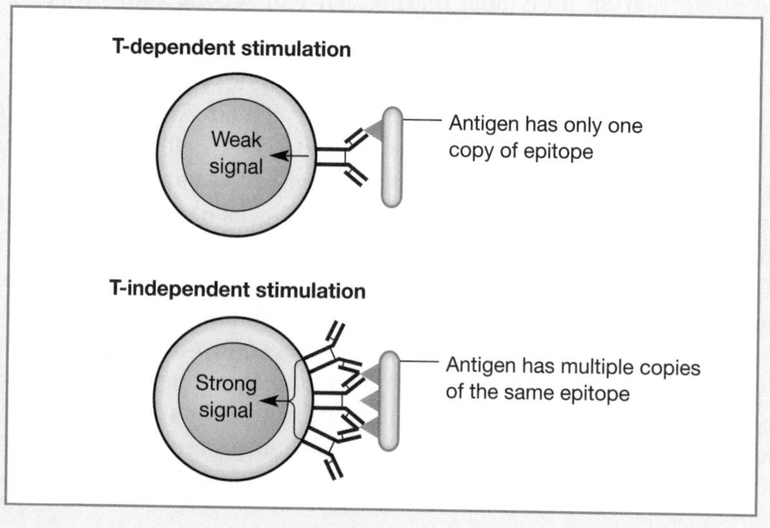

antigen are stimulated to proliferate, this process has been called **clonal selection** and the proliferation of antigen-specific cells is called **clonal proliferation** or **clonal expansion**. The principle of clonal selection and expansion in B cells is shown in Figure 7.1. T cells will undergo similar clonal selection and expansion after stimulation by antigen/MHC.

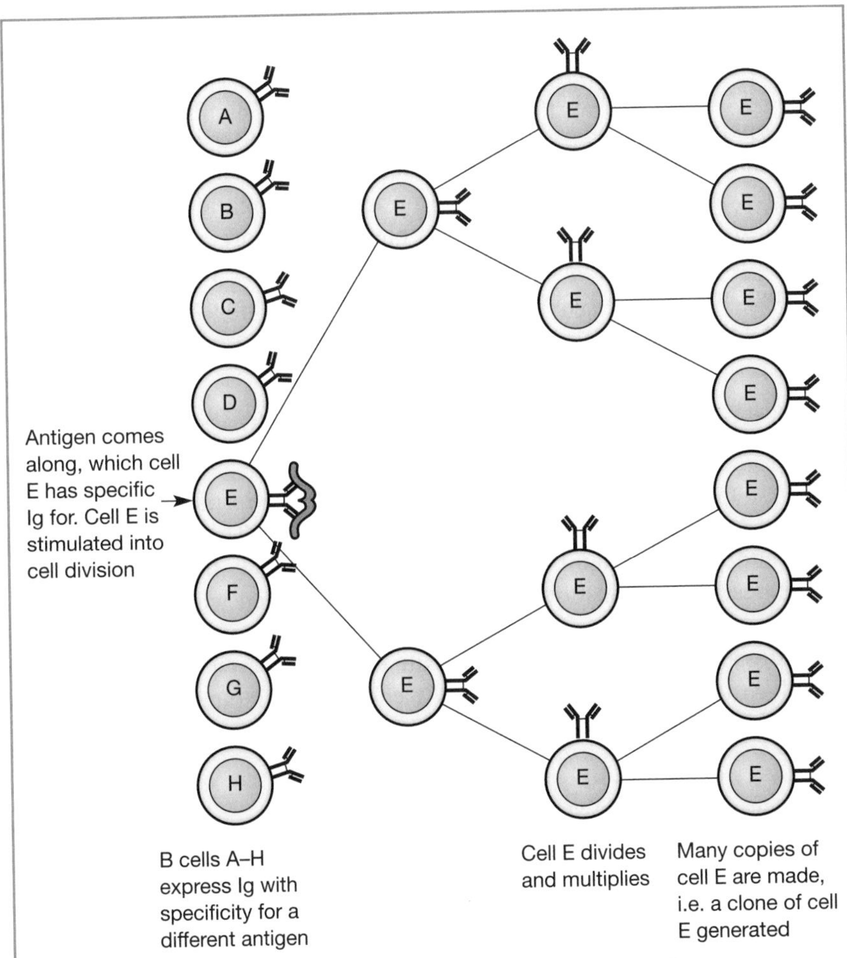
A
B
C
D
E
F
G
H
Antigen comes along, which cell E has specific Ig for. Cell E is stimulated into cell division
B cells A–H express Ig with specificity for a different antigen
Cell E divides and multiplies
Many copies of cell E are made, i.e. a clone of cell E generated

Figure 7.1 Clonal selection and expansion in B cells. The B cells labelled A–H all have Ig on their surface but the Ig on each B cell is specific for a different antigenic epitope. When antigen comes along, only B cell E has Ig that can bind to an epitope on the antigen. Binding of the antigen to the Ig on B cell E stimulates the B cell to divide many times, forming a clone of B cells derived from E.

Therefore the production of antibody can be divided into a number of stages:

1. Presentation of antigen to, and stimulation of, antigen-specific CD4 T cells to proliferate and differentiate into Th.
2. Stimulation of B cells by antigen and interaction with Th.
3. Proliferation of B cells and their differentiation into plasma or memory cells.

These events occur primarily in lymph nodes, spleen and mucosal lymphoid tissue. The fundamental ways in which the spleen and lymph nodes operate to achieve these events are quite similar although some of the details differ. A major difference between lymph nodes and spleen is that antigen is brought to the lymph nodes by lymphatic vessels, whereas antigen in the spleen comes from the blood. Therefore antigen located in tissues enters the lymphatics and stimulates responses in the lymph nodes draining the tissue. Antigen in the blood enters the spleen and stimulates immune responses there.

7.2 Activation of CD4 T cells (0–5 days)

The first, and critical, stage in any adaptive immune response is the stimulation of antigen-specific CD4 T cells to become Th. This occurs in lymph nodes and the spleen and in organised mucosal tissue, the latter of which will be described in Section 7.5. When we are infected with a microbe this occurs in various tissue sites, e.g. under the skin or sometimes in the blood. Therefore antigen has to get from the site of infection to a lymph node or to the spleen to be able to activate CD4 T cells to proliferate and differentiate into Ths. The production of Th can therefore be considered in three stages:

1. Delivery of antigen to lymph node or spleen.
2. Activation of antigen-specific CD4 T cells.
3. Proliferation of activated CD4 T cells and differentiation into Th.

7.2.1 Delivery of antigen to lymph nodes or spleen

When a tissue is infected by a microbe, the microbe will initially encounter cells of the innate immune system, such as tissue macrophages described in Chapter 2; this may initiate an inflammatory response. Another cell type found in tissues plays a vital role in stimulating CD4 T cells and initiating an immune response. This cell is called a **dendritic cell (DC)**. DCs are bone marrow-derived cells that are found in almost all tissues. DCs in non-lymphoid tissues are called tissue DCs; Langerhans cells are a specialised type of tissue DC found in the skin (Plate 12). Tissue DCs are highly efficient at taking up antigen by pinocytosis, receptor-mediated endocytosis (using the mannose receptor) or phagocytosis. Tissue DCs can also recognise microbial products through the expression of toll-like receptors (TLRs – see Chapter 2) and become activated as a consequence of this. When they are activated, DCs are stimulated to leave the tissue site and enter the draining lymphatic vessels in which they will be carried to the local lymph node. The migration of dendritic cells to the lymphatic vessels is induced by a chemokine called the Epstein–Barr virus-induced receptor ligand chemokine (ELC), which binds the chemokine receptor CCR7, which is upregulated on the activated DCs. As they migrate from the tissue to the lymph node the DCs process the antigen and express antigen fragments on their class II MHC for recognition by CD4 T cells. Once in the lymph node the dendritic cells migrate to the T cell area, the paracortex, where they can encounter CD4 T cells. Antigen derived from microbes at the site of infection can also enter the lymphatic ducts in free form and be carried to the local lymph nodes. As the antigen percolates through the lymph node it is picked up by DCs in the paracortex, where it can be processed and antigenic fragments presented to T cells on class II MHC.

Antigen can also arrive in the spleen in two ways. DCs travel from tissue via the blood to the spleen carrying antigen for presentation to CD4 T cells. Alternatively many of the central arteries in the spleen terminate near, or in, the marginal zone. Blood containing antigen is emptied into this area and some of the antigen enters the PALS area where it can be picked up and processed by dendritic cells for presentation to CD4 T cells (Figure 7.2).

7.2.2 Activation of CD4 T cells begins within hours of infection

The activation of CD4 T cells is one of the most crucial early events in initiating a specific immune response. The ideal situation is that the CD4 T cell should be able to recognise and respond to as few molecules of antigen/class II MHC as possible so that the response can be initiated soon after infection. However, it is important that the response is controlled so that only foreign antigens are responded to. One of the mechanisms by which the activation of CD4 T cells is controlled is to have a requirement that the CD4 T cell must receive two signals to be activated. The first signal is generated by the TCR on the CD4 T cell surface recognising antigen in association with class II MHC. The cell bearing the antigen on its class II MHC is called an **antigen-presenting cell**, or APC. The requirement for recognition of antigen means that CD4 T cells are activated only in the presence of antigen to which they are specific; there would be no point in activating the cells if their antigen was not present.

The second requirement for the activation of CD4 T cells is that other molecules on the T cell must bind to molecules on the APC. When these other molecules on the T cell are bound they deliver another signal to the T cell and this other signal is known as a **co-stimulus** (see Figure 7.3). The most important co-stimulatory molecule on the T cell is called CD28. This

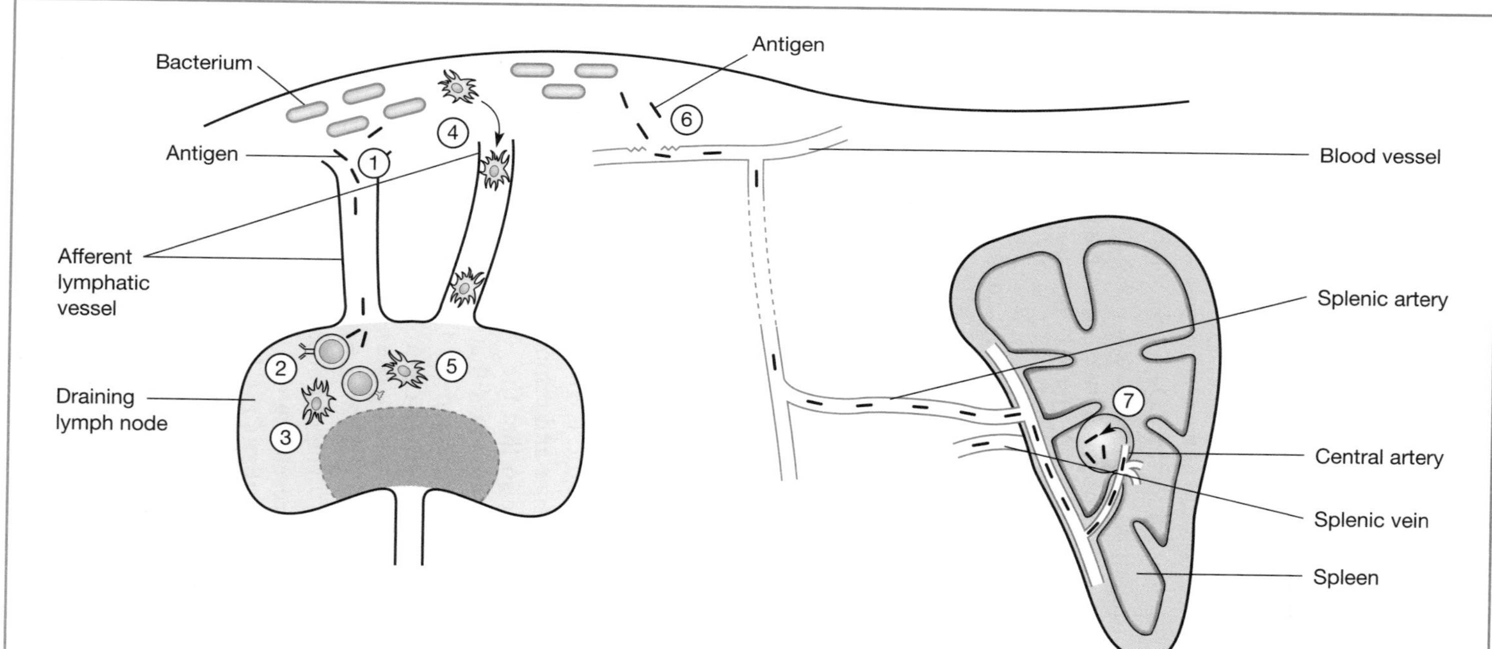

Figure 7.2 Transport of antigen to the spleen and lymph node. In this example, bacteria have infected a tissue site. Antigens shed from the bacteria can enter the draining lymphatics (1) to be transported to the draining lymph node where they can directly stimulate B cells (2) or be taken up by interdigitating dendritic cells within the paracortex of the lymph node, processed and antigenic peptides presented to CD4 T cells by class II MHC on the dendritic cells (3). Alternatively antigen in tissues can be taken up by tissue dendritic cells (4) that are stimulated to enter the lymphatics and migrate to the draining lymph node where they can present antigenic peptides to CD4 T cells (5). If there is damage to blood vessels, antigens enter the blood (6) and are transported to the spleen. Here they enter the white pulp (7) and can stimulate B cells or be taken up by interdigitating dendritic cells and antigenic peptides presented to CD4 T cells.

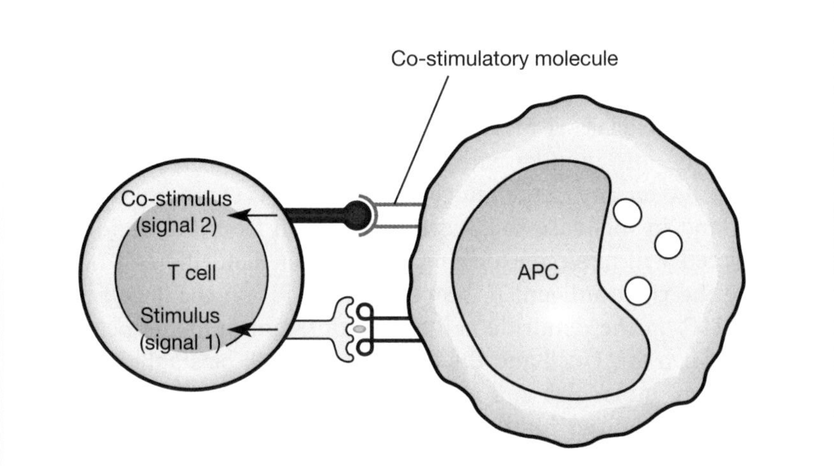

Figure 7.3 CD4 T cells require two signals to be activated. Signal 1 is provided by the TCR recognising Ag/class II MHC. The second signal, known as a co-stimulus, is provided by the cell presenting the antigen. The most important co-stimulus is provided by CD80 or CD86 on the APC, which binds to CD28 on the CD4 T cell.

can bind to either of two molecules called CD80 and CD86 (these were formerly called B7.1 and B7.2 respectively). If the TCR on a CD4 T cell binds antigen/class II MHC and CD28 binds to CD80 or CD86, the CD4 T cell will receive its two signals and be activated. In the absence of a co-stimulatory signal the CD4 T cell will not be activated and no response will occur. However, mice lacking CD28 can still generate some Th although not as effectively as normal mice, indicating that other molecules can act as co-stimuli.

Therefore an APC must express both class II MHC and CD80 or CD86 to be able to optimally activate a resting CD4 T cell. The expression of CD80 and CD86 itself is carefully controlled so that CD4 T cells only receive the co-stimulus through CD28 and are activated when it is appropriate. The tissue dendritic cells described above express some class II MHC and co-stimulatory molecules but are not very good at activating CD4 T cells. However, the DCs that have migrated to lymph nodes or the spleen following activation by microbial products have upregulated CD80 and CD86 in addition to upregulating class II MHC and have now become cells that are very good at activating CD4 T cells.

When the DCs have migrated to the T cell areas of the lymph node they secrete the chemokine ELC. ELC can bind to its receptor CCR7, which is expressed by T cells, and therefore the T cells are attracted to the dendritic cells and contact between the two cell types is promoted. In fact the interaction between T cells and dendritic cells is a very dynamic process. T cells move about very rapidly (for a cell) within the lymph node or spleen and this enables them to encounter many dendritic cells. When a T cell encounters a dendritic cell it binds to the dendritic cell in a non-antigen-specific way through adhesion molecules. The T cell then uses its TCR to check if the dendritic cell is expressing antigen/MHC for which the T cell is specific. A particular dendritic cell will express many different antigenic peptides on all its class II MHC molecules. However, most of the CD4 T cells that come into contact with the dendritic cells will not have TCRs that are specific for any of the antigen/class II MHC expressed by the dendritic cell and in this case the T cell will detach from the dendritic cell and the T cell will move on looking for another dendritic cell hopefully expressing the right antigen. If, however, the TCR on the T cell binds to antigen/MHC on the dendritic cell a very different outcome occurs. The binding between the T cell and the dendritic cell will be stabilised through other adhesion molecules and an event occurs which results in the formation of a specialised intercellular communication structure between the dendritic cell and the CD4 T cell known as the **immunological synapse** (Figure 7.4). The term 'immunological synapse' was coined because the interaction between the T cell and dendritic cell was likened to that between neurons, which occurs via neuronal synapses.

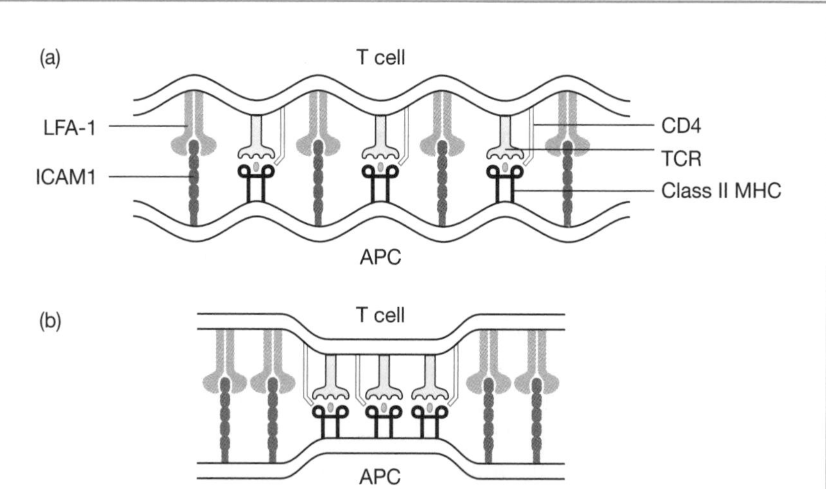

Figure 7.4 Formation of the immunological synapse. (a) Initially the TCRs on the CD4 T cell that come into contact with class II MHC/peptide on the APCs are randomly distributed on the respective cell surfaces. CD4 also binds to class II MHC. ICAM-1 on the APC binds to LFA-1 on the T cell promoting adhesion between the two cell types. (b) Following signalling from the TCR/CD3 complex, cytoskeleton events results in reorganisation of molecules on the cell membrane to form supra-molecular activation clusters (SMACs) with TCRs and associated signalling molecules located in the central region (c-SMAC) and the LFA-1/ICAM-1 located in the periphery (p-SMAC) of the synapse. Other molecules such as CD28 also locate to the c-SMAC to provide additional signalling.

Immunological synapse

The original binding of CD4 T cells to dendritic cells is through the cell surface integrin LFA-1 on the T cell binding to ICAM-1 on the dendritic cell. This binding is of low affinity and if the TCR does not recognise antigen on the dendritic cell, the binding of LFA-I to ICAM-1 will stay low affinity and the T cell will dissociate from the APC. If, however, the TCR recognises antigen being presented by the DC the TCR will bind to the antigen/class II MHC. It is also important for TCR-mediated signalling that CD4 molecules on the T cell bind to class II MHC. As a consequence of TCR recognition of antigen, signalling events occur in the T cell that result in activation of LFA-1 so it now binds ICAM-1 on the DC with high affinity. This stabilises the binding of the T cell to the dendritic cell and allows more TCR molecules on the T cell to bind to antigen/class II MHC on the dendritic cell. There is also recruitment of cytoskeletal elements within the T cell that results in redistribution of the molecules on the T cell surface into what are known as supramolecular activation clusters (SMACs). This results in SMACs where the TCRs and their associated CD3 complex (see Section 4.2), CD4 and CD28 form a central cluster surrounded by a peripheral ring of LFA-1. This clustering of the TCR/CD3 complex, CD4 and CD28 maximises the delivery of both signal 1, through the TCR/CD3 complex, and signal 2, through CD28, to the CD4 T cell, resulting in its activation.

7.2.3 Activated CD4 T cells are stimulated to proliferation and differentiate

After activation by antigen-presenting DCs the CD4 T cells are stimulated to proliferate. The proliferation of CD4 T cells is driven by cytokines. Activated CD4 T cells produce a cytokine called interleukin-2 (IL-2). IL-2 is a protein of molecular mass 15 kDa and is primarily a growth factor. It can stimulate the proliferation of CD4 T cells, CD8 T cells and B cells. Upon stimulation through the TCR, CD4 T cells are induced to express receptors for IL-2 (IL-2R). Since the same CD4 T cell can both make IL-2 and express the IL-2R, IL-2 is able to act in an autocrine manner. IL-2 can also act in a paracrine manner and bind to IL-2Rs on neighbouring cells that have been stimulated through antigen recognition to express IL-2Rs (Figure 7.5). Through the autocrine and paracrine actions of IL-2, CD-4 T cells that have recognised antigen undergo extensive proliferation so that the number of antigen-specific CD4 T cells is increased dramatically. This proliferation peaks 3–4 days after initial contact with the antigen and, although it is difficult to measure exact figures *in vivo*, eventually results in a 10 000- to 100 000-fold increase in the number of antigen-specific CD4 T cells.

Following the period of proliferation, the CD4 T cells begin to differentiate and acquire the functions they need to act as Th cells. The main function that Th cells acquire is the ability to secrete a variety of different cytokines (Figure 7.5). There are also changes in the molecules expressed on the surface of the T cells and, through the expression of these molecules

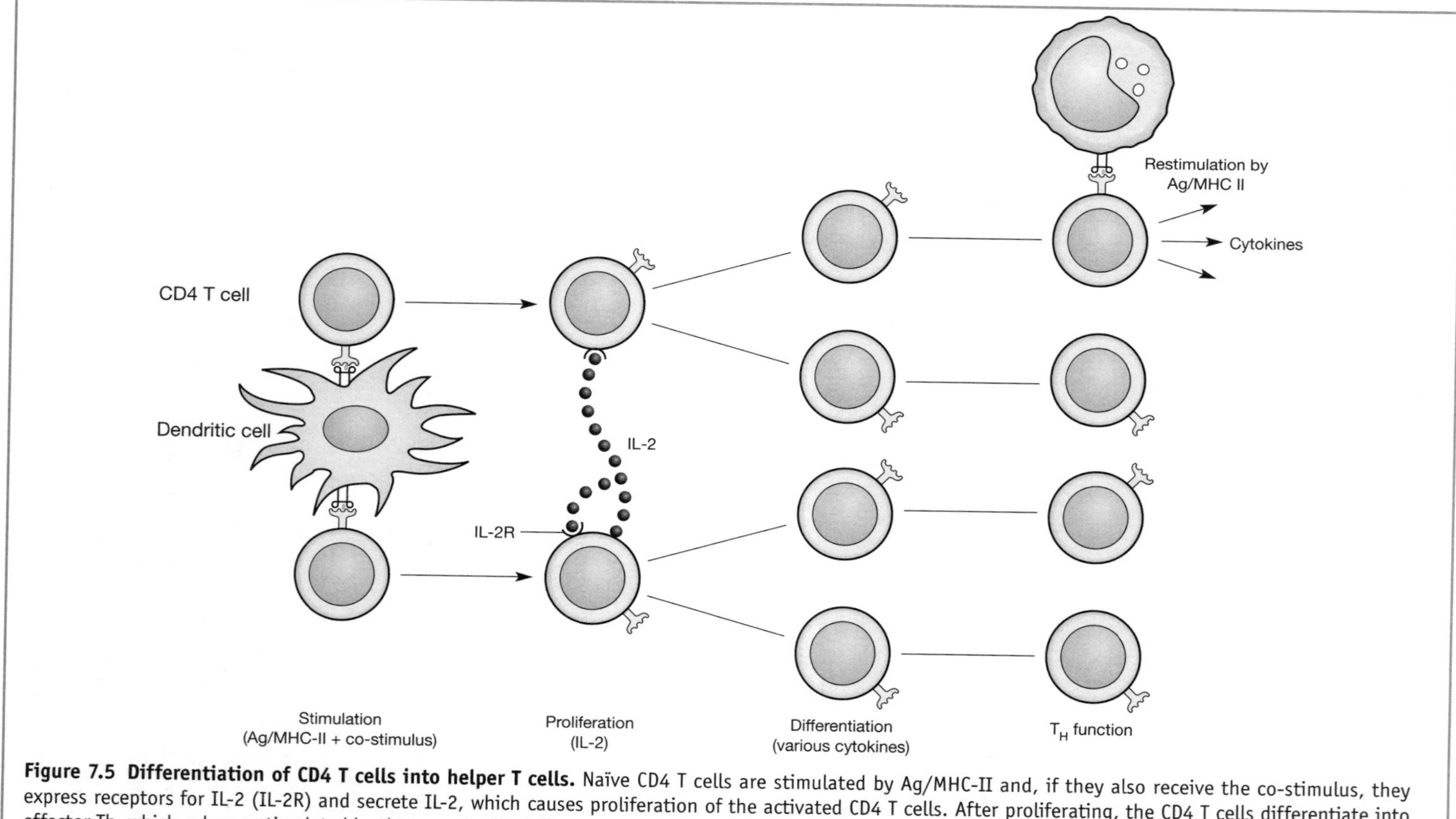

Figure 7.5 Differentiation of CD4 T cells into helper T cells. Naïve CD4 T cells are stimulated by Ag/MHC-II and, if they also receive the co-stimulus, they express receptors for IL-2 (IL-2R) and secrete IL-2, which causes proliferation of the activated CD4 T cells. After proliferating, the CD4 T cells differentiate into effector Th, which, when restimulated by the same Ag/MHC-II, secrete cytokines that affect the function of other cells.

and the secretion of cytokines, the Th cells are able to control the production of different effector responses. The Th migrate to the boundary of the paracortex and the cortex, where they can interact with B cells.

7.3 Stimulation of B cells by antigen and interaction with Th (0–5 days after antigen)

Antigen filtering through the lymph node cortex percolates through the follicles, where it can be captured by macrophages (Figure 7.6a). If the antigen on the macrophage comes into contact with B cells whose Ig is specific for the antigen, two events occur. One is that the B cells take up the antigen, process it and present antigenic peptides on their class II MHC. The B cells are also partially activated by antigen and migrate to the junction of the cortex and paracortex, where they encounter Th cells that have migrated to the edge of the paracortex. The interaction between the B and Th cells involves both direct cell–cell contact and the production of cytokines (Figure 7.7). The activated B cells express increased class II MHC on their cell surface and so are able to present antigen to the Th. The Th that are specific for the antigen/class II MHC presented by the B cells recognise the antigen/class II MHC through their TCR and are stimulated to produce cytokines. This ensures that only Th that are specific for the antigen contact the B cells. There are also interactions between other molecules on the surface of the B and Th. One interaction involves a molecule called CD40 on the B cell binding to a molecule called CD154 on the T cell. Another important interaction appears to be between a protein called inducible co-stimulator (ICOS) on the Th, which binds to a signalling protein called B7RP on B cells. Individuals with mutations in their ICOS genes have defective antibody production. As a consequence of these cell-surface interactions and cytokine production the Th and B cells are both stimulated to proliferate and eventually form a cluster, or focus, of cells at the outer edge of the paracortex. Although the exact details of this proliferation is not known, *in vitro* B cells can be stimulated to proliferate by IL-2, IL-4 or IL-5 and it assumed that one or more of these cytokines plays a similar role *in vivo*. Some of the B cells differentiate into plasma cells secreting IgM. Other B cells undergo class switch to IgG and these can then differentiate into IgG-secreting plasma cells. Class switch is described in more detail in Section 7.4.3. The foci of IgM- and IgG-secreting plasma cells are often the main source of antibody in a primary response and reach their peak about 4 days after exposure to antigen. After 4–7 days some of the B cells and Th cells migrate to primary follicles, which is where germinal centres will form.

Antigen in the marginal zone of the spleen can be picked by the many macrophages in the region and presented to B cells in the primary follicles (Figure 7.6b). The B cells migrate to the border of the PALS and primary

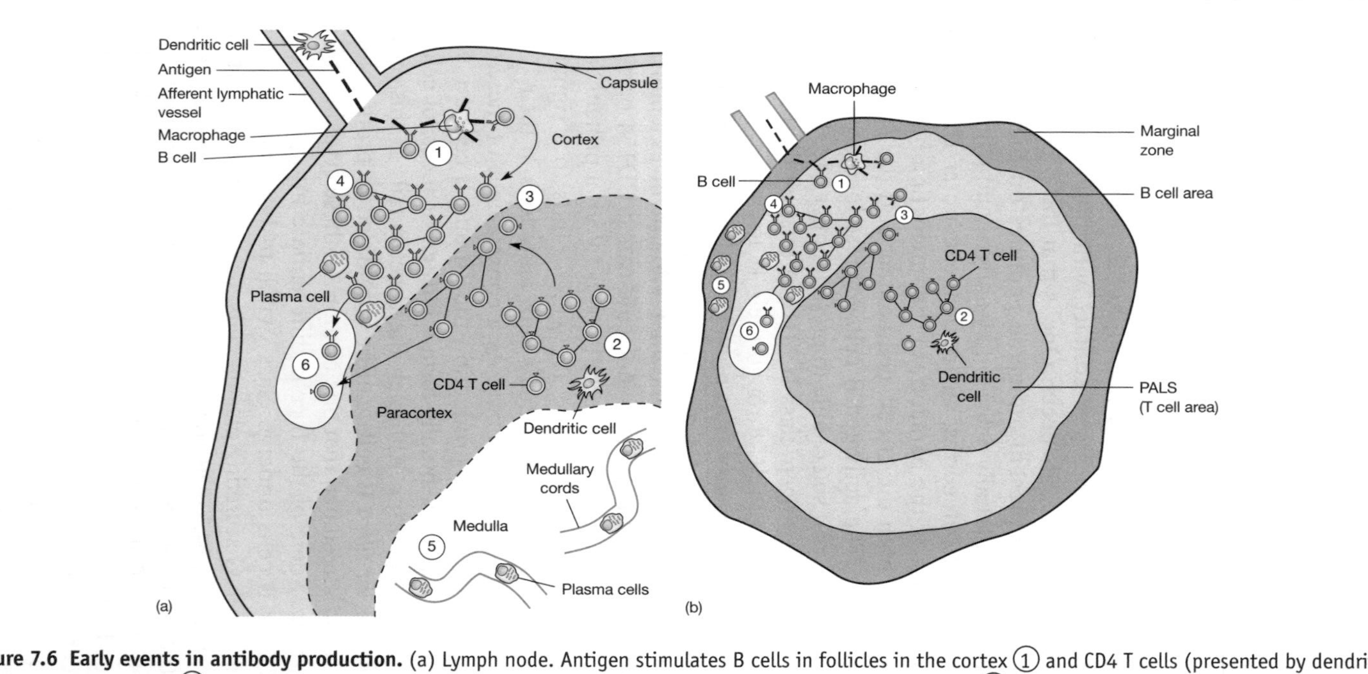

Figure 7.6 Early events in antibody production. (a) Lymph node. Antigen stimulates B cells in follicles in the cortex ① and CD4 T cells (presented by dendritic cells) in the paracortex ②. Stimulated B cells and CD4 T cells migrate to the border of the cortex and paracortex ③ where B cells present antigenic peptide on class II MHC to the CD4 T cells which in turn give signals to the B cells to proliferate and differentiate ④. Some of the B cells differentiate into plasma cells, which migrate to the medullary cords and secrete antibody ⑤. A few of the B cells and some CD4 T cells enter a primary follicle ⑥ where they will form a germinal centre. (b) Spleen. Antigen stimulates B cells at the junction of the marginal zone and primary follicle ① and CD4 T cells in the PALS ②. The CD4 T cells migrate to the B cell area ③ where CD4 T cell/B cell interactions take place, resulting in proliferation of B cells ④ and formation of some plasma cells in the same way as in the lymph node. The plasma cells migrate to the marginal zone and secrete antibody ⑤. Other B cells and CD4 T cells enter primary follicles, leading to the development of a germinal centre ⑥.

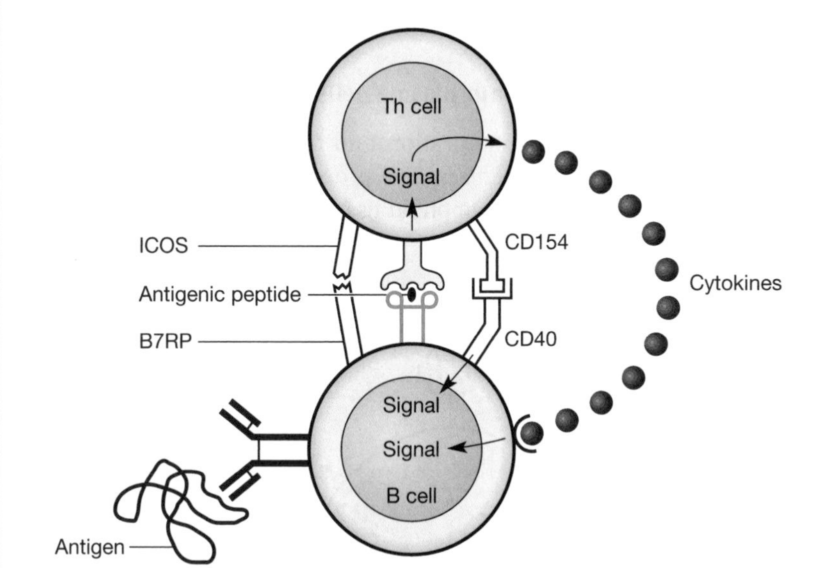

Figure 7.7 Interactions between B cells and Th. B cells bind antigen via their membrane Ig, endocytose and process it and present peptides derived from the antigen in association with class II MHC on the B cell surface. Th cells recognising Ag/MHC-II on the B cell surface are stimulated to secrete cytokines, which deliver signals to the B cells. Binding of CD154 on the Th cell to CD40 on the B cell delivers additional signals to B cells, which are essential for the B cells to switch antibody class.

follicle where they meet Th cells that have also migrated there. The same molecular interactions occur between the B cells and Th as seen in the lymph node resulting in the rapid division of the cells, the formation of foci and the production of IgM and IgG. As in the lymph node, some of the B cells and Th cells eventually migrate to primary follicles and initiate the formation of germinal centres.

7.4 Formation of germinal centres (4–14 days after antigen)

Germinal centres, also called secondary follicles, are specialised structures that form within lymphoid tissue following an encounter with antigen. Four main events occur in germinal centres:

- **Antibody class switch.** This changes the antibody on the B lymphocyte cell surface from IgM and IgD to one of the other classes of antibody – IgG, IgA or IgE.
- **Affinity maturation of antibody.** This results in antibody with high affinity for antigen, which is more effective at dealing with pathogens.

- **Differentiation of B cells into memory cells.** These are B cells that have undergone class switch and affinity maturation but have not differentiated into plasma cells.
- **Differentiation of B cells into plasma cells.**

Two of these events, affinity maturation and the production of memory B cells, occur only in germinal centres. The process of germinal centre formation is the same in all organised lymphoid tissue and is as follows (Figure 7.8).

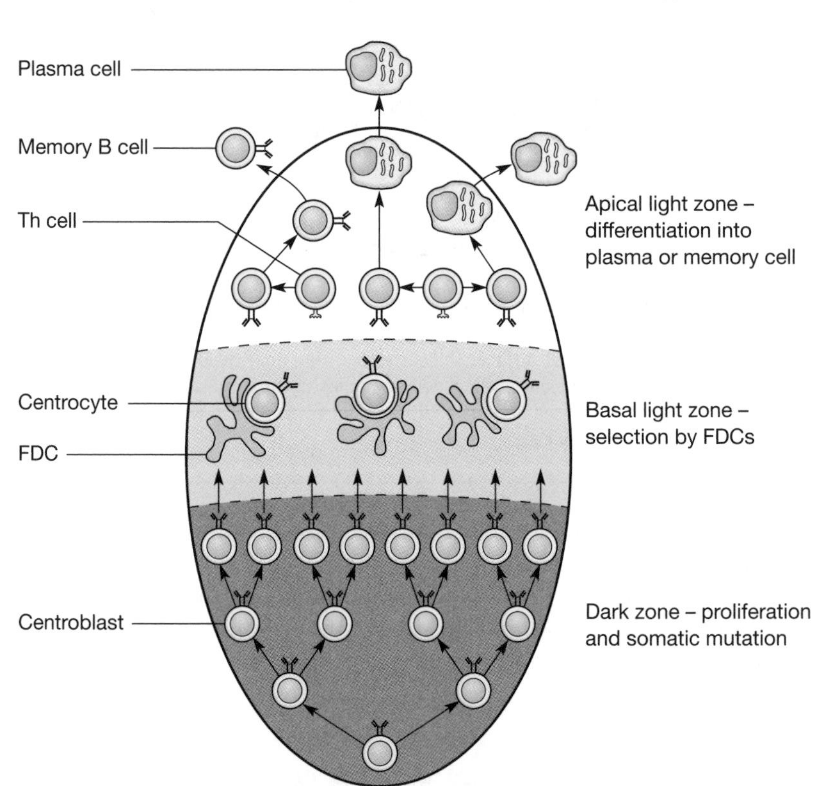

Figure 7.8 Germinal centre formation. B cells proliferate in the dark zone, where they are known as centroblasts. During this proliferation, somatic mutation of the B cell Ig takes place. The centroblasts stop dividing and enter the basal light zone, where they are known as centrocytes. The centrocytes encounter antigen being presented by follicular dendritic cells (FDC) and those centrocytes whose Ig binds antigen are rescued from cell death. Centrocytes whose Ig does not bind antigen die by apoptosis. Surviving centrocytes enter the apical light zone, where they present antigen on their class II MHC to Th cells. B cells presenting antigen that is recognised by the Th cells receive a survival signal from the Th cell and differentiate into plasma cells or memory cells. B cells presenting antigen for which there are no Th die. Note that although centroblasts and centrocytes are given special names, they are in fact B cells at special stages of differentiation.

7.4.1 Initiation of germinal centres

One, or a few, B cells together with some Th from a foci of antibody-producing cells enter a primary follicle. Primary follicles are specialised structures containing a unique type of antigen-presenting cell called a **follicular dendritic cell**. Follicular dendritic cells form a dendritic network throughout the follicle and are very good at retaining antibody–antigen complexes on their surface and presenting the antigen to B cells. The antibody for the immune complex is produced initially by the plasma cells in the extra-follicular foci.

The migration of B cells and Th cells into follicles is promoted by the chemokine B lymphocyte chemoattractant, which is produced by follicular dendritic cells and other stromal cells in the B cell area and binds to the chemokine receptor CXCR5 expressed on B cells and Th.

7.4.2 Affinity maturation increases affinity of antibody for antigen

The B cells that have entered the primary follicle down-regulate their Ig membrane receptors and undergo extensive proliferation. At this stage they are called centroblasts. During this period of proliferation the B cells undergo a process called **affinity maturation**. It is important to produce antibody with high affinity for antigen because high-affinity antibody works better. Affinity maturation occurs in the following way.

As the centroblasts in the germinal centre divide (Figure 7.8) special molecular mechanisms cause hyper-mutation of the H and L chain variable genes. Because this is happening in non-germ-line cells (germ-line cells are sperm and ova) it is called somatic mutation (Figure 7.9). The mutations in the Ig genes lead to changes in the nucleotide sequence and some of these changes will alter the amino acid sequence of the hyper-variable regions of the antibody molecule and therefore change the conformation of the antigen-binding site. Since the mutations are random, the change in the conformation of the antigen-binding site could either increase or decrease the affinity of the antibody for antigen or have no effect on the affinity. Since we want only the B cells that have mutated their Ig genes to produce high-affinity Ig, it is necessary to select for these high-affinity cells and this occurs as follows.

The centroblasts stop dividing and re-express their membrane Ig; they are now called **centrocytes** (Figure 7.8). The centrocytes must now recognise antigen on the follicular dendritic cells to receive a survival signal. Centrocytes whose surface Ig has high enough affinity for antigen will successfully bind to the antigen on the follicular dendritic cell and receive a survival signal. Centrocytes with lower affinity for antigen will not be able to bind to the antigen and do not receive the survival signal; these cells die by apoptosis and are engulfed by macrophages (Figure 7.10). As the immune response progresses, antibody will begin to remove antigen and the amount of antigen available for continued stimulation of B cells will

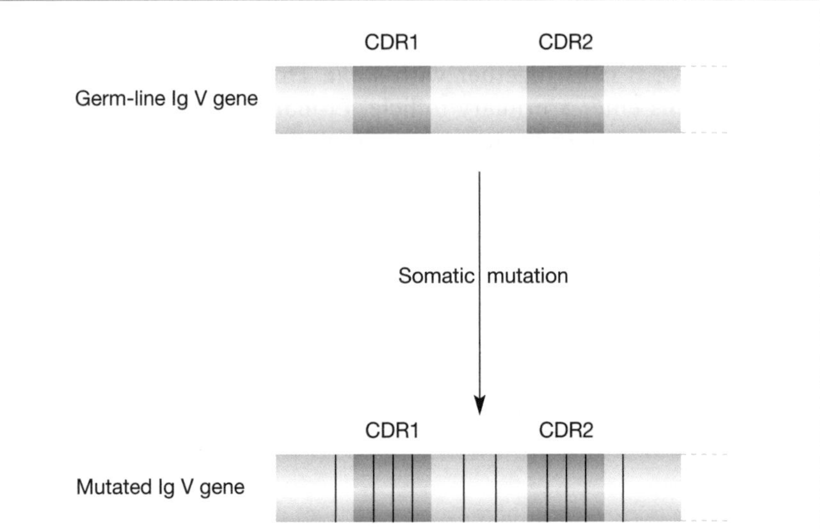

Figure 7.9 Somatic mutation of Ig genes. During somatic mutation the nucleotide sequence of V genes of the Ig H- and L-chains changes. The black lines in the lower figure indicate sites where the nucleotide sequence has changed from the original germ-line V gene sequence. Note that somatic mutation occurs in parts of the V gene coding for both CDR1 and CDR2 and the framework (non-CDR) parts of the V genes. There is, however, no somatic mutation in the C regions of the genes.

fall. Once the amount of antigen becomes limiting, the B cells will begin to compete with each other for antigen. In this competition, B cells that have mutated their antibody to a higher affinity will win the competition and be the only ones that will continue to be stimulated and undergo cell division. This will lead to the generation of many B cells with high affinity for antigen and the death of low-affinity B cells. This process of affinity maturation is extremely effective and may result in a 10 000- to 100 000-fold increase in antibody affinity as the immune response progresses.

Centrocytes and centroblasts are found in different regions of the germinal centre (Figure 7.8). Centroblasts are found in the dark zone and centrocytes in the basal light zone. The basal light zone contains many follicular dendritic cells for affinity maturation and macrophages to mop up the apoptotic B cells. It is thought that some cells can accumulate up to 20 mutations in their Ig genes, although other cells undergo less somatic mutation.

7.4.3 Class switch enables B cells to express IgG, IgA or IgE

During the centroblast/centrocyte stage B cells undergo another differentiation process called the antibody **class switch** (Figure 7.11). Using special molecular mechanisms, the B cell is able to change the heavy chain constant regions of its antibody from μ and δ to γ, α or ε while keeping the same heavy chain variable region and light chain (Box 7.2). Therefore the class switch does not change the antigen specificity of the antibody, but just the class of Ig.

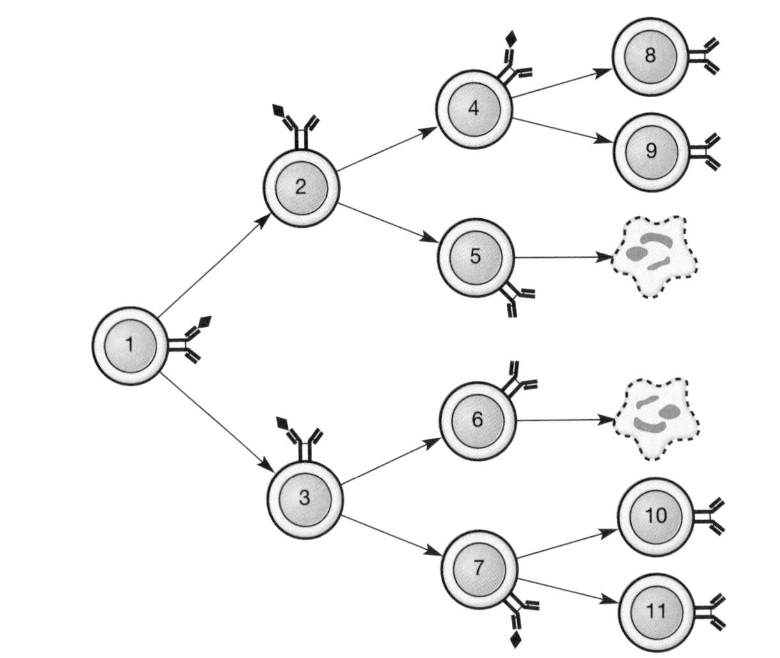

Figure 7.10 Following somatic hypermutation B cells die unless stimulated by antigen. In this example, B cell number 1 binds antigen and divides to produce daughter cells 2 and 3. Cells 2 and 3 both bind antigen and divide again to produce cells 4, 5, 6 and 7. They also undergo somatic hypermutation. Cells 4 and 7 have Ig which has mutated to bind antigen with high affinity and are stimulated to produce daughter cells 8, 9, 10 and 11. Cells 5 and 6 have mutated their Ig so that it does not bind antigen and they die by apoptosis.

Class switch is controlled by the Th cells and cytokines. As mentioned above the interaction between B cells and Th involves both cell contact and cytokines (Figure 7.7). The direct cell–cell interactions between the B cells and Th are thought to be the same as described above. B cells express class II MHC on their cell surface and present antigen to the Th that stimulates the Th to produce cytokines. CD40 on the B cell binds to CD154 on the T cell. This interaction between CD40 and CD154 is essential for the class switch and without it only IgM is produced. The importance of this interaction in humans was demonstrated in a congenital (genetically determined) immunodeficiency disease known as hyper-IgM syndrome. Individuals with this syndrome have high levels of circulating IgM but little or no IgG, IgA or IgE. It has been demonstrated that these individuals have mutations in their CD154 gene so that they either do not make CD154 or they make a mutated form of the protein that binds poorly, if at all, to CD40. Because they cannot switch antibody class, affected individuals give especially poor secondary antibody responses and are prone to certain types of infections. Additionally ICOS on the Th interacts with B7RP on

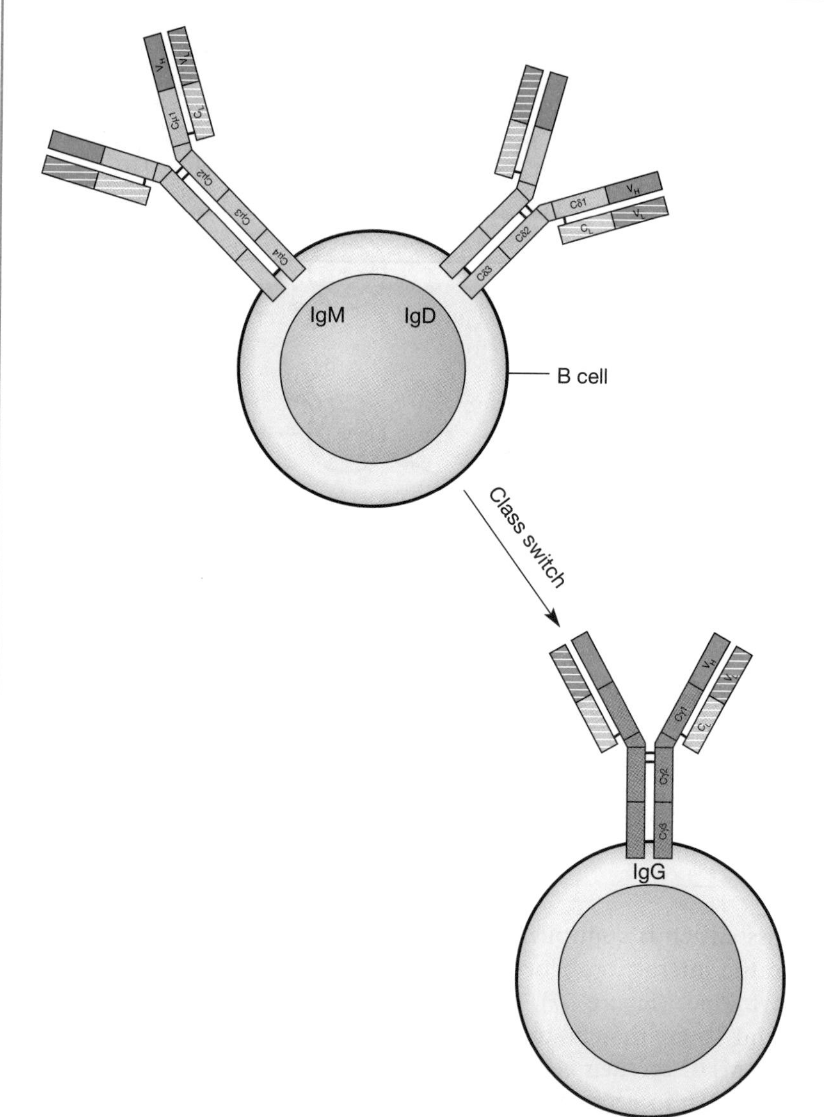

Figure 7.11 B cell antibody class switch. During antibody class switch, B cells stop expressing IgM and IgD on their cell surface and express another class of antibody, in this example IgG. Note that the V_H and the L-chain stay the same; only the C regions of the H-chain change.

the B cells. Although the CD154/CD40 and ICOS/B7RP interactions are essential for the class switch to occur, they do not appear to influence the class of Ig that a B cell switches to. This appears to be influenced more by cytokines produced by the Th when it interacts with the B cell.

Cytokines affect the proliferation, differentiation and class switching of B cells and also the level of antibody production by plasma cells. Some cytokines, such as IL-6 and IL-21 promote the general differentiation of B cells into plasma cells. Other cytokines promote the switch to particular classes. Table 7.1 and Figure 7.12 show the cytokines involved in plasma cell generation. The control of switching to the IgG subclasses is least well understood; IL-4 and IL-21 promote the switch to IgG1 and IgG3 while IgG4 production is increased by IL-13. A number of cytokines, including IL-10, IL-21 and transforming growth factor-β (TGFβ) promote IgA production. Finally IL-4 and IL-13 promote the switch to IgE.

BOX 7.2: MOLECULAR ASPECTS OF CLASS SWITCH

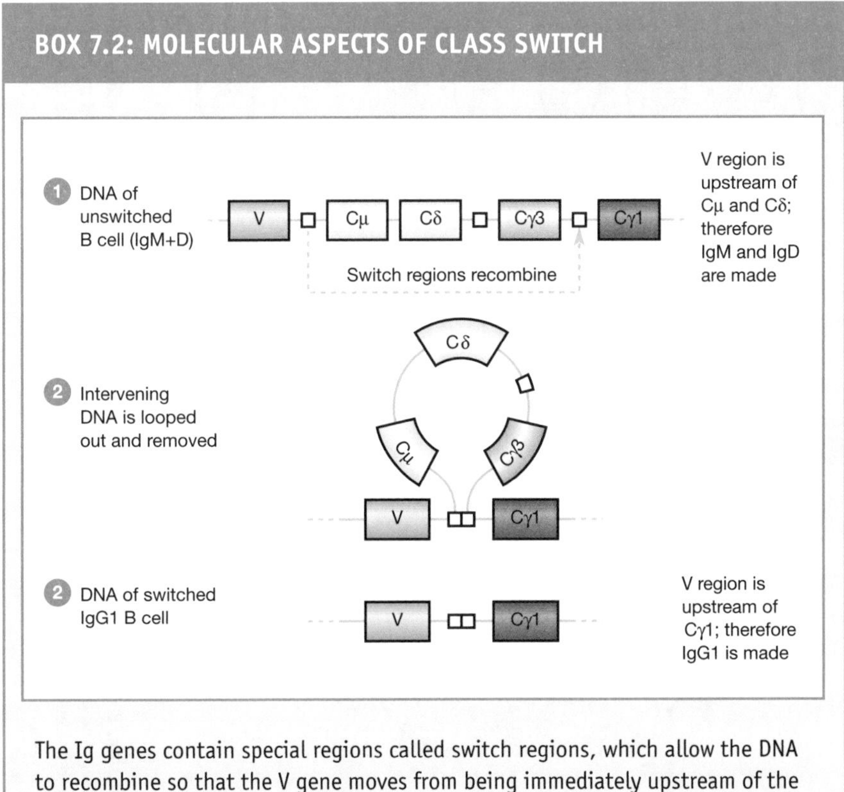

The Ig genes contain special regions called switch regions, which allow the DNA to recombine so that the V gene moves from being immediately upstream of the $C\mu$ gene to being upstream of another C-region gene; in the diagram the V region switches to being upstream of $C\gamma 1$.

7.4.4 B cells differentiate into antibody-secreting plasma cells or memory B cells

The final stage of B cell differentiation is the differentiation of B cells into plasma cells or memory B cells (Figure 7.8). Plasma cells secrete large amounts of antibody (Figure 7.13) to deal with the current threat while memory B cells provide protection against future infection. Memory B cells are long-lived and can survive long after an initial infection has been

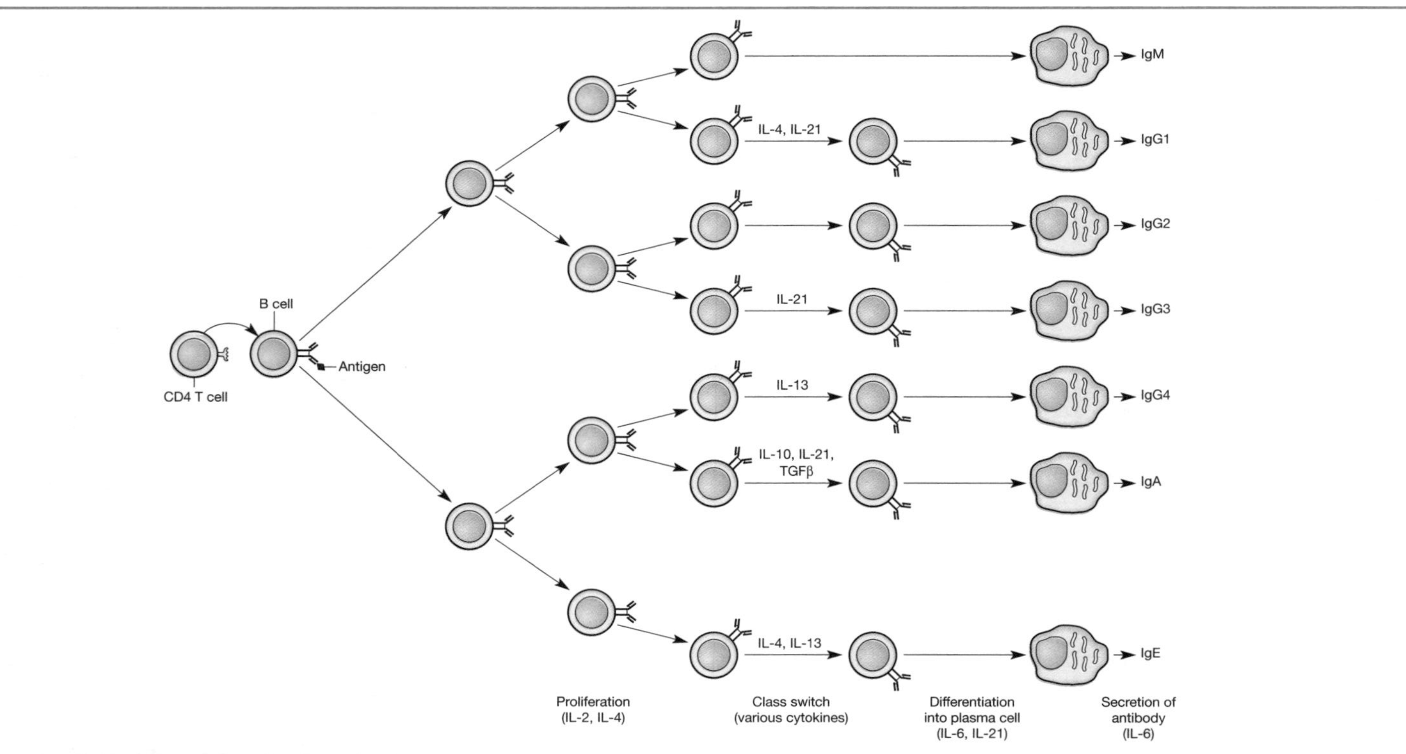

Figure 7.12 Cytokines and antibody production. B cells that are stimulated by antigen interact with CD4 T cells and proliferate under the influence initially of IL-2 and later IL-4 and IL-5. After proliferating, the B cells undergo class switch controlled by various cytokines secreted by Th cells. Finally, differentiation into plasma cells is influenced by IL-4 and IL-5 and antibody secretion is stimulated by IL-6. Some cytokines, e.g. IL-6, may be secreted by cells other than Th. In reality not all classes of antibody are produced in response to a single pathogen.

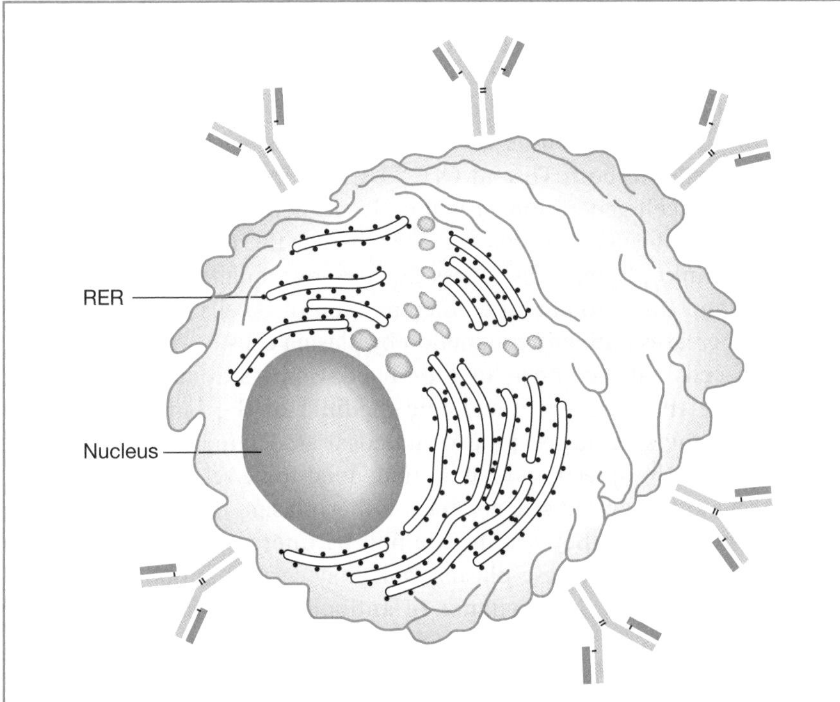

Figure 7.13 Diagram of plasma cell. Note the extensive RER typical of a cell synthesising large amount of protein for secretion. In this case the secreted protein is antibody.

eliminated. Memory B cells have undergone affinity maturation and class switch but retain surface expression of the class of Ig they have switched to. They enter the recirculating lymphocyte pool and migrate through lymph nodes, spleen and some other lymphoid tissue. If they encounter antigen in the future they are able to very rapidly develop into plasma cells secreting high-affinity IgG, IgA or IgE because they do not need to undergo affinity maturation or class switch again. Therefore upon second encounter with an antigen you get a faster and bigger IgG, IgA or IgE antibody response (see Chapter 10).

Table 7.1 Influence of cytokines on antibody production

Antibody class	*Cytokines promoting switching/production*
IgG1	IL-4, IL-21
IgG2	Not known
IgG3	IL-21
IgG4	IL-13γ
IgA	IL-10, IL-21,TGFβ
IgE	IL-4, IL-13

Each B cell in the apical light zone can differentiate into either a plasma cell or a memory B cell (Figure 7.8). The factors that affect whether a B cell becomes a plasma cell or a memory cell are not fully understood. There is good evidence that if the B cell interacts with a Th, the binding of CD154 on the Th to CD40 on the B cell stimulates the B cell to become a memory B cell and both Th and CD154/CD40 interactions are essential for memory B cell generation. It is not clear whether other signals stimulate the cells to become plasma cells or whether the B cells automatically differentiate into plasma cells unless they are restimulated through CD40. Both IL-4 and IL-5 promote the differentiation of B cells into plasma cells and IL-6 increases antibody production by plasma cells.

Plasma cells can be short- or long-lived. Some stay in the lymphoid tissue where they are produced, the medulla of the lymph node, the sinusoids of the splenic red pulp or mucosal associated lymphoid tissue. Plasma cells in the spleen and lymph node tend to be relatively short-lived and secrete antibody for a few weeks. Other plasma cells migrate to the bone marrow where they are maintained by interactions with the bone marrow stromal cells. These plasma cells may secrete antibody for many months or years. The maintenance of antibody production can be truly remarkable; a vaccine for the yellow fever virus was tested in a trial in the USA and some individuals who had received the vaccine still had antibody against the virus 70 years later with no deliberate reinfection.

Figure 7.14 summarises the different fates that can befall B cells in germinal centres during the antibody production process.

7.4.5 The combination of class switch and affinity maturation significantly reduces the time taken to make antibody

Class switch and affinity maturation appear to be other examples of the immune system being unnecessarily complicated. If you need to produce different classes of antibody with high affinity for antigen, surely the easiest way is to produce B lymphocytes already expressing different classes of high-affinity antibody and just select those B lymphocytes with the class of antibody required to become plasma cells secreting the same high-affinity antibody of the required class. In fact, having class switch and affinity maturation is more efficient for the following reasons.

Remember that each B cell has only one antigen specificity and that different B cells have different antigen specificities. The immune system must provide B cells, which as a population have as many antigen specificities as possible. It is not possible to escape this responsibility because that would leave you open to infection by pathogens whose antigens you did not have B cells specific for. However, your body only has room for a fixed number of B cells. Therefore you can only have a certain number of B cells that are specific for any one antigen. The more B cells you have that are specific for an antigen, the quicker you will generate enough plasma cells to produce

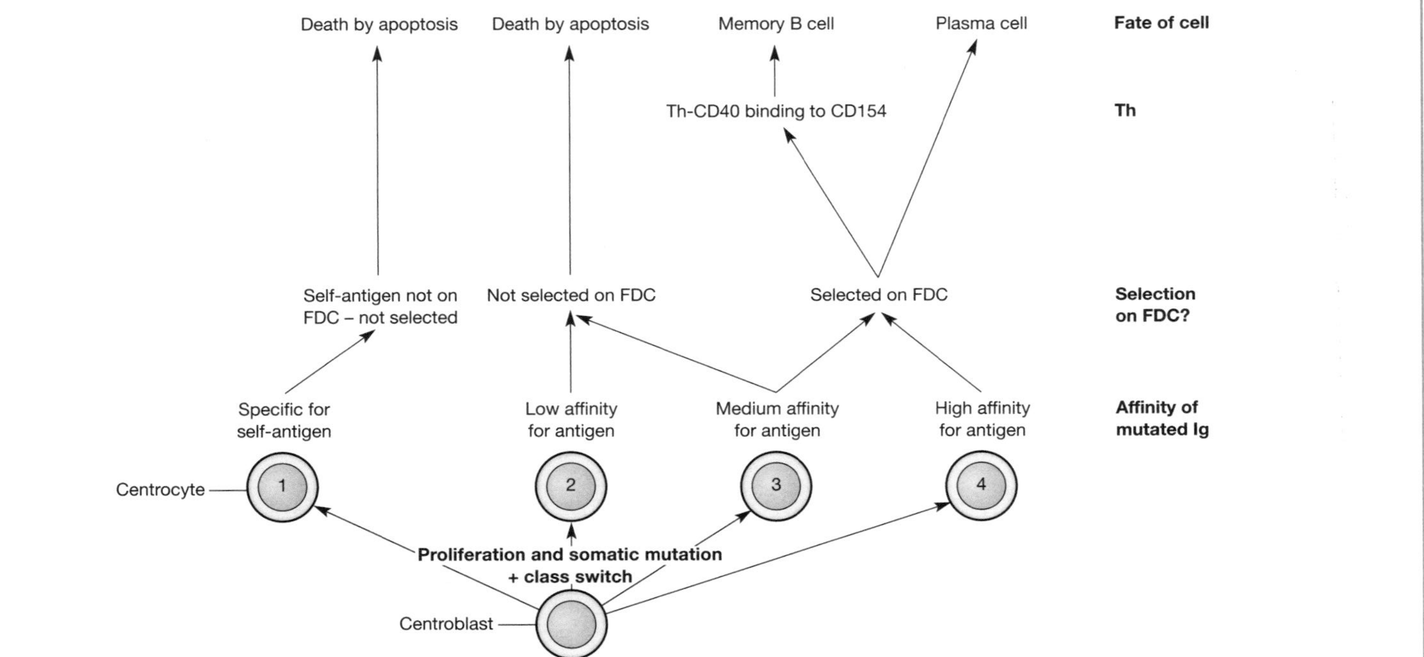

Figure 7.14 Fate of B cells in germinal centres. Centroblasts in the dark zone proliferate and their Ig V genes undergo somatic mutation. This can result in cells with Ig that has high (cell 4), medium (cell 3) or low (cell 2) affinity for the foreign antigen stimulating the response. Some centroblasts may mutate their Ig so that it is specific for self-antigen (cell 1). The centroblasts stop dividing and become centrocytes, which undergo selection by antigen on FDCs in the basal light zone. Cells with low affinity for the foreign antigen (cell 2) will not be selected and will die. Cells with medium affinity for the foreign antigen (cell 3) may or may not be selected. Those cells that are selected, along with high-affinity cells (cell 4), proceed to the apical light zone, where they differentiate into plasma cells or, if they encounter CD154 on Th, memory B cells. Centrocytes with Ig specific for self-antigen will die because the self-antigen is not on FDCs in the form of Ab–Ag complexes.

the amount of antibody needed to protect you against the pathogen. Class switch and affinity maturation maximise the number of B cells that can be initially stimulated by a single antigen to become high-affinity antibody producing plasma cells in the following ways.

Class switch If class switch did not occur, you would need a separate B cell with the same antigen specificity for each class of antibody. Since there are eight classes of antibody in humans you would need eight times as many B cells to cover the same range of antigen specificities. Because you cannot reduce the number of antigen specificities covered by your B cell population as a whole, and you cannot increase the total number of B cells that your body can accommodate, the only solution to this problem would be to have eight times fewer B cells specific for any particular antigen. This would not be a problem if you always made all eight different classes of antibody. However in most infections a few classes of antibody predominate and you do not make some classes at all. Therefore B cells of the class you do not want are wasted (Figure 7.15).

Affinity maturation Affinity maturation enables more B cells to be initially stimulated by antigen in the following way. B cells must bind antigen with a certain affinity to be stimulated to eventually become plasma cells. For any particular antigen some B cells will have Ig on their surface with high affinity for the antigen but there will be many more B cells whose Ig binds the antigen with much lower affinity. It is difficult to estimate exactly how many more low-affinity B cells there would be compared with high-affinity cells but an estimate of 1000-fold is realistic. If affinity maturation did not occur, the only way to make high-affinity antibody would be to select the very small number of high-affinity B cells for clonal expansion. This would require more rounds of cell division to generate the required number of plasma cells. By having affinity maturation you can also select the low-affinity B cells for clonal expansion and allow them to affinity-mature, thereby initially stimulating 1000 more antigen-specific B cells.

The combination of affinity maturation and class switch means that when a new antigen comes along you may have 8000 times as many B cells to select for clonal expansion compared with the situation where affinity maturation and class switch did not occur. In other words, **without** class switch and affinity maturation you have to generate 8000 B cells from a single B cell to be in the same starting position as you are **with** class switch and affinity maturation (Figure 7.15). Generating 8000 cells from one cell involves 13 cell divisions. Assuming that B cells can divide once every 10 hours, 13 generations would take 130 hours, or 5.5 days. Even allowing that affinity maturation is not 100% efficient and that you may want to make more than one class of antibody (but not all eight classes), you still save many days in generating antibody by having class switch and affinity maturation. Given the rate of multiplication of some pathogens, these days may be crucial to whether you survive an infection or not and from an evolutionary point of view would be a powerful selection force for class switch and affinity maturation.

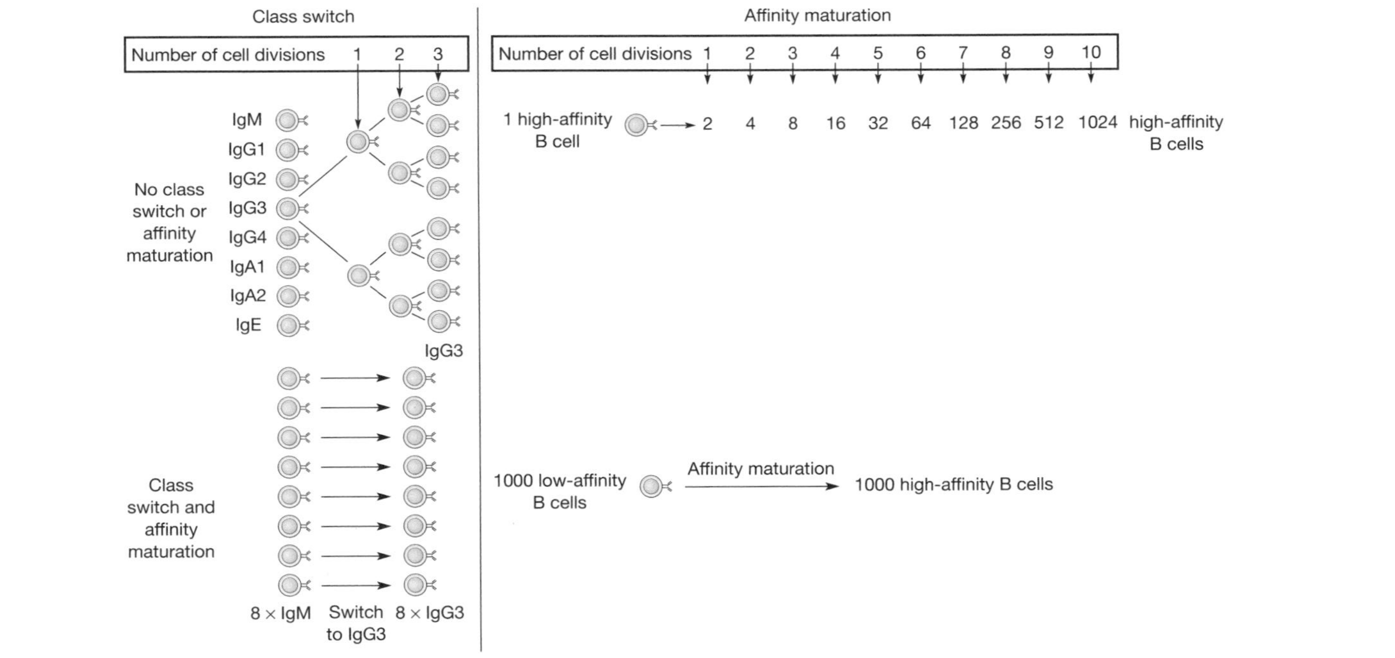

Figure 7.15 Benefits of class switch and affinity maturation. Without class switch it would be necessary to have a separate B cell for each class of antibody. To make just IgG3, as in this example, only one of the eight B cells is utilised – the others are wasted. With class switch, it is possible to have eight times as many cells specific for the antigen that can all switch to IgG3, therefore saving three rounds of cell division. Without affinity maturation the only way to generate high-affinity antibody would be to select B cells with high affinity for antigen – these would be quite rare. With affinity maturation it is possible to select low-affinity B cells, which may be present at 1000-fold higher frequency than the high-affinity cells – this saves ten rounds of cell division. Overall, if they were 100% efficient, the combination of class switch and affinity maturation would save 13 rounds of cell division, which would take about 5.5 days. Even if the processes were only 5% efficient, they would save between eight and nine divisions, or about 4 days.

7.5 MALT and the production of IgA

MALT contains many lymphoid follicles, which serve the same purpose as those seen in the spleen and lymph nodes. As might be expected, in areas that are exposed to many environmental antigens, most of the follicles contain germinal centres where plasma cells and memory B cells are generated. The area surrounding the follicles is rich in T cells, which will provide the Th for antibody generation.

The epithelium overlying the follicles in the gut is specialised to enable it to transport antigen into the lymphoid follicle from the lumen of the intestine. The epithelium above lymphoid follicles is flattened and does not contain villi even in mucosa, such as the ileum, which normally have many villi (see Figure 7.16). Special epithelial cells called **M cells** overlie the follicles. The internal side of the M cells contains deep invaginations that surround clusters of B cells, T cells, macrophages and dendritic cells. The M cells endocytose antigen present in the mucosal lumen and the endocytic vesicle crosses the M cell before fusing with the basal cell membrane, thereby releasing the endocytosed antigen into the invagination containing the lymphocytes.

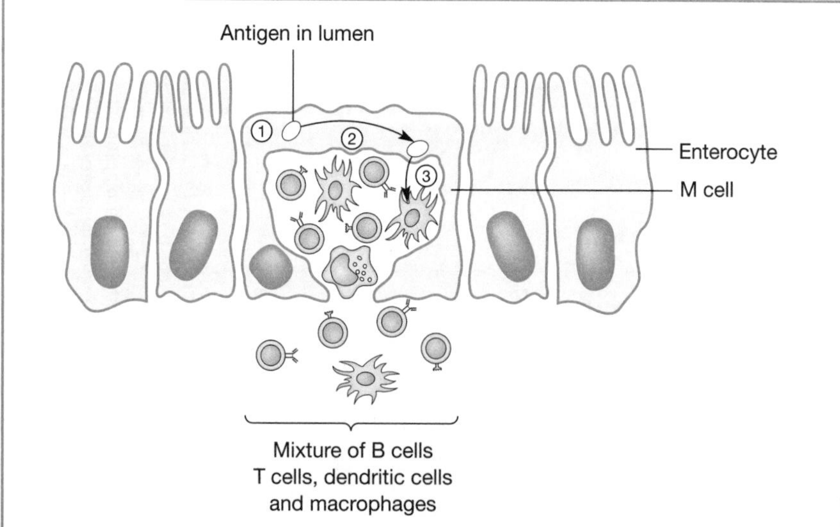

Figure 7.16 Diagram of mucosal lymphoid tissue. Specialised epithelial cells called M cells have a flattened surface and are specialised for capturing antigen in the lumen and transporting it to the mucosal lymphoid tissue where antigen can stimulate B cells and be processed by dendritic cells for presentation to CD4 T cells.

The antigen can then be taken up by dendritic cells for presentation to T cells or recognised by B cells in the follicles. The sequence of events that follows leading to the formation of germinal centres and production of plasma and memory B cells is the same as that occurring in lymph nodes or the spleen. Activated B cells become committed to IgA, leave the follicles in the Peyer's patches and enter lymphatic vessels, eventually reaching the blood. The vascular system delivers the B cells to the lamina propria at any mucosal site in the body. The B cell now becomes an IgA-secreting plasma cell. In this way the IgA-secreting cells can be distributed far from the site of original antigenic stimulation and provide protection against infection by that pathogen at all mucosal sites.

The IgA produced by the plasma cells in the lamina propria is taken up by the epithelial cells of the mucosa. The epithelial cells add the secretory piece to the IgA before releasing the IgA into the lumen (see Figure 7.17). The secretory piece inhibits the degradation of the IgA by enzymes in the mucosal secretions.

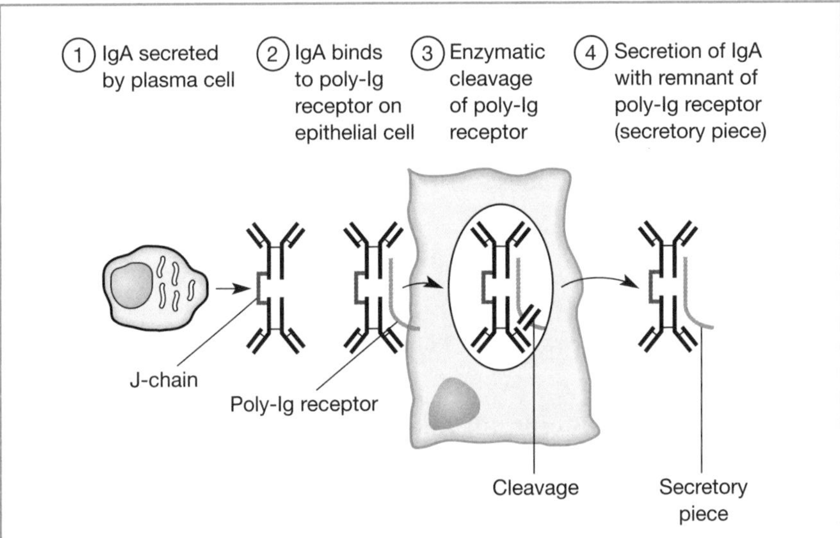

Figure 7.17 Secretion of IgA. IgA secreted by plasma cells in the mucosa binds to a receptor, called the poly-Ig receptor, on mucosal epithelial cells. The poly-Ig receptor, together with the bound IgA, is internalised by the epithelial cells and the poly-Ig receptor is cleaved. Finally the IgA is secreted into the lumen with part of the poly-Ig receptor, now known as the secretory piece, attached.

7.6 Summary

- Antigen entering tissue can enter the MALT or it can enter lymphatic ducts draining the tissue site, where it will go to the local lymph nodes. Antigen in the blood goes to the spleen.
- Antigen is taken up by dendritic cells in tissue or lymphoid tissue, and dendritic cells present antigen to, and activate, CD4 T cells to become Th cells.

- Th cells migrate to the edge of follicles, where they encounter B cells that have been stimulated by antigen. The B cells differentiate into antibody producing extra-follicular foci.
- Some B cells migrate into primary follicles accompanied by Th cells, resulting in germinal centre formation. Here plasma cells or memory B cells are produced.
- In the germinal centre B cells undergo two unique events. They switch the class of antibody they express on their cell surface from IgM and IgD to IgG, IgA or IgE, a process known as class switch. The B cells also randomly mutate their Ig V genes, therefore altering the nucleotide, and hence amino acid, sequence of the variable region. B cells whose mutated Ig has higher affinity for antigen compete favourably for antigen and continue to be stimulated by antigen. This results in selection of higher-affinity B cells – a process known as affinity maturation.
- After class switch and affinity maturation, B cells complete their differentiation and become either plasma cells, which produce large amounts of antibody, or memory B cells. Memory B cells are long-lived and able to respond rapidly to a second exposure to the same antigen. This gives a bigger, better and longer response to the antigen and is known as a secondary antibody response.

7.7 Questions

1) Listed in random order are 11 events that occur during antibody production.

   (i) Proliferation of B cells, (ii) Antibody class switch, (iii) Processing of antigen by dendritic cells, (iv) Affinity maturation, (v) Presentation of antigen to CD4 T cells by dendritic cells, (vi) Differentiation of CD4 T cells into helper T cells, (vii) Differentiation of B cells into plasma cells, (viii) Proliferation of CD4 T cells, (ix) Stimulation of B cells by antigen, (x) Secretion of antibody, (xi) Interaction of Th with B cells.

   Arrange them in the order in which they occur. N.B. Some events can occur simultaneously.

2) Cytokines control proliferation, class switch and antibody secretion. For each cytokine note whether they PRIMARILY effect (A) proliferation, (B) class switch, (C) secretion of antibody.

   Cytokine
   - IL-2
   - IL-4
   - IL-5
   - IL-6
   - IFNγ
   - TGFβ

3) What factors promote the interaction of different cell types (APCs, T cells, B cells) in a lymph node?
4) In the following diagram of a germinal centre, what are the cell types indicated by letters A–F? Choose from the following: (i) centroblast, (ii) centrocyte, (iii) follicular dendritic cell, (iv) helper T cell, (v) memory B cell, (vi) plasma cell.

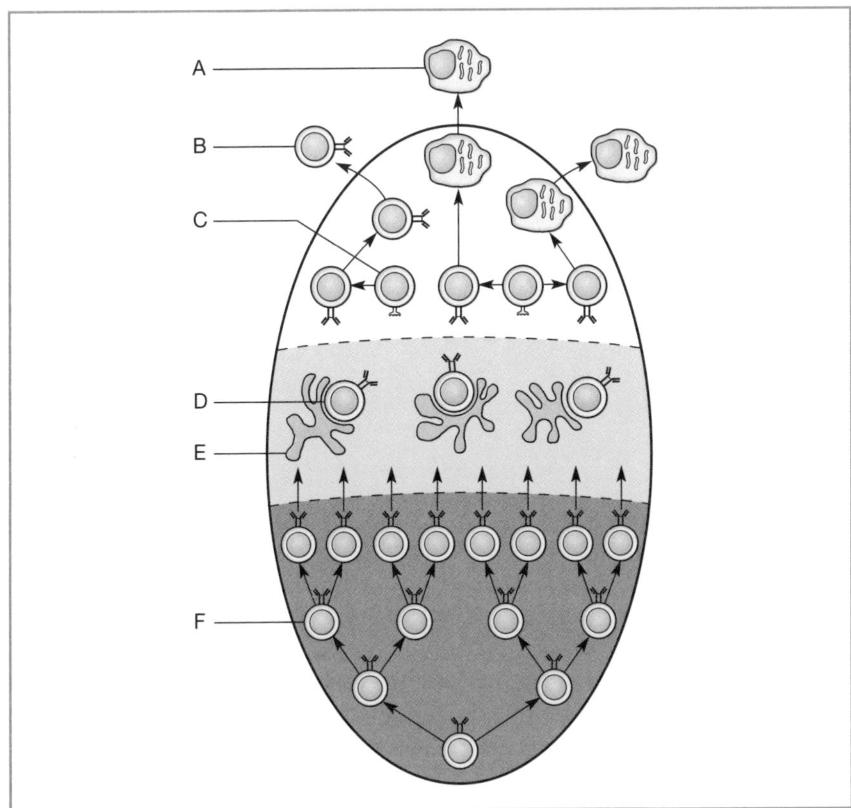

5) What are the special features that are involved with the production of secretory Ig?

The answers to these questions can be found on page 337.

7.8 Further reading

1) Honjo T, Kinoshita K, Muramatsu M. (2002) Molecular mechanism of class switch recombination: linkage with somatic hypermutation. *Annual Review of Immunology* 20:165–196.

2) Mills DM, Cambier JC. (2003) B lymphocyte interactions during cognate interactions with CD4+ T lymphocytes: molecular dynamics and immunological consequences. *Seminars in Immunology* 15:325–329.

3) Schwickert TA, Lindquist RL, Shakhar G, Livshits G, Skokos D, Kosco-Vilbois MH, Dustin ML, Nussenzweig MC. (2007) *In vivo* imaging of germinal centres reveals a dynamic open structure. *Nature* 446:83–87.

CHAPTER EIGHT

Effector mechanisms: dealing with pathogens *in vivo*

(1) Antibody-mediated responses

Learning objectives

To know about the different ways the specific immune response aids in neutralising or eliminating pathogens. To understand the different ways in which antibody provides protection from infectious agents. To learn about the components and biology of the complement system.

Key topics

- Antibody-mediated effector mechanisms
  - Neutralisation
  - Agglutination
  - Opsonisation
  - Antibody-dependent cell-mediated cytotoxicity (ADCC)
- Killing by phagocytes
- Complement
  - Components of the complement system
  - Three pathways of complement activation
- Biology of the complement system

8.1 Humoral and cell-mediated immunity

The ways in which effector responses can protect the individual are by first of all neutralising any immediate threat posed by the pathogen (e.g. toxins produced by the pathogen) and then bringing about the destruction and/or elimination of the pathogenic organism so that it no longer has the potential to cause disease. The types of specific effector response were mentioned in Chapter 6; these are antibody responses, cytotoxic CD8 T cell responses, delayed-type hypersensitivity responses and neutrophil-mediated inflammatory responses. Antibody responses have also been called humoral immunity because, historically, the immunity could be transferred with 'humour'; that is, liquid (this was serum and not funny jokes). Cell-mediated immunity was so-called because it could only be transferred with cells and not with humour – it is now clear that cell-mediated immunity is particularly referring to immune responses involving CD8 T cells or delayed-type hypersensitivity responses, neither of which involve antibody.

8.2 Antibodies provide protection in many different ways

Antibody can protect against infectious agents in a number of ways, many of which involve interaction with components of the innate immune system. These can be summarised as:

- neutralisation;
- agglutination;
- opsonisation;
- activation of complement;
- antibody-dependent cell-mediated cytotoxicity.

As described in Chapter 3, there are many different classes and subclasses of antibody. Each class and subclass of Ig has a different Fc portion of the IgH chain and these different Fc portions endow the antibodies with different functions.

8.3 Neutralisation by antibody

This is the simplest way in which antibody can act as an effector molecule and a variety of pathogens or their products can be neutralised by antibody (Figure 8.1).

8.3.1 Toxins

Antibody can neutralise toxins simply by binding to them and thereby inhibiting their action. This is particularly important with diseases such as tetanus, diphtheria and botulism (a severe type of food poisoning), where the pathology is due totally to production of powerful toxins.

8.3.2 Virus

For a virus to infect a host cell, specific molecules on the viral surface must bind to molecules on the surface of the host cell (see Box 8.1). Antibody can bind to the viral receptor molecule and stop it binding to the cell, thereby preventing infection of the cell. Antibody present in mucosal secretions, particularly IgA, can prevent viral entry into the body and stop the infection from being established. Many viruses spread through the

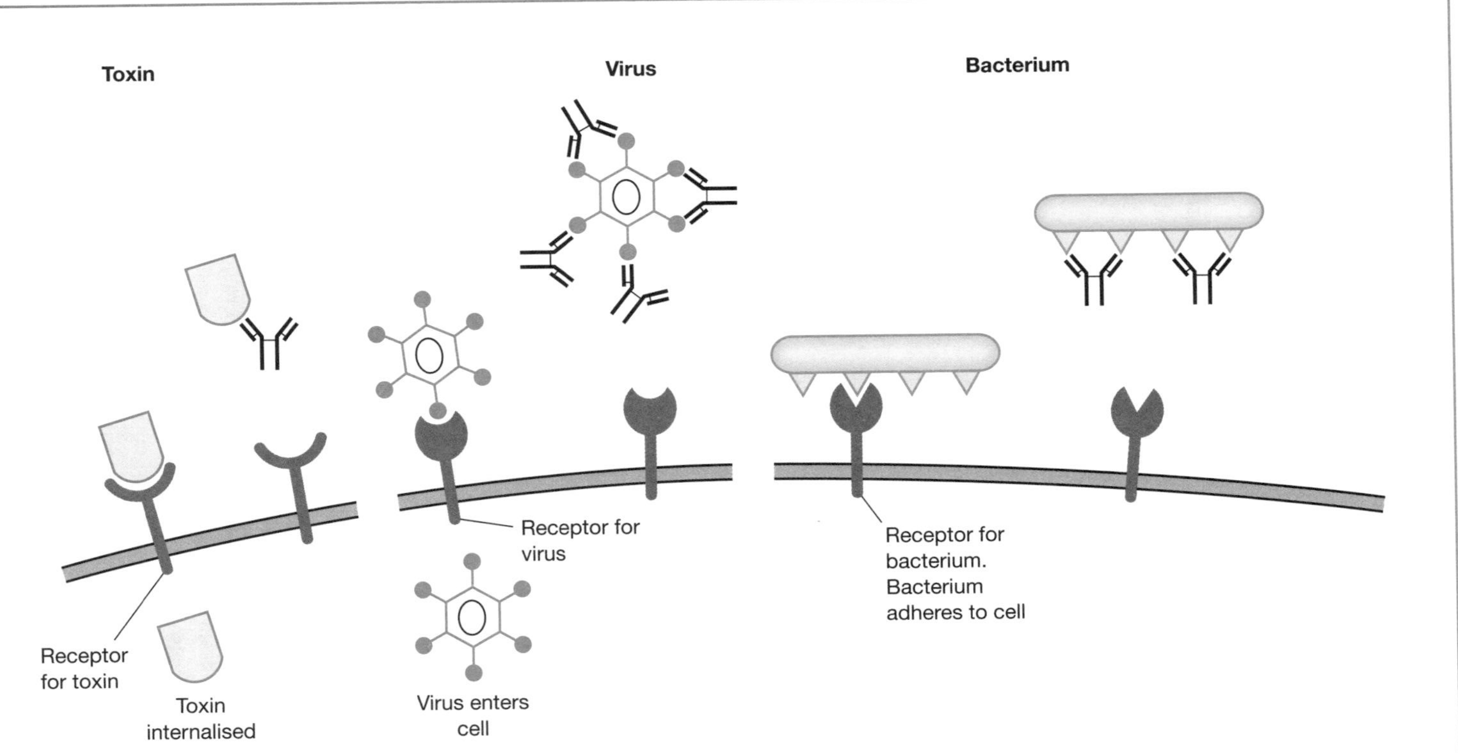

Figure 8.1 Neutralisation by antibody. Antibody binding to toxins or molecules on the surfaces of viruses or bacteria prevents their binding to cellular receptors. This prevents internalisation of toxins or viruses and prevents adhesion of bacteria to cell surface.

bloodstream to their target organs and during this extracellular phase they are exposed to circulating antibody that can bind to the virus and prevent it from infecting the target cells.

BOX 8.1: VIRAL AND BACTERIAL ATTACHMENT TO HOST CELLS

Viruses have to infect host cells in order to replicate. The first stage in infecting the cell is the attachment of the virus to a molecule on the host cell. In some cases, such as 'flu or cold viruses, the molecules that the virus attaches to are expressed on many cell types. In other cases the molecule has a very restricted cellular distribution, which determines the tissue specificity (tropism) of the virus.

Virus	Disease	Host cell attachment molecule
Influenza	'Flu	Sialic acid on glycoproteins
Rhinovirus	Cold	ICAMs
Rabies virus	Rabies	ACh-R
HIV	AIDS	CD4 (+CCR5 or CXCR-4)
Epstein–Barr Virus	Infectious mononucleosis	C3 receptor 2 on B cells
Herpes simplex 1 Virus	Cold sores	Fibroblast growth factor receptor (FGF-R)
Rotavirus	Infantile diarrhoea	β-Adrenergic receptor

Bacterial adhesion is often of a much less cell-specific nature because the molecule on the host cell is found in many cell types. A very important role for bacterial adhesion is to allow the bacteria to colonise mucosal surfaces in the GI, respiratory and GU tracts.

Gram-negative bacteria (e.g. *E. coli, Vibrio cholera and Neisseria gonorrhoeae*) can produce pili – small hair-like extensions from the bacterial cell surface that are smaller than flagella. These pili can enable the bacteria to adhere to epithelial cells. Many bacteria have pili that bind to α-mannosides.

Both Gram-positive and Gram-negative bacteria have non-pili-associated adhesion molecules. *Bordatella pertusis*, the cause of whooping cough, produces a number of adhesins. One, called filamentous haemagglutinin (FHA), binds to ciliated cells. It also contains the RGD motif (arginine-glycine-aspartate) that enables it to bind to CR3 (CD11a/CD18 integrin) on macrophages. *Neisseria gonorrhoeae* makes proteins called opacity proteins that are involved in binding of the bacteria to epithelial cells. *Staphylococcus aureus* produces a protein that can bind to fibronectin on epithelial cell surfaces.

Antibody can also prevent the virus from replicating after it has entered the cell. In ways that are not understood, antibody prevents the virus from uncoating, a necessary prerequisite to replication; this has been suggested to occur for measles and influenza viruses. Other antibodies inhibit viral replication in ways that are poorly understood.

8.3.3 Bacteria

In addition to neutralising bacterial toxins, antibody can also affect bacterial adherence to cells and inhibit bacterial metabolism. Mucosal IgA can bind cholera bacteria and prevent their adherence to the intestinal epithelium. Other antibodies have been found to inhibit bacterial metabolism by binding to pores in the bacterial cell wall that are involved in nutrient transport.

8.4 Antibodes can cause agglutination of microbes

Because antibodies are multivalent they are able to bind to more than one microbial particle and can form complexes of microbe and antibody (Figure 8.2), a process known as agglutination. Agglutination can limit the spread of pathogens by retaining them in clumps; larger complexes of antigen are more likely to be phagocytosed and killed (see Section 8.5).

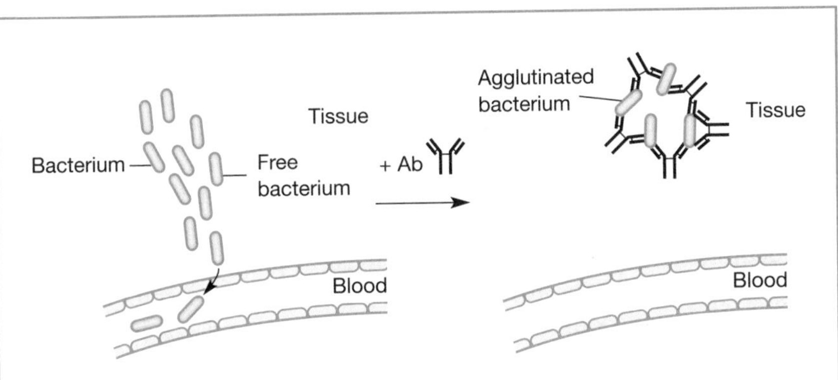

Figure 8.2 Agglutination. Antibody binds to antigens on bacteria, causing them to agglutinate or clump, which inhibits the bacteria from leaving the site and also makes bacteria more susceptible to phagocytosis.

8.5 Antibodies can act as opsonins and promote phagocytosis

Phagocytosis involves the engulfment and ingestion of a particle. Usually these particles are microorganisms, but damaged cells or tissue components can also be phagocytosed in the process of tissue cleansing. Phagocytosed microbes will normally be killed by the phagocyte, although some (e.g. mycobacteria) have developed resistance to killing by phagocytes.

Phagocytosis and killing can be divided into four stages (see Section 2.2.4):

1. Recognition and attachment.
2. Ingestion.
3. Killing.
4. Degradation.

8.5.1 Recognition and attachment – antibodies can act as opsonins

The first stage of phagocytosis involves recognition of the microorganism by the phagocyte. As mentioned in Chapter 2, phagocytes can recognise microbes directly using specific receptors on their cell surface that can recognise (bind to) various molecules on microorganisms. However, many pathogens have evolved ways of avoiding direct recognition by phagocytes and the phagocytes need help in recognising pathogens. Opsonins are molecules that bind to pathogens and to phagocytes and promote phagocytosis (see Chapter 2). Some opsonins belong to the innate immune system but are limited in which pathogens they can recognise. Because they are specific for antigens on the pathogen, antibodies can be very effective opsonins and recognise pathogens that opsonins of the innate immune system cannot.

Antibodies can act directly as opsonins by binding to the pathogen using their Fab antigen-binding sites and then binding to receptors on the phagocyte that are specific for the Fc part of the antibody molecule (Figure 8.3). The receptor on the phagocyte that binds the Fc portion of the antibody is called, not surprisingly, an Fc receptor, or FcR. There are a number of different FcRs that bind different classes of antibody and their main properties are listed in Table 8.1. It should be noted that not all the Fc receptors listed in Table 8.1 are involved in phagocytosis. Fcγ RI and Fcγ RIII promote phagocytosis. Fcγ RII is an inhibitory receptor and inhibits macrophage and other cell functions. The IgE- and IgA-binding FcRs

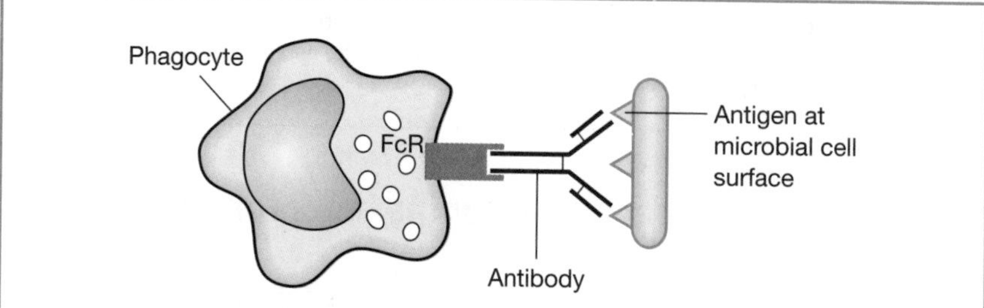

Figure 8.3 Antibody as an opsonin. Antibody binds to antigen on microbial surfaces via the antigen-binding site. Phagocytes have FcRs that bind to the Fc part of the antibody molecule and the antibody/microbe complex is phagocytosed.

are involved in specialised activities of IgA and IgE that do not include phagocytosis. When antibody binds to antigen, the Fc portion of the antibody undergoes a conformational change that enables it to bind to FcR on phagocytes. The binding of antibody to FcRs triggers the phagocyte to phagocytose and, it is hoped, to kill the ingested pathogen. The FcRs therefore play an active role in triggering phagocytosis and do not simply bind the antibody–antigen complex. Opsonisation by antibody can also be promoted by complement (see Section 8.6).

Table 8.1 Fc receptors

Receptor	*Ig bound*	*Distribution*
Fcγ RI	IgG1, 3	Macrophages, dendritic cells Inducible on neutrophils, eosinophils
FcγRII	IgG1, 3	Different forms with different intracellular regions expressed on all leukocytes except T cells
FcγRIII	IgG1, 3	Monocytes, neutrophils, T cells, NK cells, eosinophils, mast cells, FDCs
FcεRI	IgE	Mast cells, basophils, FDCs Inducible on eosinophils
FcεRII	IgE	B cells, macrophages, eosinophils
FcaRI	IgA1, 2	Macrophages, neutrophils Inducible on eosinophils

8.5.2 Phagocytes ingest bound particles

Once phagocytosis has been triggered the phagocyte extends its membrane around the particle, forming structures called pseudopodia. Eventually the particle becomes completely surrounded by the phagocytic cell and is engulfed in a phagocytic vacuole. This completes the process of ingestion and the phagocyte will now normally proceed to kill the ingested organism.

8.5.3 Phagocytes have different ways of killing microbes

When the microbe is inside a phagocytic vacuole the phagocyte will attempt to kill it. Phagocytes have a number of mechanisms for killing ingested microorganisms. Some of these can be performed within the vacuole by mechanisms that are activated by the formation of the vacuole and by microbial products. These do not require fusion of the phagosome with a lysosome and hence are termed lysosome-independent. However, not all microorganisms are killed by these lysosome-independent mechanisms. The phagosome may, and usually does, fuse with a lysosome, resulting in exposure of the phagocytosed microorganism to lysosomal-dependent killing mechanisms (Figure 8.4).

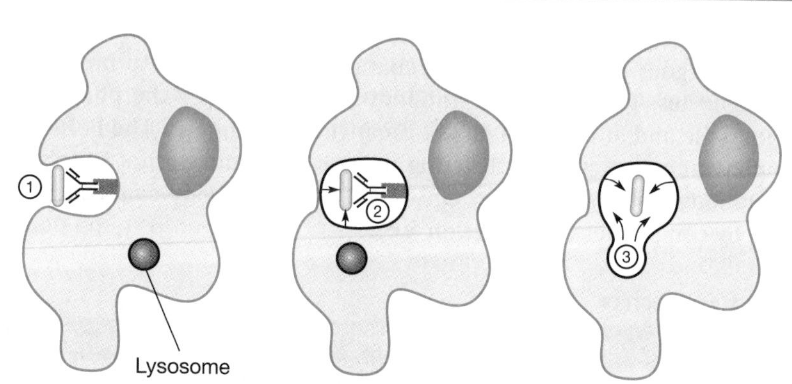

Figure 8.4 Phagocytosis and killing. Phagocytes that have bound opsonised microbes extend their membrane round the microbe ① eventually taking the microbe into a phagocytic vacuole ② where the microbe is exposed to lysosomal-independent killing mechanisms. The phagosome may fuse with a lysosome ③, forming a phagolysosome where the microbe is exposed to lysosomal-dependent killing mechanisms in addition to the lysosomal-independent mechanisms.

Lysosome-independent killing mechanisms

There are two main microbicidal pathways that can occur without lysosomal fusion with the phagosome. These are the generation of **oxygen radicals** and the production of **nitric oxide**.

- **Oxygen radicals**. Oxygen radicals are highly reactive chemicals with bactericidal activity. They can bind to and damage a variety of microbial products such as membranes, proteins and DNA, leading to killing of the microbe. Oxygen radicals involved in killing by phagocytes include superoxide anion (O_2^-), hydrogen peroxide (H_2O_2), singlet oxygen ($^1O_2$) and free hydroxyl radicals (·OH). The reactions generating oxygen radicals are shown in Box 8.2.
- **Nitric oxide pathway**. The other lysosomal-independent pathway involves the generation of nitric oxide (NO), which is highly toxic to bacteria and can also inhibit viral replication. NO can bind iron and deprive bacteria of this essential growth element. NO is produced by the combination of oxygen with nitrogen derived from the amino acid L-arginine in a reaction catalysed by the enzyme nitric oxide synthase (NOS) (see Box 8.2).

Lysosomal-dependent killing mechanisms

Although lysosomal-independent mechanisms of killing exist, under normal circumstances phagosomes containing ingested microbes fuse with lysosomes to form a phagolysosome. This process exposes the contents of the phagosome to lysosomal products, which have a variety of microbicidal activities:

- **Generation of chlorine products.** Lysosomes contain an enzyme called myeloperoxidase (MPO). This catalyses the production of hypochlorous acid (HOCl) from hydrogen peroxide and chloride; the hypochlorous acid is then converted to hypochlorite (OCl^-) and chlorine (Cl_2) (see Box 8.2). As might be expected, these chlorine-containing products are toxic to many microbes (OCl^- is used as a bleach) and contribute to phagocytic killing.
- **Defensins.** These are cationic proteins that are able to form ion pores in membranes and can kill a variety of microbes, including bacteria, fungi and some viruses.
- **Proteolytic enzymes.** Lysosomes contain many proteolytic enzymes that may degrade microbial products although it is not clear how important they are in attacking intact microbes. Another lysosomal product is lysozyme; as its name implies, lysozyme is able to degrade the peptidoglycan layer of Gram-positive bacteria.

BOX 8.2: GENERATION OF BACTERICIDAL RADICALS AND NITRIC OXIDE

Oxygen radicals

The first step in this process involves the generation of the superoxide anion, O_2^-:

$$NADPH + O_2 \xrightarrow{\text{Cytochrome b-245}} NADP + O_2^-$$

The O_2^- then spontaneously reacts with other molecules, generating a number of other oxygen radicals: these are hydrogen peroxide (H_2O_2), singlet oxygen ($^1O_2$) and free hydroxyl radicals (•OH), which are produced as follows:

$$2O_2^- + 2H^+ \longrightarrow H_2O_2 + {}^1O_2$$
$$O_2^- + H_2O_2 \longrightarrow \bullet OH + OH^- + {}^1O_2$$

Nitric oxide

The production of nitric oxide is catalysed by the enzyme nitric oxide synthase (NOS), of which there are two types: endothelial NOS (eNOS) is constitutively expressed in many cell types; inducible NOS (iNOS) is, as its name implies, induced and is the main enzyme used by phagocytes.

The reaction is as follows:

$$O_2 + \text{L-arginine} \xrightarrow{\text{NOS}} NO + \text{citrulline}$$

Generation of chlorine products

Initially hypochlorous acid (HOCL) is generated from hydrogen peroxide (H_2O_2) and chloride (Cl^-):

$$H_2O_2 + Cl^- + H + \longrightarrow HOCl + H_2O$$

The HOCl generates hypochlorite (OCl^-) and combines with more chloride to generate chlorine:

$$HOCl \rightleftharpoons H^+ + OCl^-$$
$$HOCl + Cl^- \longrightarrow Cl_2 + OH^-$$

8.5.4 Killed microbes are degraded

An important role of the proteolytic enzymes found in lysosomes is to degrade microbial products so they can be excreted by the phagocyte. Additionally degradation of microbial proteins by macrophages can generate antigenic peptides that can be presented on the surface of the macrophage in association with class II MHC and can therefore stimulate CD4 T cells, thereby contributing towards the specific immune response against the pathogen.

8.6 Complement is a protein cascade with antimicrobial functions

Another important function of some classes of antibody is to activate the complement system. Complement is not a single molecule but a cascade of proteins whose closest parallel is the clotting system, although the functions of complement are very different from those of the clotting system. Like the clotting system, complement consists of a series of inactive precursor proteins that are activated and then activate the next protein in the sequence.

Three complement pathways exist: the classical, alternative and lectin pathways. All three pathways have a common final stage but the earlier components in the pathways differ and the ways in which the pathways are activated are also different. When describing complement components a convention is adopted because many complement components are inactive pro-enzymes. Therefore the active forms of the protein are indicated by the suffix **.

8.6.1 The classical complement pathway

The main components of the classical complement pathway are proteins called C1–C9. They were numbered in the order in which they were discovered and fortunately, with one exception, they become involved in the chain of events in the same numerical order. Therefore the initiation of the classical complement pathway involves the first component of complement, C1.

Plate 1. Elephantiasis. Obstruction of lymphatic vessels in chronic filiarisis caused by the nematode worm *Wucheria bancrofti* can lead to massive oedema and other complications, resulting in elephantiasis. (Source: Courtesy of Richard Suswillo, Department of Infectious Diseases, School of Medicine, Imperial College, London, UK.)

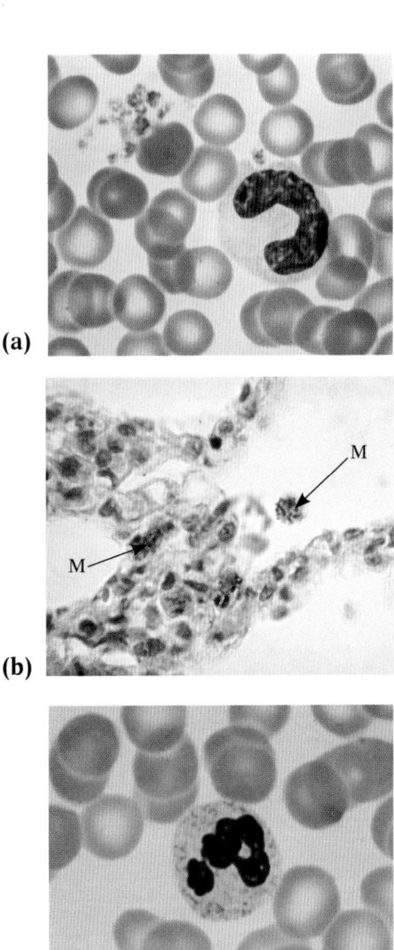

Plate 2. Cells of the phagocyte lineage. (a) Monocytes circulate in the bloodstream, where they are not phagocytic. Monocytes leave the bloodstream and enter sites of inflammation, where, under the influence of inflammatory mediators, they differentiate into inflammatory macrophages (H&E × 1000). (b) Tissue macrophages are weakly phagocytic unless stimulated by microbial products or inflammatory mediators. The macrophages (M) have phagocytosed black carbon particles (× 400). (c) Neutrophils have a characteristic multi-lobed nucleus. They are usually the first cells to be recruited from the bloodstream to sites of inflammation. Once they have left the bloodstream they perform a 'kamikaze' role, rapidly phagocytosing and killing bacteria before dying themselves in a few hours. The mixture of dead neutrophils and bacteria products make up pus (H&E × 800). (Source: Courtesy of Mike Mahon, John S. Dixon and Philip F. Harris, University of Manchester.)

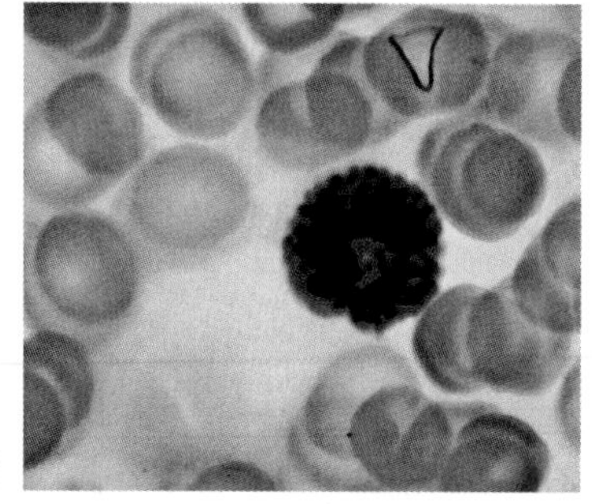
(a)

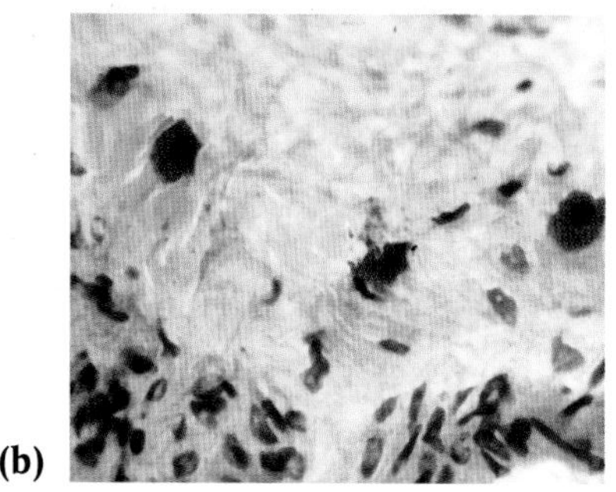
(b)

Plate 3. Mast cells and basophils. These cells have many similarities and are characterised by numerous large granules. (a) Basophils are white blood cells and can enter some inflammatory sites (H&E × 800). (b) Mast cells are found in connective and mucosal tissue. This picture is of the bladder wall and the mast cell granules are stained red. The mast cell third from the right is partially degranulated (pararosanaline alpha napththyl stain × 400). Despite their similarities there is no convincing evidence that mast cells are derived from basophils. (Source: Courtesy of Mike Mahon, John S. Dixon and Philip F. Harris, University of Manchester.)

Plate 4. Lymphocyte and plasma cell. (a) The lymphocyte consists of a darkly staining nucleus with a narrow layer of cytoplasm (H&E × 1000). (b) Plasma cells are bigger than lymphocytes and have much more cytoplasm (H&E × 4000). (Source: Courtesy of Mike Mahon, John S. Dixon and Philip F. Harris, University of Manchester.)

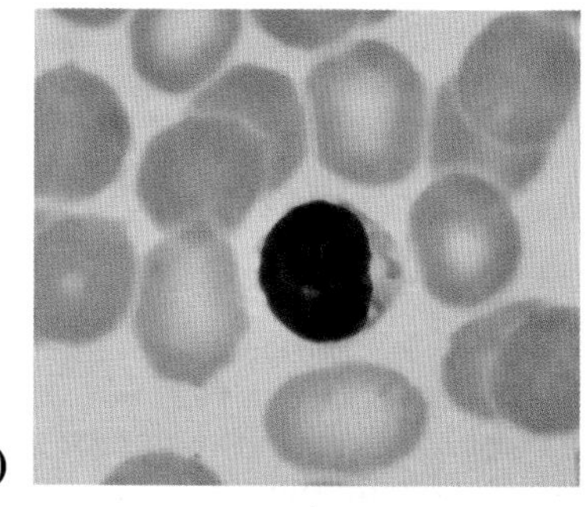
(a)

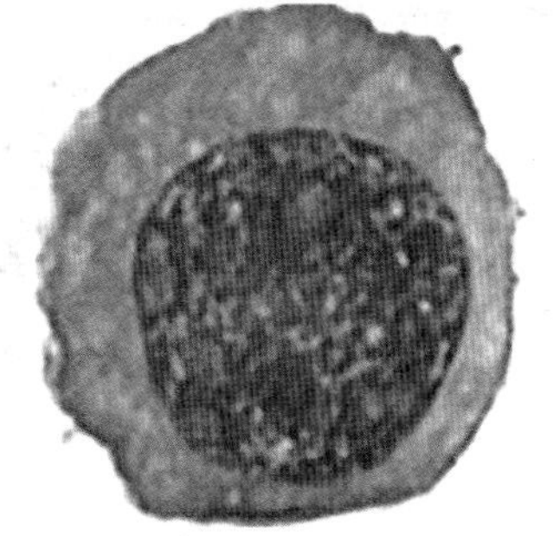
(b)

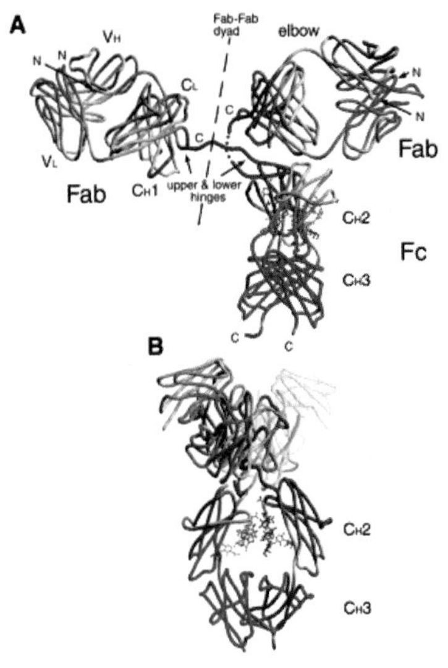

Plate 5. Antibody structure. This plate shows the structure of an antibody directed against the gp120 envelope protein of HIV. The antibody shows a globular Y-shaped structure with asymmetric arrangement of the variable regions due to conformation of the hinge regions. (Source: From Saphire, E.O. *et al.* (2001), 'Crystal structure of a neutralising human IgG against HIV-1: A template for vaccine design'. *Science*, Vol. 293, pp. 1155–59. Copyright © 2001 American Association for the Advancement of Science (AAAS). Reprinted with permission.)

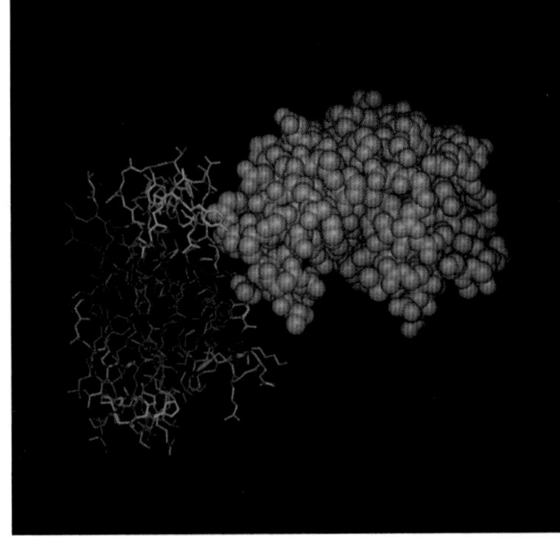

Plate 6. Antibody–antigen interaction. The antibody (in stick form) is shown interacting with antigen, hen-egg lysozyme, in purple. The interface between antibody and antigen is flattish with dents and bumps formed by the amino acid side chains. (Source: Courtesy of Dr Annemarie Honegger and Prof. Dr. Andreas Plueckthun, Biochemisches Institut Der Universitat Zurich, Winterthurerstrasse 190, CH-8057 Zurich, Switzerland. Reprinted with permission.)

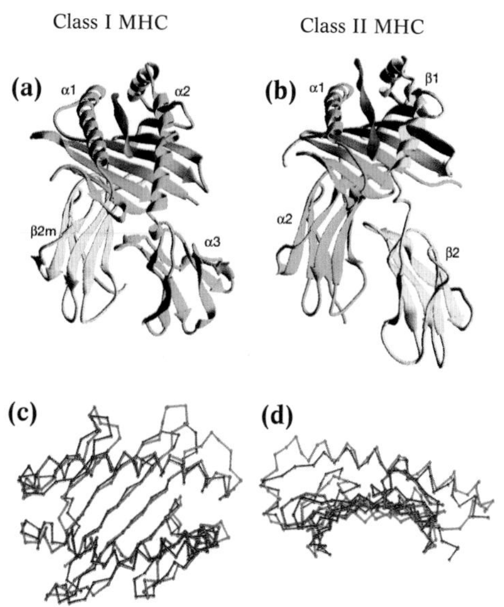

Plate 7. C1 I and C1 II MHC. The top figures show class I(a) and II(b) MHC with the peptides binding to the MHC shown in blue. The bottom figures show an overlay of class I (blue) and class II (red) MHC from the top c) and side d) showing the overall similarity in structure of the two molecules. (Source: From Stern, L.J. and Wiley, D.C., 'Antigenic peptide binding class I and class II histocompatibility proteins'. *Structure*, Vol. 2, pp. 245–51. Copyright © 1994 Elsevier. Reprinted with permission.)

Plate 8. Interaction between the T cell receptor (TcR) and antigen/class I MHC. The class I MHC molecule is at the bottom of the picture with the α-chain in green and β_2-microglobulin in yellow. The peptide being presented by class I MHC is in red. The Vα chain of the TcR is in purple and the TCR β-chain in blue; for simplicity the Cα part of the TCR is not shown. (Source: Courtesy of Dr Annemarie Honegger and Prof. Dr. Andreas Plueckthun, Biochemisches Institut Der Universitat Zurich, Winterthurerstrasse 190, CH-8057 Zurich, Switzerland. Reprinted with permission.)

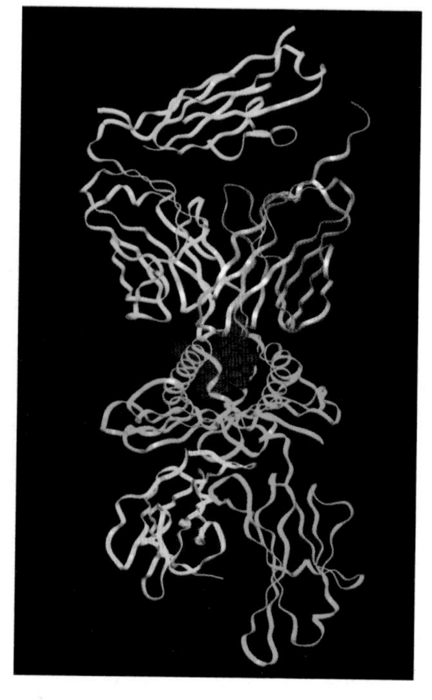

Plate 9. Histological appearance of thymus. The darker staining outer area of each lobule is the cortex and the lighter inner staining area the medulla (H&E × 15). (Source: Courtesy of Mike Mahon, John S. Dixon and Philip F. Harris, University of Manchester.)

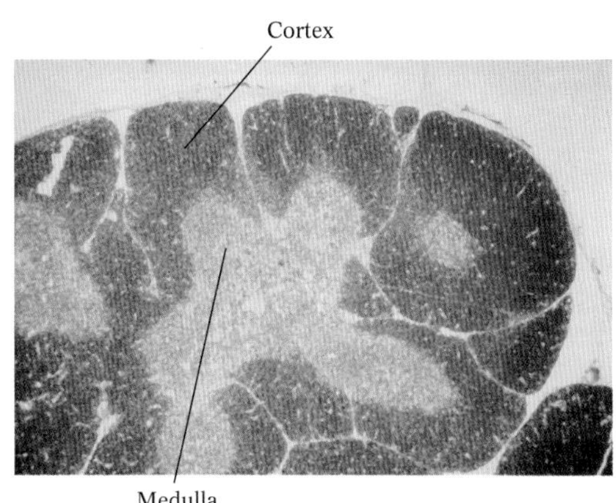

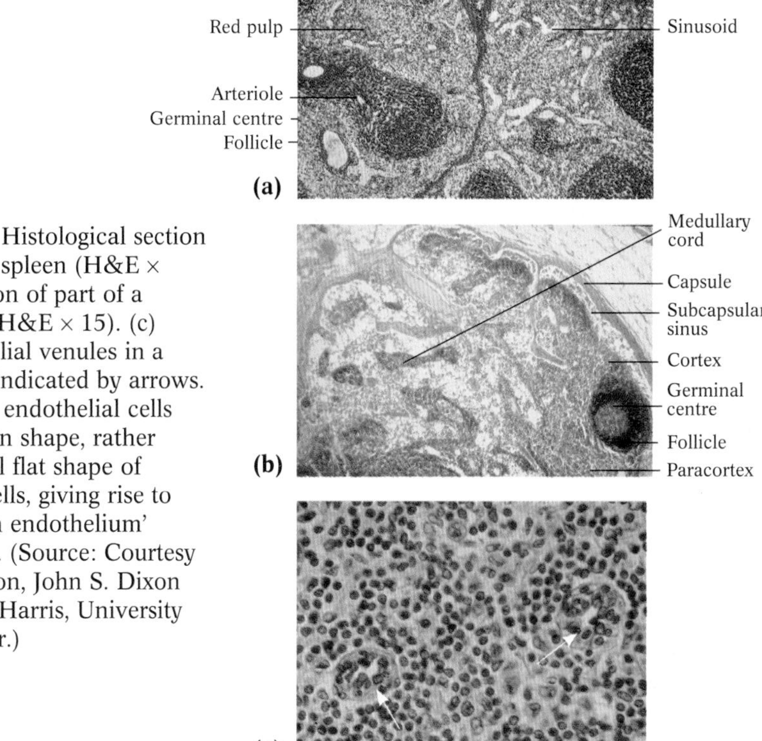

Plate 10. (a) Histological section of part of the spleen (H&E × 20). (b) Section of part of a lymph node (H&E × 15). (c) High endothelial venules in a lymph node, indicated by arrows. Note that the endothelial cells are cuboidal in shape, rather than the usual flat shape of endothelial cells, giving rise to the term 'high endothelium' (H&E × 200). (Source: Courtesy of Mike Mahon, John S. Dixon and Philip F. Harris, University of Manchester.)

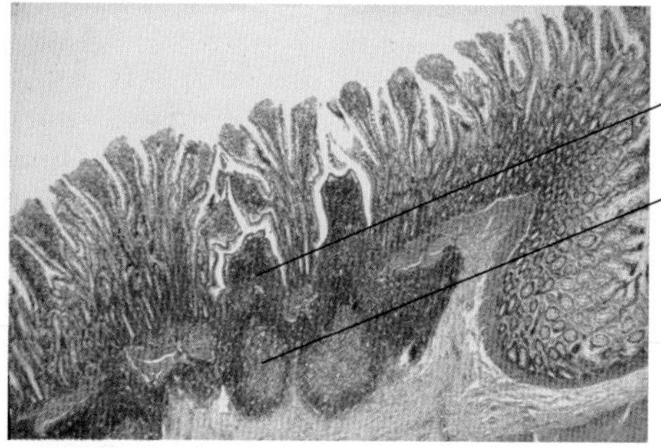

(a)

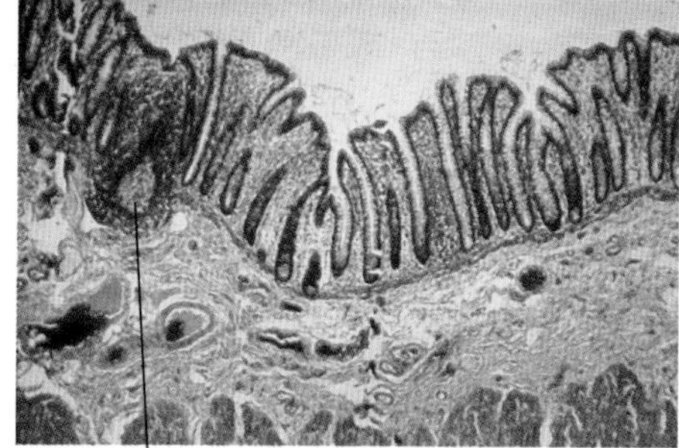

(b)

Plate 11. Histological appearance of mucosal associated lymphoid tissue (MALT). (a) Section of ileum with Peyer's patches. Note the flattened appearance of the epithelium above the Peyer's patches (H&E × 10). (b) Section of large intestine with lymphoid nodule (H&E × 10). (Source: Courtesy of Mike Mahon, John S. Dixon and Philip F. Harris, University of Manchester.)

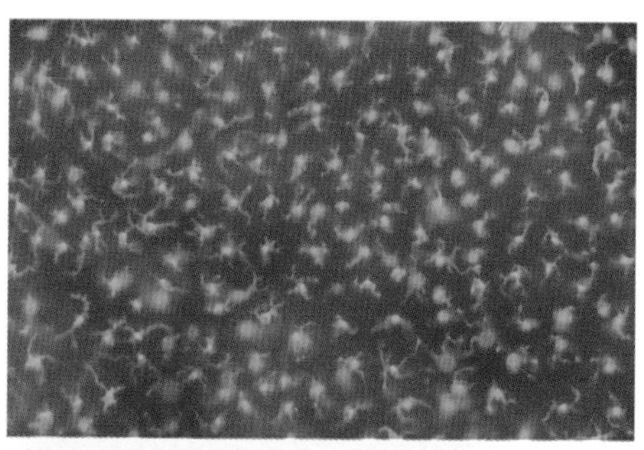

(a)

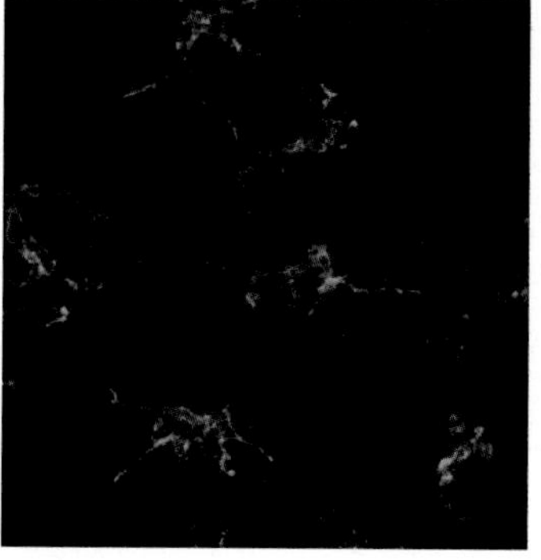

(b)

Plate 12. Langerhans cells and dendritic cells. (a) The bright green cells are Langerhans cells in the skin stained for immunofluorescence (with antibodies to CD1a). The dendritic processes of the Langerhans cells form a web that enables the cells to take up antigen in the skin (× 100). (b) Confocal image of dendritic cells (× 400). ((b) (Source: Courtesy of Dr Pedram Hamrah and Dr Reza Dana, Schepens Eye Research Institute, Harvard Medical School, Boston, Massachusetts, USA.)

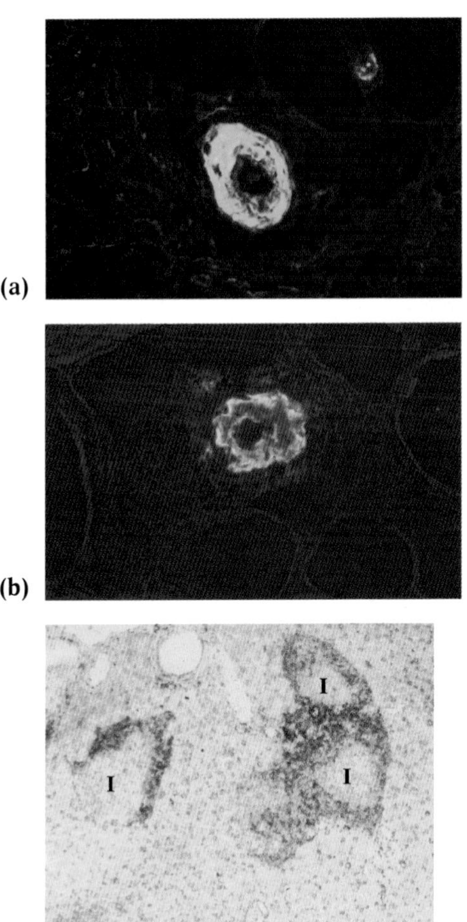

Plate 13. Appearance of immunofluorescence and immunoperoxidase. (a) Section of skin from an individual with polyarteritis nodosa, a disease in which the blood vessel walls are weakened and IgG is deposited. The IgG has been detected by incubating the tissue with fluorescent anti-human Ig (× 480). (b) Complement deposition can be detected in the same disease, polyarteritis nodosa, with fluorescent antibody to C3 (× 480). (c) Immuno-cytochemical staining of pancreas with diabetes. The section was stained with antibody to CD4 followed by peroxidase-conjugated anti-Ig and a substrate that is turned brown by the peroxidase. The section is lightly stained with haematoxylin and the brown colouration indicates the presence of many CD4 T cells in the islets (indicated by I) (× 100).

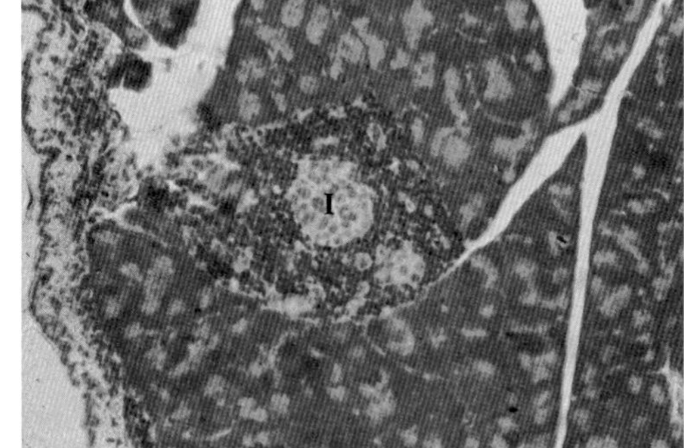

Plate 14. Infiltrating cells in a diabetic pancreas. The islets of Langerhans are indicated by the letter I and this diabetic pancreas has extensive infiltration of mononuclear cells (small dark staining cells) into the islets (H&E × 100).

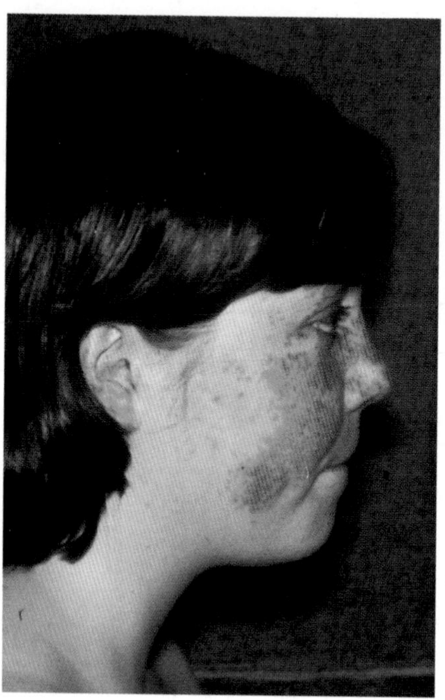

Plate 15. Butterfly rash of SLE. The characteristic rash is seen most frequently on areas of skin exposed to the sun.

Plate 16. Allergic reactions in skin and lung. (a) Wheal and flare skin reaction to applied allergen. (b) Lung of chronic asthmatic during symptom free phase. The bronchiole and alveoli are clear (van Gieson's elastica stain × 15). (c) High-power picture of the lung of a chronic asthmatic patient during an asthmatic attack. The basement membrane is thickened and there is extensive infiltration of eosinophils into the respiratory mucosa (H&E × 80).

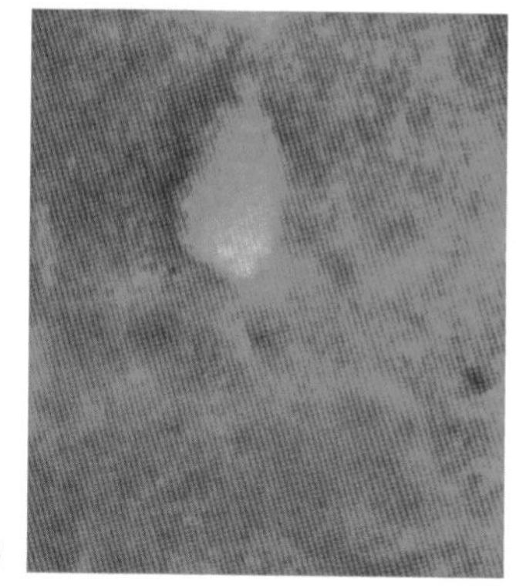

(a)

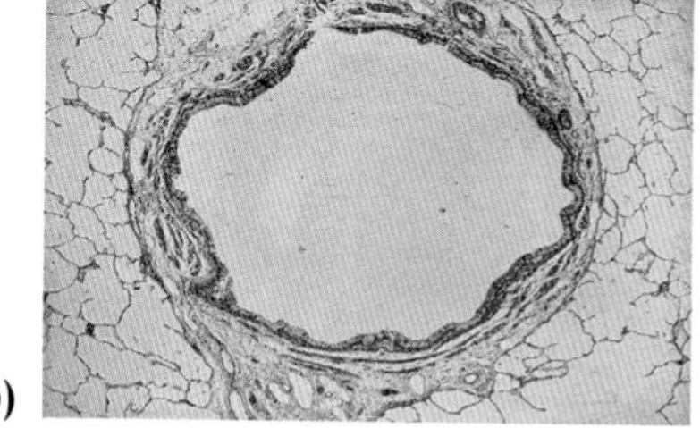

(b)

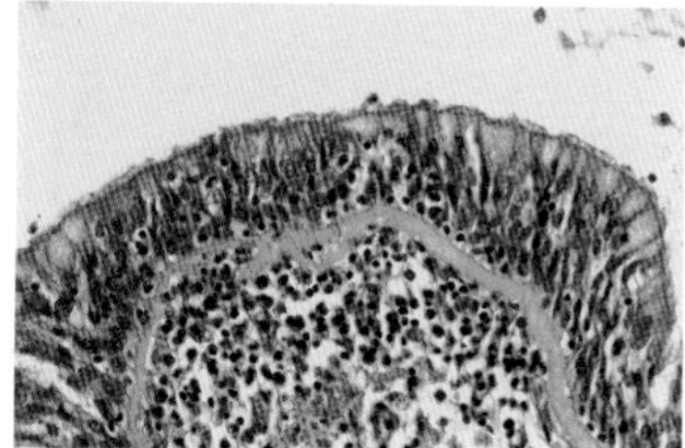

(c)

(1) Binding of C1 to antibody-antigen complexes initiates the classical pathway

For the complement pathway to be activated antibody must bind to antigens on a solid surface, for example the membrane of a bacterium. When an antibody binds to an antigen, the Fc portion of the antibody alters conformation so that it can bind the first component of complement, C1. This process is also known as **complement fixation**. Not all antibody classes can bind C1: IgM is the most efficient and, in humans, IgG1, IgG2 and IgG3, but not IgG4, can bind C1.

C1 has three components called C1q, C1r and C1s and the complex has the appearance of a bunch of tulips, with six stalks composed of C1q bound to two molecules each of C1r and C1s (see Figure 8.5, and Box 8.3 for more details). Each C1q molecule has a globular protein 'flower head' that is capable of binding to the Fc part of the antibody molecule. At least two of the C1q globular heads must bind to an Fc binding site for C1 to be activated. For IgG molecules, which have only one Fc portion, this means that two or more IgG molecules must be bound close together on a particulate surface, such as a bacterium, to activate C1. IgM, which is a pentamer, has five Fc regions and therefore one IgM molecule is capable of activating C1 (① and ②in Figure 8.5). When two or more of the C1q globular heads have bound to Ig Fc regions the C1q undergoes a conformational change, which results in activation of C1r. C1r is an inactive serine protease, which, after activation, cleaves itself to form more active C1r**. C1r** can also cleave C1s, another serine protease, leading to activation of C1s**.

(2) Activated C1 generates products from C4 and C2

C1s** then cleaves another complement component, C4, into a small fragment called C4a and a larger fragment called C4b. The C4b can then attach to the cell membrane or particle to which the Ab/C1 complex is bound. The C4a diffuses away but has important biological properties (see Section 8.6.4). The membrane-bound C4b can now bind the C2 component of complement and when C2 is bound to C4b it in turn is cleaved by C1s** into C2a and C2b. The C2b diffuses away but the C2a remains bound to C4b and the C4b2a forms an active protease complex C4b2a**③ in Figure 8.5).

(3) C4b2a is a C3 convertase

The substrate for C4b2a** is C3, which is a central and key component of all the complement pathways. Because C4b2a** converts inactive C3 into its active form it was originally called **C3 convertase**. The importance of C3 is demonstrated by the fact that individuals who lack C3 suffer from recurrent, life-threatening bacterial infections. C3 binds to C4b2a and is cleaved into C3a and C3b④ in Figure 8.5). C4b2a** is an enzyme and one molecule of C4b2a** can generate more than 200 molecules of C3b. C3a diffuses away but has important functions (see Section 8.6.4). The C3b

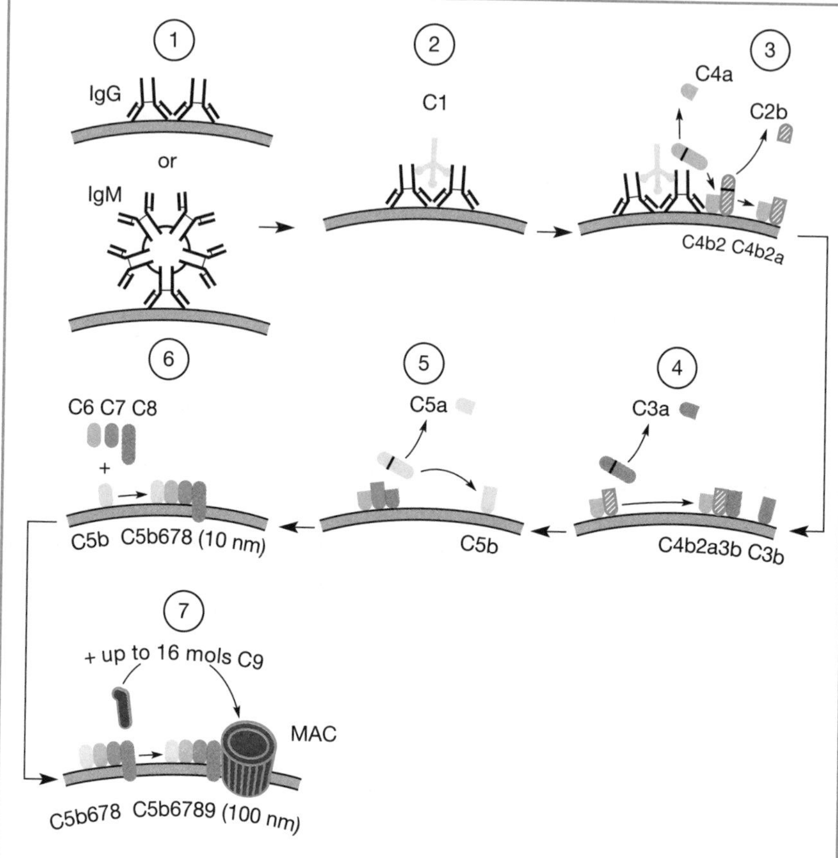

Figure 8.5 The classical complement pathway. Two molecules of IgG are needed to bind (fix) C1. Because IgM is a pentamer of Ig, one molecule of IgM can bind C1.

generated has two important functions. Most of the C3b binds to the cell to which the antibody is bound and acts as an opsonin, promoting the phagocytosis of the cell. The ability of C3b to act as an opsonin is one of the most important functions of complement. Some of the C3b binds to the C4b2a**, forming C4b2a3b** ④ in Figure 8.5).

BOX 8.3: STRUCTURE OF C1

C1 consists of 18 polypeptide chains. The three subunits are called C1q, C1r and C1s and there are six C1q, two C1r and two C1s subunits in a C1 macromolecule. Each C1q subunit actually consists of three different polypeptide chains called A, B and C, which twist together in a collagen-like triple helix topped with a globular head.

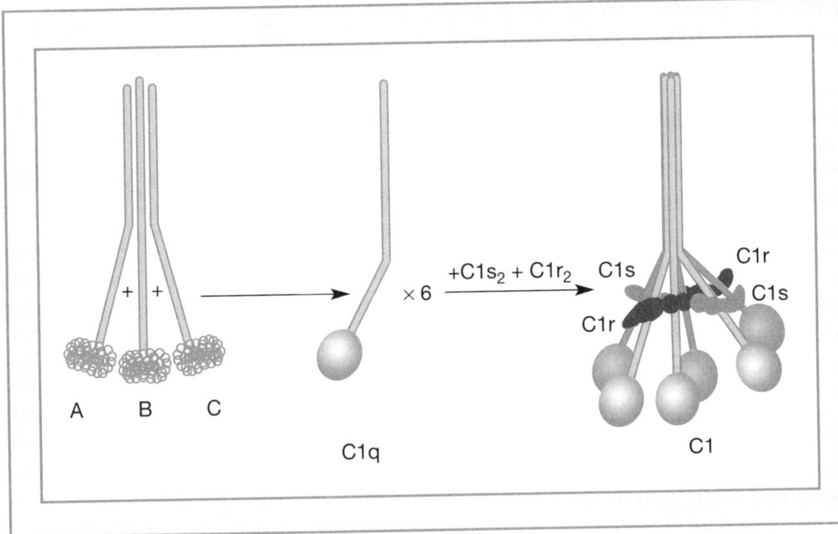

(4) C4b2a3b is a C5 convertase

C4b2a3b is a C5 convertase and binds to and cleaves C5 into C5a and C5b⑤ in Figure 8.5). C5a diffuses away but has important biological activities (see Section 8.6.4) and the C5b binds to the cell surface. The cleavage of C5 is the last enzymatic step in the complement pathway. The later steps are involved in generating pores in the membrane that will result in lysis of the cell. These pores are known as the **membrane attack complex**.

The generation of the C5 convertase is also the last step at which the three pathways differ; the subsequent stages are the same for all the pathways.

(5) C6–9 are also involved in the formation of the membrane attack complex

C5b is quite labile and has about two minutes to bind the next complement component, C6, before it is inactivated⑥ in Figure 8.5). The C5b6 complex binds C7 and C8 to form a C5b678 complex. During the binding of these molecules some of them undergo conformational changes that result in highly lipophilic structures in C7 and C8 inserting into the membrane. These C5b678 complexes can form small pores (10 Å in diameter) ⑥ in Figure 8.5) in the membrane, which can result in the lysis of some microorganisms and therefore act as a membrane attack complex.

Most cells are not lysed by C5b678 but require the final complement component C9 to be added. Up to 16 C9 molecules can be added to one C5b678 complex and the C9 molecules assemble to form a pore some 100 Å in diameter ⑦ in Figure 8.5). This is usually what is meant by the membrane attack complex (MAC).

(6) Lysis occurs due to osmotic imbalance

The pores formed by the MAC are large enough to allow the passage of ions, small molecules and water but too small for proteins to pass through.

There is therefore an influx of water into the cells and a loss of ion balance, which leads to lysis by 'bursting' of the cell.

8.6.2 The lectin pathway of complement activation

This is very similar to the classical pathway except for the very first steps (Figure 8.6). Instead of antibody binding to an antigen on a pathogen, mannose binding protein (MBP), which is always present in the serum, binds to mannose residues on the surface of pathogens. These mannose residues may be components of glycoproteins or polysaccharides. When MBP has bound to the pathogen, a protease called MBL-associated serine protease (MASP) binds to the MBP. The complex of MBP and MASP cleaves C4 into C4a and C4b and C2 into C2a and C2b in the same way that C1s** performs these functions in the classical pathway. Once C4b is generated it binds to the cell surface and the rest of the sequence is the same as for the classical pathway.

8.6.3 Alternative pathway of complement activation

The important differences between the classical and lectin pathways and the alternative pathway of complement are that C1, C4 and C2 are not involved in the alternative pathway. However, the components C3 and C5–C9 are involved in all pathways. The alternative pathway uses different components to generate C3 and C5 convertases, as described below (Figure 8.6).

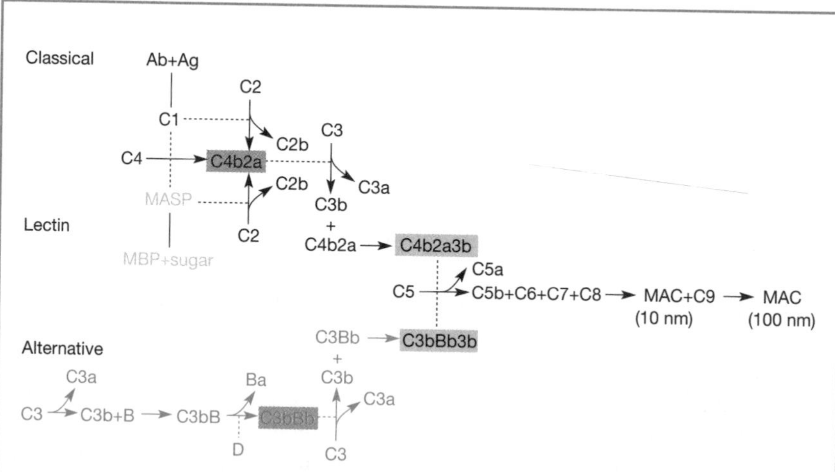

Figure 8.6 Complement pathways. The classical and lectin pathways result in the generation of the C3 convertase C4b2a and the C5 convertase C4b2a3b. The alternative pathway results in the formation of the C3 convertase C3bBb and the C5 convertase C3bBb3b. Once C5b has bound to the cell surface, the later steps are identical for all three pathways.

(1) Formation of C3 convertase

C3 is quite labile and undergoes a low level of spontaneous hydrolysis to produce C3a and C3b. Most of the C3b generated is itself hydrolysed to an inactive form in the serum within milliseconds of formation. Some of the C3b binds to the body's own cells where it is inactivated by surface membrane regulatory proteins (see Section 8.6.5). Many microbes lack these regulatory proteins and therefore C3b bound to a microbial membrane is not inactivated. This system provides an elegant, innovative mechanism by which the innate immune system can distinguish self from non-self. Self-tissue is able to inactivate C3b but many microbes are not and go on to be opsonised or lysed by the alternate complement pathway.

If C3b is not inactivated it binds another component of the alternative pathway, factor B. The factor B bound to C3b is then cleaved by another alternative pathway serum protein, factor D, to yield Ba and Bb. Ba diffuses away but Bb remains bound to C3b, forming an active C3 convertase called C3bBb**, which is equivalent to C4b2a of the classical pathway. The C3bBb** is stabilised by another protein called properdin. The C3bBb** can cleave more C3, generating more C3b that can bind to the microbial membrane, which provides a powerful amplification loop. It has been estimated that in this way upwards of a million C3b molecules can be bound to the membrane within five minutes.

This C3b can also perform one of two functions. One of these is to act as an opsonin and just as in the classical pathway the opsonic activity of C3b plays a very important role in eliminating microorganisms. The other role of C3b is as part of the C5 convertase of the alternative complement pathway.

(2) Formation of C5 convertase

Some of the C3b generated by the C3bBb** actually binds to the C3bBb to form C3bBb3b, which is a C5 convertase equivalent to C4b2a3b in the classical pathway. C3bBb3b cleaves C5 into the same C5a and C5b components as C4b2a3b. Once C5a and C5b have been generated the latter stages of the complement pathway involving formation of the MAC proceed in exactly the same way as for the classical pathway (see above).

8.6.4 Different components of the complement system have many biological activities

The complement cascade, whether it is activated by the classical, lectin or alternative pathway, generates a number of molecules that have a variety of biological activities (Figure 8.7). The main activities of the complement components are as follows:

- **Cell lysis.** This is mediated by the components C5–9 forming the MAC, as described above.
- **Opsonisation.** C3b is a very important opsonin. Monocytes, macrophages and neutrophils all have receptors for C3b on their surface.

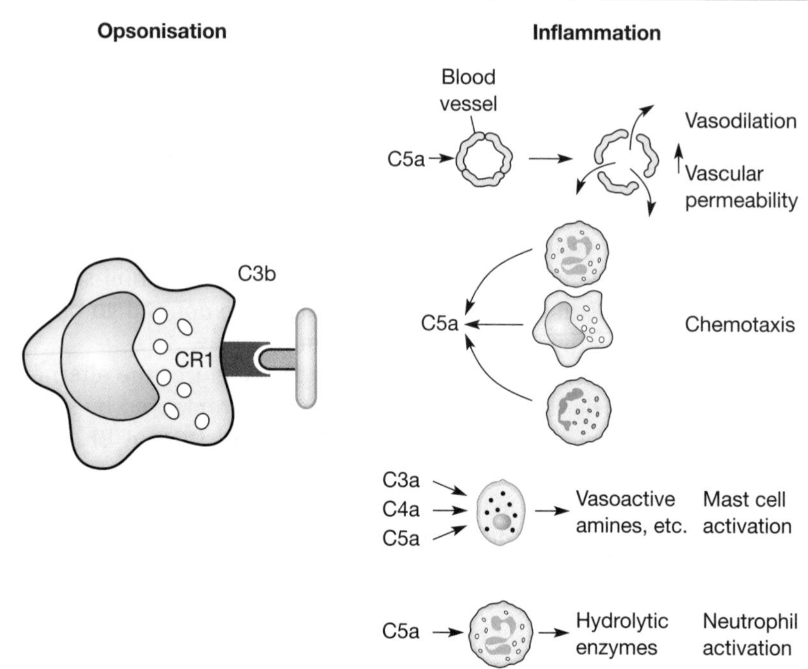

Figure 8.7 Functions of complement components. Complement has many functions in addition to lysis of cells. C3b is an opsonin and when bound to a microbial cell surface can bind to complement receptors (CR1) on phagocytes. Complement components also play a number of roles in inflammation, including causing vasodilation and an increase in vascular permeability (C5a), chemotaxis of neutrophils, monocytes and eosinophils (C5a), and activation of mast cells (C3a, C4a and C5a) and neutrophils (C5a).

These receptors, called complement receptors type 1 (CR1), can bind C3b on the surface of microbes, thereby acting as opsonins and promoting the phagocytosis of the microbe.

- **Chemotaxis.** Many complement components produced by the splitting of inactive precursors have chemotactic activity and therefore play a role in recruiting cells to the site of an inflammatory response. The most active is C5a, which is chemotactic for neutrophils, monocytes and, to a lesser extent, eosinophils. C3a is also chemotactic for eosinophils.
- **Inflammatory mediators.** C3a, C5a and C4a can all act as anaphylatoxins – these are molecules that can activate mast cells and cause their degranulation. C5a can cause release of hydrolytic enzymes from neutrophils. C3a and C5a can cause degranulation of eosinophils.
- **Clearance of immune complexes.** When antibody binds to a soluble antigen, such as a toxin protein secreted by a bacterium, there is the possibility that insoluble antigen–antibody complexes could be formed. These can become trapped in small capillaries and deposited on the capillary walls, leading to a local inflammatory response resulting in immune complex disease (see Chapter 13). Complement has two important roles in preventing this from happening. C3b is able to interact with

antigen–antibody and limit the size of the lattice formed by the complex (Figure 8.8). This prevents the complexes from reaching a size where they become insoluble and could be deposited in the blood vessels.

Erythrocytes and complement can also collaborate in the removal of immune complexes (Figure 8.8). Erythrocytes have CR1 on their surface. Immune complexes with C3b bound to them are able to bind to the CR1 on erythrocytes, which also prevents the immune complexes being deposited in blood vessels. When the erythrocyte with the bound immune complex travels through the liver or spleen, macrophages bind the erythrocyte/immune complex pair through Fc receptors on the

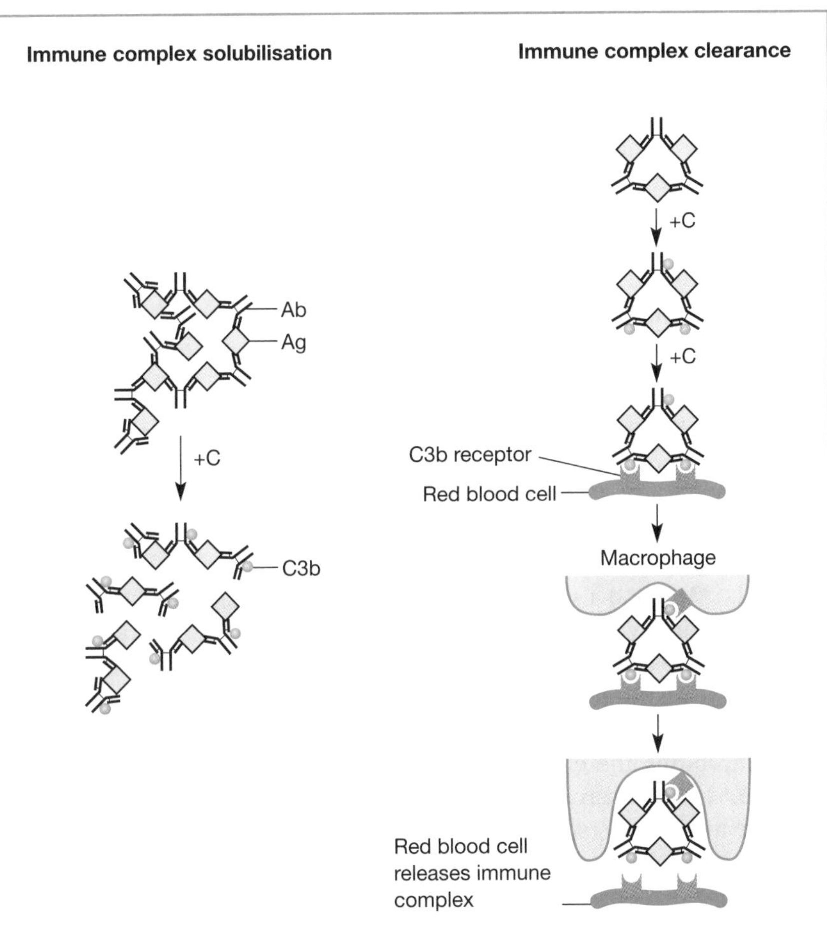

Figure 8.8 Immune complex solubilisation and clearance. Immune complexes can result in the fixation of C1 leading, via the classical pathway, to the generation of C3b, which dissociates immune complexes and prevents them getting bigger. The C3b in immune complexes can also contribute to immune complex clearance by binding to CR1 on the surface of erythrocytes. The erythrocytes transport the immune complexes to the liver or spleen where macrophages take up the immune complex and remove it, leaving the erythrocytes to continue circulating in the blood.

macrophages binding to the Ab in the immune complex. The erythrocyte dissociates from the immune complex and returns to the circulation. The immune complex is phagocytosed and destroyed by the macrophage.

8.6.5 Complement has to be regulated to avoid damage to cells and tissues

It is important that a system with the amplification and potentially tissue-damaging properties of the complement system is very tightly regulated. Regulation occurs at two levels. One is to regulate the amount of complement activity and the other is to regulate the site of complement activity. If the site of complement activation is the surface of a pathogen it is important that active complement components do not spread to host cells in the vicinity of the pathogen. Both soluble inhibitors in the plasma or extracellular fluid and membrane-bound inhibitors are important in regulating complement activity.

There are remarkable parallels between regulation of the classical and alternative complement pathways and many regulatory proteins are involved in inhibition of both pathways (Figure 8.9).

Regulation of C3 convertases

Various proteins either inhibit the assembly of C3 convertases or accelerate their dissociation (Figure 8.9):

- **Regulators of C3 convertase assembly.** The classical C3 convertase is C4b2a. Two membrane proteins, complement receptor 1 (CR1) and membrane co-factor protein (MCP), and one soluble protein, C4b binding protein, are able to bind C4b and stop it binding to C2a, thereby inhibiting the formation of C4b2a. Another regulatory protein, factor I, then cleaves C4b, resulting in irreversible inhibition of C4b2a assembly.

  CR1, MCP and a soluble protein, factor H, are able to bind C3b and stop it associating with factor B to form the alternate C3 convertase, C3bBb. Factor I then cleaves C3b, irreversibly inhibiting assembly of C3bBb.
- **Factors causing decay of C3 convertases.** A membrane protein, decay accelerating factor, causes dissociation of C2a from C4b or C3b from Bb, inhibiting the classical and alternate C3 convertases respectively. Factor I then cleaves the membrane-bound C4b or C3b, making the inactivation irreversible.

Regulation of the membrane attack complex (MAC)

A soluble protein called S-protein (vitronectin) binds the C5b-7 complex and stops it inserting into the cell membrane. A cell surface protein, CD59, inhibits MAC formation on the cell surface by blocking the binding of C8 and C9 to the C5b-7 complex.

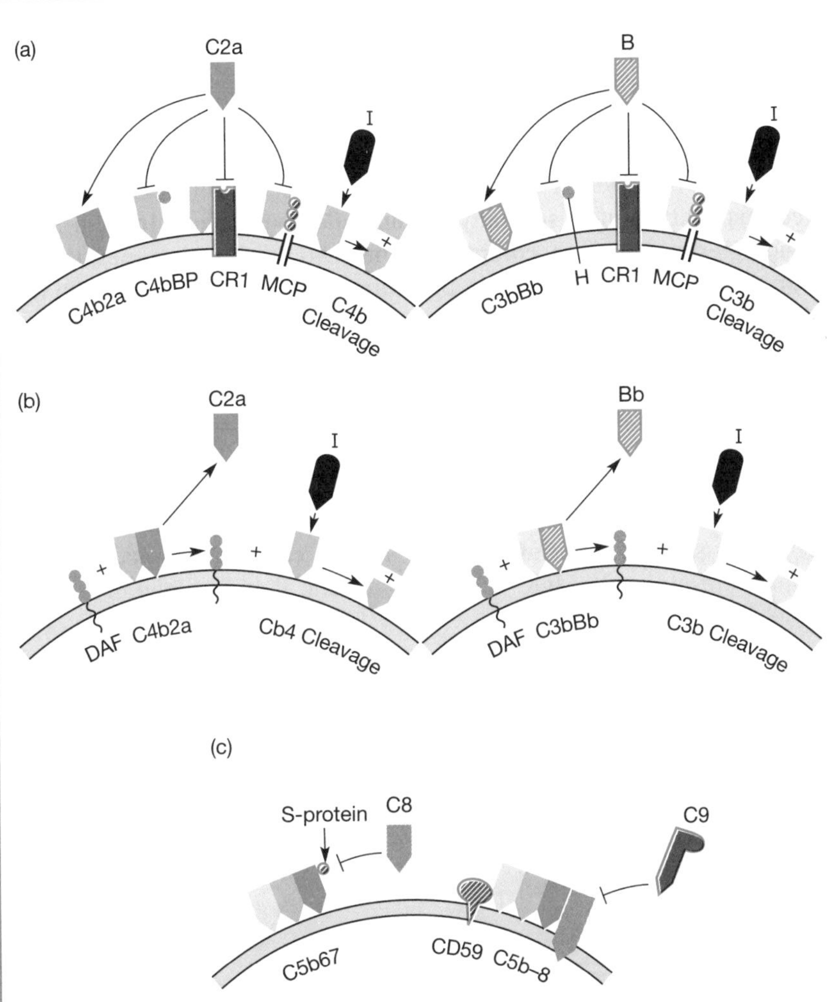

Figure 8.9 Regulation of complement. (a) Prevention of formation of C3 convertase. C4bBP, CR1 and MCP are able to bind to C4b and prevent C2a from binding to C4b in the classical pathway; Factor H, CR1 and MCP bind to C3b and prevent factor B binding to C3b in the alternative pathway, thereby preventing the formation of the C3 convertases C4b2a and C3bBb respectively. C4b and C3b are irreversibly cleaved by factor I. (b) Decay of C3 convertase. Decay accelerating factor (DAF) causes dissociation of C2a from C4b2a or dissociation of Bb from C3bBb and the C4b and C3b are cleaved by factor I. (c) Prevention of MAC (membrane attack complex) formation. S-protein stops the addition of C8 to C5–7 and CD59 stops the addition of C9 to C5–8.

8.7 Antibody and complement synergise to promote the opsonisation of microbes

Although antibody and complement can act independently of each other, the ability of antibody to activate the complement cascade provides the antibody response with a range of additional effector functions with which to combat pathogens, especially extracellular pathogens. In addition to providing extra functions, complement has two other important attributes. It can synergise with antibody in promoting opsonisation, one of the most important functions of these molecules. Phagocytes have receptors for both the Fc portion of antibody (FcR) and C3b (CR1). If a bacterium or other organism is coated with both antibody and C3b, the binding and activation of the phagocyte are increased synergistically (Figure 8.10); that is, the combination is greater than can be achieved by either component alone. This results in much more effective clearance of microbes.

Secondly, complement provides a powerful amplification system so that a relatively small amount of antibody–antigen complex can activate many complement molecules. IgM is a pentamer and therefore one molecule of

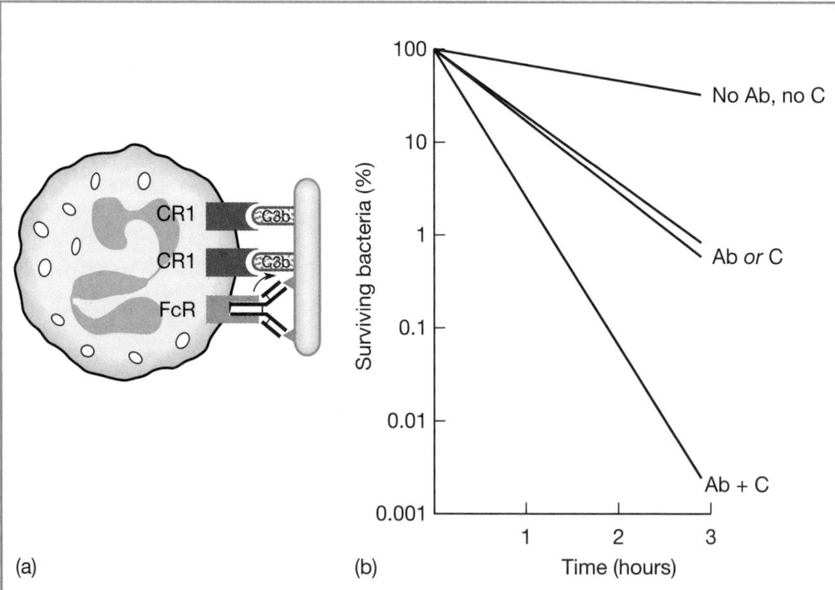

Figure 8.10 Antibody and complement synergise to opsonise microbes. (a) Antibody bound to a particle (in this case a bacterium) can fix complement, resulting in the attachment of C3b to the surface of the particle. Phagocytes have receptors for both antibody (FcR) and C3b (CR1) and therefore there is very efficient opsonisation and phagocytosis of the antibody/C3b-coated particle. (b) The kinetics of bacterial clearance *in vivo* show the synergistic effects of antibody and complement. It has been estimated that, in the presence of complement, one molecule of IgM may be enough to cause phagocytosis.

IgM can fix complement. Because of this *a single molecule of IgM*, in the presence of complement, can cause the opsonisation and phagocytosis of a bacterium. This is a good illustration of what a powerful combination antibody and complement can be.

8.8 Antibody-dependent cell-mediated cytotoxicity (ADCC)

There is one final set of effector mechanisms involving antibody that collectively are called antibody-dependent cell-mediated cytotoxicity, or ADCC for short. Cytotoxicity literally means killing of cells and there are many kinds of killer cells including neutrophils, macrophage/monocytes, eosinophils and natural killer cells. These cells have Fc receptors (FcRs) on their surface and are therefore able to bind to the Fc part of antibody. Antibody can bind to antigen on a cell surface; this might be a host cell that is infected with a pathogen and is expressing antigens derived from the pathogen on its cell surface, or it may be an antigen on a larger pathogen such as a worm. The ADCC killer cells can bind to the Fc part of the antibody and kill the antigen-bearing cell or worm (Figure 8.11). It is important to realise that killing by ADCC is not the same as phagocytosis. The targets for ADCC are generally too big to be phagocytosed and the killing is extracellular. In phagocytosis the microorganism is ingested and killed intracellularly.

The role of ADCC *in vivo* has been difficult to elucidate and has been presumed from *in vitro* studies. Host cells that have been infected by viruses or intracellular dwelling bacteria or parasites may express antigens derived from the pathogen on their cell surface and therefore be recognised by antibody and be susceptible to ADCC primarily performed by macrophages and NK cells. Additionally eosinophils have been shown to kill helminthic worms coated with antibody *in vitro*. Both IgG and IgE can promote this type of ADCC (Figure 8.11).

8.9 Summary

- Antibody is able to provide protection by both neutralising toxins produced by pathogens and promoting the destruction and elimination of pathogens. Antibody can agglutinate microorganisms promoting their phagocytosis or act as an opsonin, also promoting phagocytosis.
- Phagocytes kill phagocytosed microorganisms through lysosome-independent and lysosome-dependent mechanisms. The former involves the generation of nitric oxide and oxygen radicals and the latter involves the generation of chlorine products and the action of defensins and proteolytic enzymes.

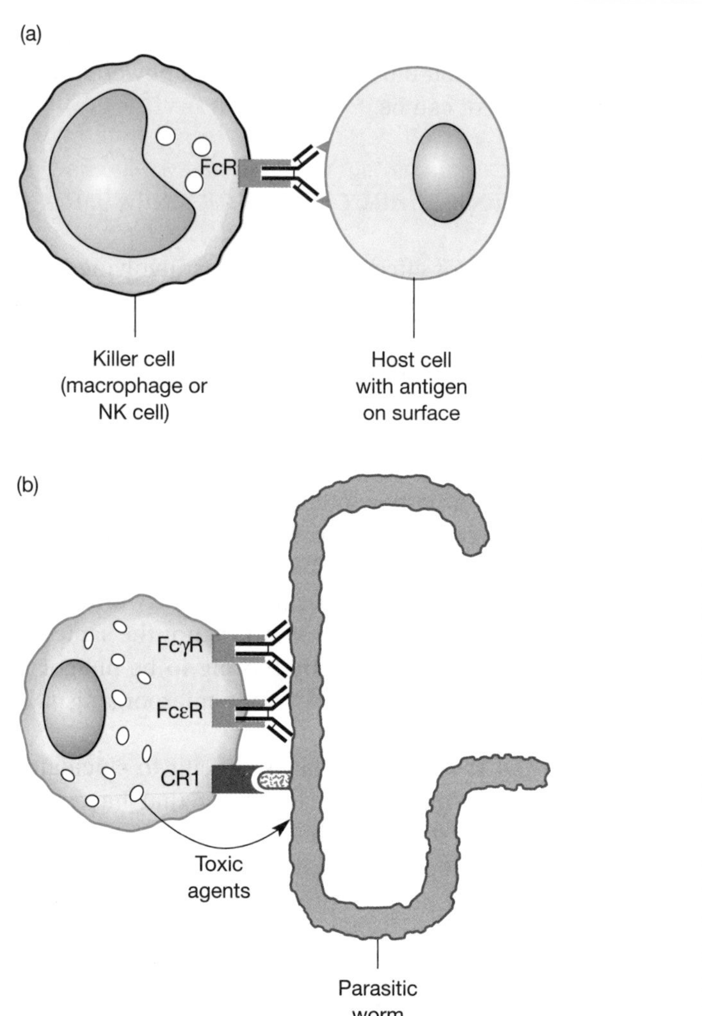

Figure 8.11 Antibody-dependent cell-mediated cytotoxicity (ADCC). (a) Host cells with antigens on their surface may bind antibody. These antigens may be derived from intracellular pathogens or may be tumour-specific antigens (see Section 15.3). Killer cells with FcRs bind the antibody and kill the host cell. The killer cells in this situation are macrophages and natural killer (NK) cells; the killing is extracellular and not by phagocytosis since often the host cell is too big to be phagocytosed. (b) A type of killing of parasitic worms has been demonstrated *in vitro*. Eosinophils have FcRs for IgE and IgG and therefore are able to attach to worms that have antibody specific for parasite antigens on their surface. The eosinophils release toxic agents, especially major basic protein, peroxidase and eosinophil cationic protein, which kill the worm.

- The complement system is a series of serum proteins that have a number of biological functions. There are three pathways of complement activation: the classical pathway, the alternative pathway and the lectin pathway.

- The many products generated as a result of complement activation have a number of different biological activities. Complement can generate a membrane attack complex that punches holes in cells, thereby killing them. Complement can also act as a powerful opsonin, acting independently or in concert with antibody to promote phagocytosis. Many complement components act as inflammatory mediators. Complement also acts to solubilise immune complexes and promote their clearance from the blood.
- Antibody can also promote ADCC in which different cell types kill other cells. These may be host cells infected with bacteria or viruses or helminths, which can be killed by eosinophils via ADCC.

8.10 Questions

1) The diagram below shows neutralisation of a virus by antibody. Match the following labels with the arrows: (i) Antibody, (ii) Attachment molecule for virus, (iii) Cell receptor for virus, (iv) Target cell for virus, (v) Virus.

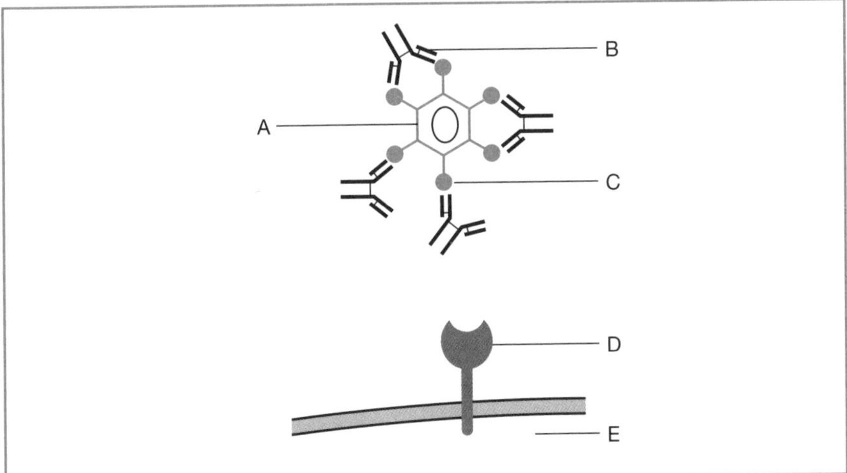

2) What mechanisms do phagocytes use to kill ingested microbes?
3) Name the three pathways of complement.
4) Bacteria were injected into normal mice or mutants lacking C3 (C3–) or mutants lacking C3 and antibody (C3–Ab–). The clearance of the bacteria was measured over the 24 hours after injection and the results are shown in the figure overleaf.

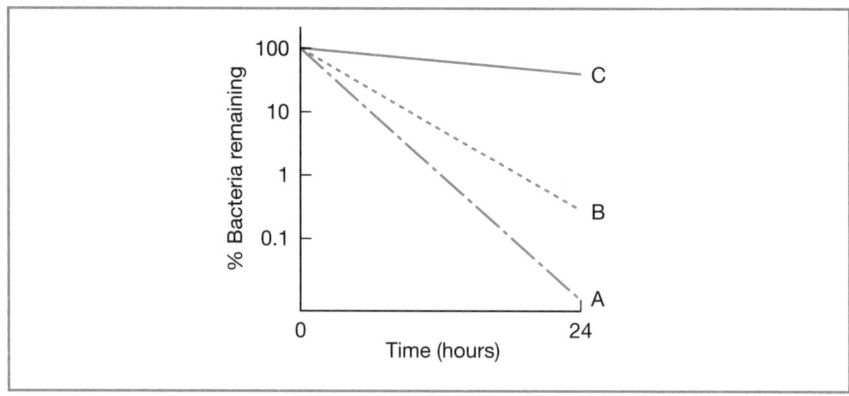

Which of the mice A), B) and C) are the normal, C– and C–Ab– mice?

5) What are the biological functions of complement?

The answers to these questions can be found on page 337.

8.11 Further reading

1) Schroeder HW Jr., Cavacini L. (2010) Structure and function of immunoglobulins. *Journal of Allergy and Clinical Immunology* 125:S41–S52

2) Walport MJ. (2001) Complement. *New England Journal of Medicine* 344:1058–1066

4) Walport MJ. (2001) Complement. *New England Journal of Medicine* 344:1141–1144.

5) Fagarasan S. (2008) Evolution, development, mechanism and function of IgA in the gut. *Current Opinion in Immunology* 20:170–177.

CHAPTER NINE

Effector mechanisms: dealing with pathogens *in vivo*

(2) Cell-mediated immunity

Learning objectives

To know how cytotoxic CD8 T cells are generated and kill infected host cells. To understand the events contributing to a delayed-type hypersensitivity reaction. To appreciate the differing cost to the host of the different types of effector response.

Key topics

- Different type of helper T cells
- Cytotoxic CD8 T cells
  - Generation of cytotoxic T cells
  - Granule-mediated killing
  - Fas-mediated killing
- Delayed-type hypersensitivity
  - Cellular features
  - Migration of effector Th cells
  - Recruitment of monocytes
  - Activation of monocytes/macrophages
- Neutrophil mediated immunity
- Consequences to the host of different effector mechanisms

9.1 Introduction

Antibody is essential for protection against a wide variety of bacteria, viruses and parasites. There are a number of immunodeficiency diseases in which babies are born with genetic mutations that render them unable to make normal antibody responses, and some are totally unable to make antibody. Before treatment became available, these babies invariably died of overwhelming infection, especially with pyogenic (causing fever) bacteria, such as *Staphylococcus aureus, Haemophilus influenzae* and *Streptococcus pneumoniae*, which have a polysaccharide coat that makes them resistant to phagocytosis in the absence of antibody. These immunodeficiencies illustrate the importance of antibody in fighting infection.

However there are some situations with infectious agents where antibody is not effective. These are where the pathogens live inside our own cells, the so-called intracellular pathogens. Viruses have to live inside cells because they lack the machinery for protein synthesis and other processes necessary for their replication. Many bacteria and protozoan parasites also live inside our cells (Table 9.1). When the microbes are located intracellularly they are not accessible to antibody. This does not mean that antibody plays no role with these infections because when the pathogens spread from cell to cell they may spend a period outside of the cell where they are potentially exposed to antibody and can be susceptible to antibody-mediated effects described in Chapter. 8. Antibody can also provide protection against future infection by binding to intracellular pathogens when they first arrive in the body, thereby preventing them from infecting cells. Mucosal IgA is particularly effective at preventing viruses from infecting through mucosa.

Table 9.1 Intracellular pathogens requiring cell-mediated immune responses for elimination

Pathogen	*Disease*	*Response*
Bacteria		
Mycobacterium spp.	Tuberculosis, leprosy	DTH
Legionella pneumophilia	Legionnaires' disease	DTH
Listeria monocytogenes	Meningitis	Tc
Chlamydia psittaci	Psittacosis	Tc
Fungi		
Pneumocystis carinii	Pneumonia	DTH
Cryptococcus neoformans	Meningitis	DTH
Viruses		
Herpes simplex virus	Cold sores	DTH
Measles virus	Measles	Tc
Influenza virus	Influenza	Tc
Protozoa		
Plasmodium berghei	Malaria	DTH, Tc
Trypanosoma	Sleeping sickness	Tc

Tc, CD8 cytotoxic T lymphocyte; DTH, delayed-type hypersensitivity.

The problem is that upon first infection with these pathogens you do not have antibody against them and by the time antibody is made the pathogens have had time to infect cells and hide in them so that antibody cannot get at them. To deal with this problem other types of immune responses are generated. These may involve CD8 cytotoxic T cells (Tcs) of the specific immune system and cells of the innate immune system such as natural killer cells, monocytes, neutrophils, eosinophils and basophils. These responses do not involve antibody and have been called cell-mediated responses or cell-mediated immunity. Parasitic worms, which are the largest infectious pathogens, pose special problems for the immune system and responses to these involve antibody and other effector mechanisms.

9.2 CD4 T cells develop into different types of helper T cells during immune responses

As with antibody responses most cell-mediated immune responses require the involvement of CD4 T cells and the first stage of the response is the stimulation of CD4 T cells by antigen peptide/class II MHC. The CD4 T cells then proliferate and differentiate into helper T cells (Th). It is now clear that CD4 T cells can differentiate into different types of Th which secrete different patterns of cytokines and therefore have different regulatory functions. The three main types of Th that have been identified so far are called Th1, Th2 and Th17 and the main cytokines secreted by different Th are shown in Figure 9.1. The three types of Th are involved in different types of cell-mediated immune responses and also promote different classes of antibody production.

9.2.1 Factors controlling Th differentiation

The type of Th generated in response to infection is crucial and different pathogens require different types of Th responses for effective protection. The initial stimulation of CD4 T cells occurs in secondary lymphoid tissue – the lymph nodes, spleen and organised mucosal associated lymphoid tissue (see Chapter 6). In the same way that occurs with antibody responses (Chapter 7), either antigen travels to the T cell area of the lymphoid tissue and is taken up by dendritic cells (DCs), or antigen (even whole microorganisms) is taken up by tissue DCs which transport the antigen to the lymphoid tissue. The DCs process the antigen and present antigenic-peptide on their Cl II MHC for recognition by CD4 T cells whose TCRs are specific for the peptide/Cl II MHC. The DCs also secrete cytokines which influence the type of Th response the CD4 T cells make. DCs recognise pathogens through their pattern recognition receptors (PRRs) and which PRRs on a DC are stimulated determines the cytokines the DC will produce. For example the fungal form of the yeast

Candida albicans stimulates IL-12 production by DCs which promotes a Th1 response whereas the hyphal form stimulates IL-23 production by DCs driving the response down the Th17 pathway. Other cells of the innate immune system, such as macrophages and natural killer cells, can also secrete cytokines that influence the Th response.

The cytokines that are important in the generation of different types of Th are shown in Figure 9.1. The most important cytokine determining Th1 development is called interleukin-12 (IL-12). IL-12 can bind directly to IL-12 receptors on CD4 T cells and signal them to differentiate down the Th1 pathway. It can also stimulate the production of IFNγ by T cells or natural killer cells and the IFNγ promotes Th1 differentiation. Interleukin-4 is

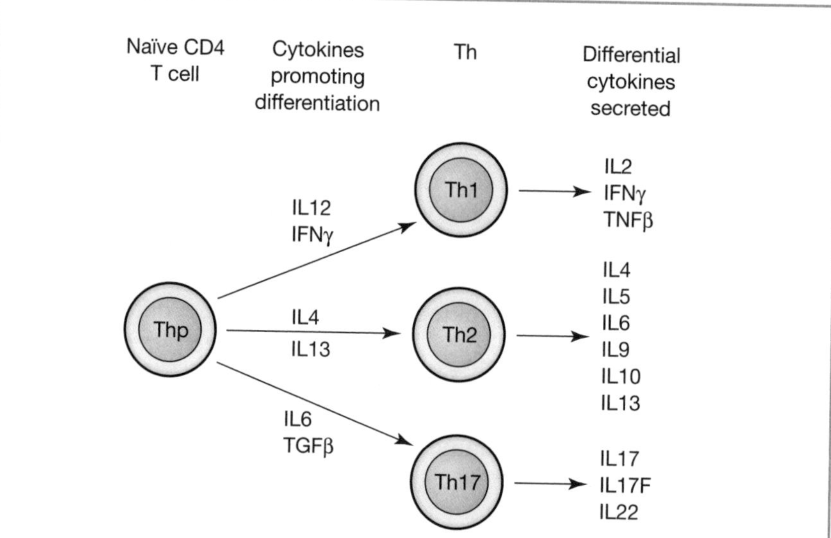

Figure 9.1 Differentiation of CD4 T cells. Following stimulation by antigen/class II MHC, CD4 T cells can differentiate into different types of helper T cells – Th1, Th2 and Th17 – which secrete different combinations of cytokines. This differentiation is controlled by the cytokines the CD4 T cells are exposed to.

the major cytokine for Th2 differentiation. Although once Th2 cells develop they can produce IL-4, the source of IL-4 driving the initial differentiation of CD4 T cells to Th2 cells is unknown. For Th17 cells the cytokines IL-6 and transforming growth factor β (TGFβ) are required for development of Th17 and IL-23 is important in maintaining Th17 responses.

The CD4 T cells proliferate under the influence of IL-2 and differentiate into Th1, Th2 or Th17 cells depending on the cytokines present. The Th can now participate in immune responses and often they do this by going to the site of infection and regulating the behaviour of other cell types which will hopefully destroy the pathogen. Because these Th are acting more directly at the site of infection they are also be referred to as effector CD4 T cells.

In real life, most pathogens probably stimulate a mixture of responses. Although responses may initially be mixed, in many cases, especially in chronic infections, one particular type of Th response will eventually predominate.

As mentioned above, different Th types are involved in responses to different pathogens. Th1 responses are particularly effective against intracellular pathogens, Th2 responses are mounted against various parasites, especially worms, and Th17 responses are important in fungal and extracellular bacterial infections where a large inflammatory type response is required to deal with the pathogen.

9.3 CD8 cytotoxic T cells are important in intracellular infections

Cytotoxic T cells (Tcs) are CD8 T cells that are able to recognise antigenic peptides presented by class I MHC on the surface of a cell. As their name implies, Tcs are able to kill the cells expressing antigen on their class I MHC. The Tcs are sometimes called killer cells and the cell being killed is called the target cell (Figure 9.2). As might be imagined, cytotoxic T cells are important in viral infections, especially if the virus is non-lytic. Viruses can be lytic or non-lytic. Lytic viruses, as their name implies, replicate within cells and then lyse the cell, causing it to burst open and release all the viral particles which can go on to infect other cells. Other viruses are non-lytic: they replicate within cells and are continually released from the cells but do not kill the cell. Tcs, by killing the cell, stop further viral replication and provide an important mechanism for dealing with these viral infections.

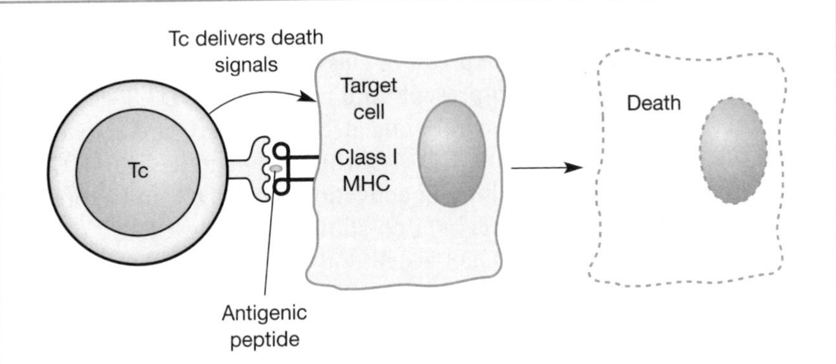

Figure 9.2 CD8 T cell-mediated cytotoxicity. The TCR of the cytotoxic T cell (Tc) recognises antigenic peptide (e.g. from a virus) being presented by a target cell. The Tc is stimulated to deliver death signals to the target cell, which is killed.

9.3.1 Cytotoxic T cells can be generated with or without CD4 T cell help

CD8 T cells that have not been stimulated by antigen before are not cytotoxic. Just as B cells have to proliferate and differentiate after their first encounter with antigen to become antibody-producing plasma cells, so CD8 T cells also have to proliferate and differentiate after their first encounter with antigenic peptide/class I MHC to become Tcs.

The generation of Tcs in response to viral infection usually requires help from CD4 T cells. However some viral infections can result in the generation of Tcs in the absence of CD4 T cells. Both of these pathways involve activation of naïve CD8 T cells by dendritic cells. Dendritic cells have two useful features that make them efficient at stimulating CD8 T cells. They express high levels of class I MHC and are therefore able to present antigenic peptides to CD8 T cells. Furthermore, dendritic cells can take exogenous antigen and present it on class I MHC; usually exogenous antigen goes to class II MHC – see Section 4.5.2. This means that the dendritic cell does not actually have to be infected with the virus to present antigen to CD8 T cells, which is a very useful feature since not all viruses infect dendritic cells.

Dendritic cells have to be activated to express co-stimulatory molecules, such as CD80 and CD86, to be able to stimulate CD8 T cells to proliferate and differentiate into Tcs, and it is how the dendritic cells are activated that determines whether Tc generation is CD4 T cell-dependent or -independent (Figure 9.3). Some viruses are able to activate dendritic cells directly through stimulation of toll-like receptors on DCs or through the induction of an inflammatory response. These viruses can therefore stimulate Tc generation in a CD4-independent manner. The initial activation of DCs is thought to take place at the site of infection. The activated DCs migrate to the T-cell area of the draining lymph node or spleen, where they can encounter CD8 T cells, and if the TCR on the CD8 T cell is specific for the viral antigen/class I MHC the dendritic cell can activate the CD8 T cell to become a Tc.

Other viruses are not able to directly activate DCs and require CD4 T cells to do so. In addition to expressing class I MHC, dendritic cells also express class II MHC and can present viral antigens to CD4 T cells. CD4 T cells that are specific for the viral antigen/class II MHC are activated to express the cell surface molecule CD154, which binds to its ligand, CD40, on the surface of the DC, resulting in activation of the DC, and expression of CD80 and CD86, which acts as a co-stimulus for the CD8 T cell and results in proliferation of the CD8 T cell and its differentiation into a Tc. The CD4 T cell itself is stimulated to produce IL-2, which promotes the proliferation of both the CD8 T cells and CD4 T cells, which both express IL-2 receptors after antigen stimulation. The CD4 T cells will differentiate into Th cells, which secrete other cytokines, such as IFNγ, which can further promote the differentiation of CD8 T cell into Tcs. It can be seen

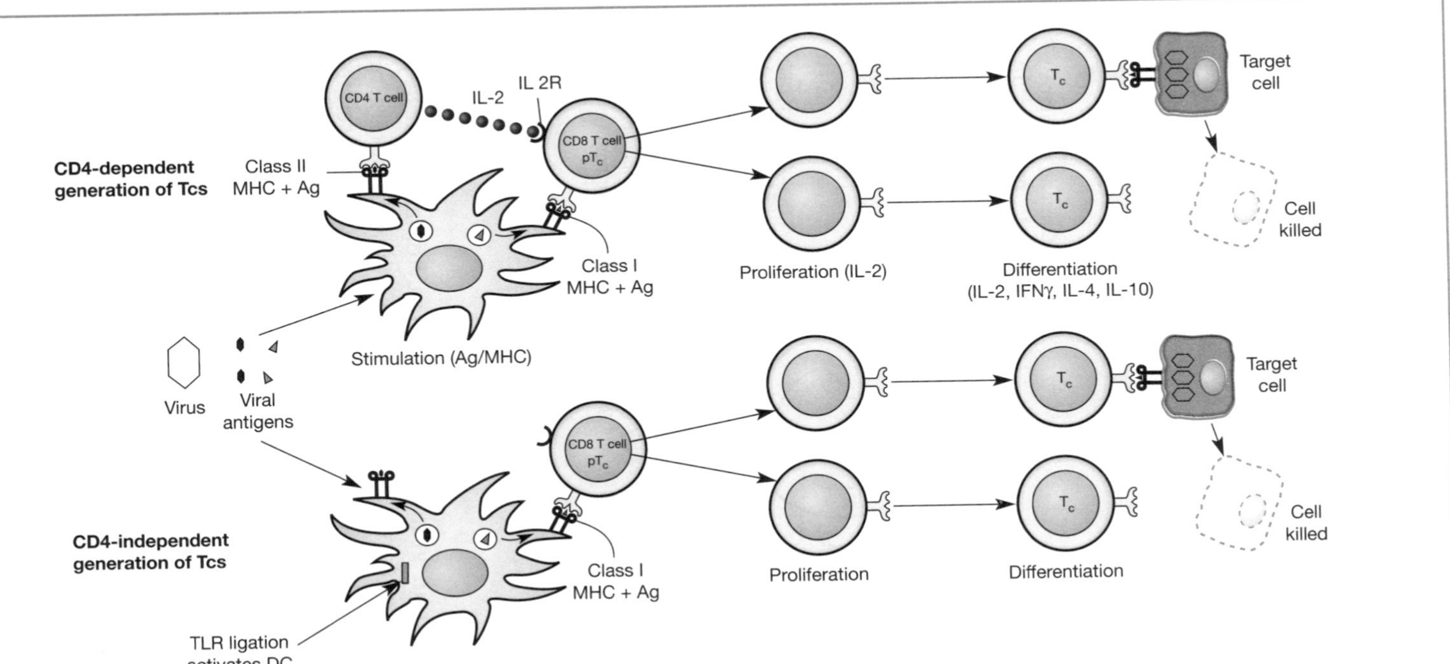

Figure 9.3 Differentiation of CD8 T cells into Tcs. Tcs can be generated in a CD4-dependent fashion (top). Dendritic cells can take up viral antigens and process them so that viral antigen peptides are expressed on the DC cell surface in association with both class I and class II MHC. Ag/C1 I MHC stimulates CD8 T cells to express receptors for IL-2 and Ag C1 II MHC stimulates CD4 T cells to produce IL-2 which causes proliferation of CD8 T cells. Eventually the CD8 T cells differentiate into Tcs, a process that is promoted by IFNγ, IL-4 and IL-10 although these are not essential. Tcs can also be generated in the absence of CD4 T cells (bottom). This requires that the DCs are activated through recognition of pathogen products, e.g. through toll-like receptors (TLRs). The activated DCs express co-stimulatory molecules and can stimulate CD8 T cells by expression of antigen peptides on their class I MHC to differentiate into Tcs.

that the ability of the same DC to present antigen to, and stimulate, both CD4 and CD8 T cells provides an efficient process for interaction between cells via contact and cytokines. Even with viruses that can stimulate CD4-independent responses, the normal immune response is thought to involve CD4 T cells.

9.3.2 Killing by cytotoxic T cells: more than one way of death

CD8 T cells can kill by two different mechanisms. The first mechanism is via granule exocytosis and the second involves the Fas pathway (Figure 9.4).

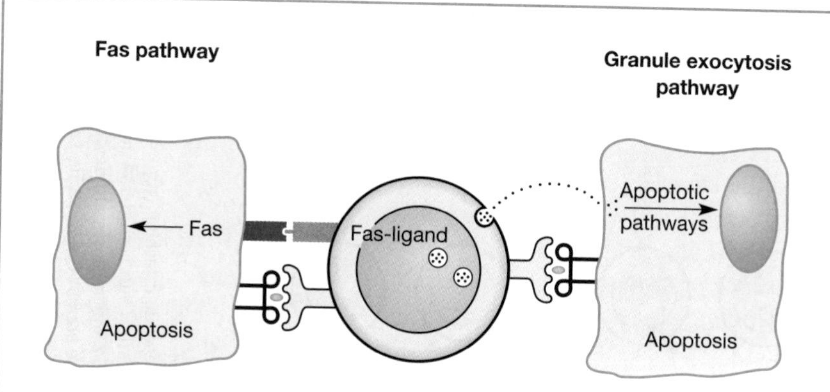

Figure 9.4 Two pathways of CD8 T cell cytotoxicity. In the Fas pathway, Fas-ligand on the cytotoxic T cell cross-links Fas on the target cell activating the Fas death pathway, leading to apoptosis of the target cell. The granule exocytosis pathway involves degranulation of the Tc and release of granule contents (perforin and granzymes), which activate apoptosis in the target cell.

Granule exocytosis

CD8 T cells contain granules within their cytoplasm. These granules contain a number of proteins that can cause the lysis of target cells. Important proteins found in granules are perforin, a number of serine proteases called granzymes, and a small protein called granulysin.

- **Perforin.** This is a 65 kDa protein with structural homology to the complement component C9. Like C9 it is able to form pores in cell membranes.
- **Granzymes.** These are a series of serine esterases that cleave proteins at serine residues. There are many granzymes in cytotoxic T cell granules, although the specificity of their enzymatic activity appears similar.
- **Granulysin.** This is a 9kDa protein that can kill mammalian cells, especially tumour cells, and also kills a variety of microbes such as bacteria, fungi and protozoa. There is particular interest in its role in tuberculosis because it can kill the causative agent *Mycobacterium tuberculosis*.

Cytotoxicity through granule exocytosis can be divided into four stages:

1. **Recognition and binding of the target cell.** For cytotoxicity to be triggered in the cytotoxic T cell, the cytotoxic T cell must recognise antigen/class I MHC on the target through specific binding of its TCR to antigen/MHC. The requirement that the TCR must bind its specific antigen ensures that only target cells bearing antigen for which the cytotoxic T cell is specific will be killed. The binding of the cytotoxic T cell to the target is promoted by binding of other molecules on the cytotoxic T cell molecules on the target cell. The most important of these other molecular interactions involve CD2 on the T cell binding to CD58 (LFA-3) on the target cell and CD11/18 (LFA-1) on the T cell binding to CD54 (ICAM-1) on the target cell.
2. **Delivery of the lethal hit.** Following triggering of the cytotoxic T cell, the granules in the T cell move towards the site of attachment to the target cell. The granules fuse with the membrane of the cytotoxic T cell and the contents of the granules are released into the intracellular space between the cytotoxic T cell and the target cell.
3. **Death of the target cell.** Perforin is released as monomers but in the presence of calcium the monomers insert into the target cell membrane and polymerise to form pores of about 50 Å in diameter. For some target cells this formation of pores is enough to cause osmotic lysis in a manner analogous to that produced by complement. However, many target cells are not killed by perforin alone and killing requires the presence of granzymes, which are thought to enter the target cell through the perforin pores. Granzymes cleave proteins in the target cells, which, amongst other things, results in activation of apoptosis pathways in the target cell. The target cell then undergoes programmed cell death. The exact role of granulysin in this killing process is not clear but it enhances killing through the perforin pathway and granulolysin can also induce apoptosis independently of granzymes.
4. **Recycling of the cytotoxic T cell.** Once the cytotoxic T cell has delivered the lethal hit it can detach from the target cell and is then capable of killing other target cells bearing the specific antigen/class I MHC. The cytotoxic T cell remains bound to the target cell for 10–15 minutes and the death of the target cell can take place at any time from 5 minutes to 3 hours from the time of detachment of the cytotoxic T cell.

Fas-mediated cytotoxicity

Fas is a death molecule that can be expressed on many cell types. Its expression is also stimulated by a variety of signals. If Fas on a cell interacts with Fas-ligand (Fas-L) the Fas is cross-linked and this activates pathways leading to apoptosis of the cell. Binding of the TCR of CD8 cytotoxic T cells to its specific Ag/MHC class I causes expression of Fas-L. Fas on the target cell will be cross-linked by the Fas-L on the cytotoxic T cell and apoptosis and death of the target cell will be triggered (Figure 9.4). Not all cells express Fas, so this mechanism of cytotoxicity is not effective against all cells.

Cytotoxic T cell responses provide important protection against many viral infections and the magnitude of the cytotoxic T cell response can be huge. It is now possible to measure the number of CD8 T cells that are specific for a particular viral antigen in association with class I MHC. These studies have shown that during some viral infections, up to 20% of all the CD8 T cells in a person's bloodstream are specific for antigens on the virus. This is a truly remarkable response but sometimes even this is not enough to eliminate all intracellular pathogens. In these cases a delayed-type hypersensitivity response may be required.

9.4 Delayed-type hypersensitivity and the activation of macrophages

Delayed-type hypersensitivity responses are often generated against pathogens that live inside macrophages themselves. It might seem that the last place a pathogen would want to live is inside a phagocytic cell, one of whose functions is to phagocytose and kill microbes. However, many pathogens have evolved a variety of mechanisms to avoid being killed by macrophages (see Box 9.1). Not all pathogens that stimulate delayed-type hypersensitivity responses live in macrophages and many live in other cell types (Table 9.2).

Table 9.2 Cells types occupied by intracellular pathogens stimulating DTH responses

Pathogen	*Disease*	*Cell type occupied*
Bacteria		
Mycobacterium tuberculosis	Tuberculosis	Macrophages
Mycobacterium leprae	Leprosy	Endothelial cells, Schwann cells
Legionella pneumophila	Legionnaires' disease	Macrophages
Rickettsia prowazekii	Typhus fever	Endothelial cells
Protozoan parasites		
Leishmania spp.	Leishmaniasis	Macrophages

The aims of a delayed hypersensitivity response are really quite simple. They are to:

- recruit monocytes to the site of infection;
- keep monocytes and tissue macrophages at the site of infection;
- activate the monocytes and macrophages to kill the intracellular organisms.

BOX 9.1: HOW PATHOGENS AVOID BEING KILLED BY PHAGOCYTES

Pathogens have evolved many ways of avoiding being killed by phagocytes. *Listeria monocytogenes*, which can cause meningitis in newborns, has evolved the ability to escape from the phagosome into the cytoplasm of the cell where it can multiply and infect other cells. Other bacteria are able to limit the involvement of lysosomal contents in bacterial killing in a number of ways. *Salmonella typhi* and *Mycobacterium tuberculosis* prevent the phagosome fusing with the lysosome, thereby preventing the bacteria from being exposed to the lysosomal contents. *Mycobacterium* spp. also produce NH_4^+, which neutralises the acid pH of lysosomes, and they can neutralise the proton pump that normally is involved in lowering the pH of the phagolysosome. Some bacteria are resistant to killing by lysosomal contents.

Other pathogens produce factors that interfere with killing by reactive oxygen products. Two enzymes produced by phagocytes are superoxide dismutase (SOD) and catalase. SOD converts superoxide to H_2O_2 in the following reaction:

$$2O_2^- + 2H^+ \xrightarrow{\text{SOD}} H_2O_2 + O_2$$

Catalase then converts hydrogen peroxide to water and oxygen:

$$2H_2O_2 \xrightarrow{\text{Catalase}} 2H_2O + O_2$$

Although normally produced by the phagocyte, many microorganisms produce these enzymes, thereby speeding up the inactivation of O_2 and H_2O_2 and protecting the bacteria.

Another approach to resistance to oxygen radicals is used by *Mycobacterium leprae*, which coats itself with phenolic glycolipid, a molecule that scavenges reactive oxygen radicals, thereby inactivating them.

Finally some organisms, such as *Mycobacterium tuberculosis*, can kill macrophages. This is a mixed blessing since the organism prefers to live and multiply within macrophages. The bacterium is protected from phagocytosis but has lost its home and may be exposed to extracellular defence mechanisms such as antibody.

Th1 cells that promote DTH reaction are usually generated in the lymph nodes draining the site of infection (Figure 9.5). When they have differentiated into effector Th1 cells they leave the lymph node via the efferent lymphatic vessels, eventually entering the bloodstream via the major lymphatic ducts. It is important that the effector Th cells can return to the site of the infection to carry out their effector functions. This is achieved by the expression of new adhesion molecules on the effector Th cells. The lymphoid tissue in which the Th cells are activated affects which adhesion molecules are induced and which site the effector Th cell will go to (Figure 9.6). If the Th cell is activated in a cutaneous lymph node draining the skin, the Th cell expresses an adhesion molecule called cutaneous lymphocyte-activation antigen (CLA). Th cells that are activated in a mucosal lymph node express different adhesion molecules called LPAM-1 and L-selectin.

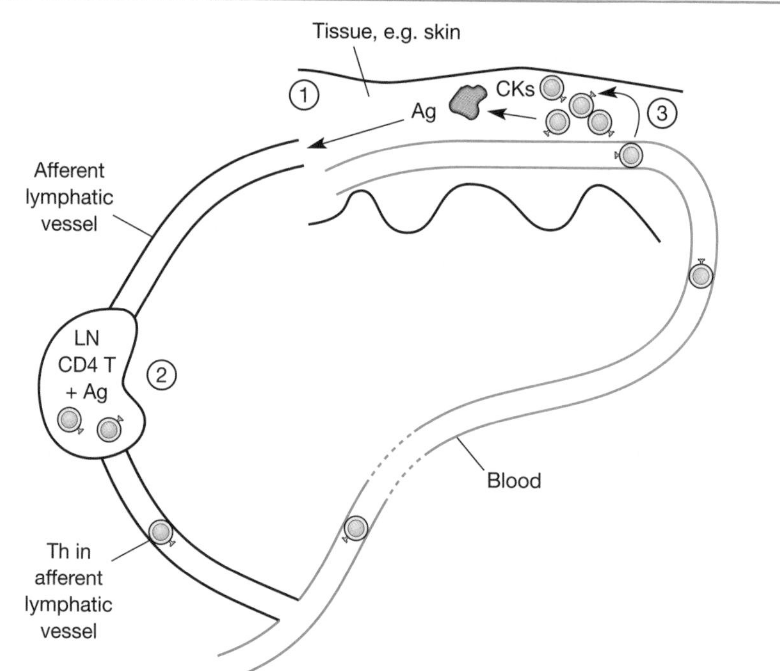

Figure 9.5 CD4 T cells and delayed-type hypersensitivity. ① Antigen in the tissue enters afferent lymphatic vessels and is transported to a lymph node. ② DCs present the antigen to CD4 T cells, which are stimulated to become Th. Th cells leave blood vessels at sites of inflammation and enter the tissue. ③ The Th cells recognise antigen being presented on the surface of monocytes in association with class II MHC, and the Th cells release TNFα, which acts on the endothelium to increase the recruitment of monocytes and Th cells from the blood. The Th cells also secrete IFNγ and IL-2, which, together with TNF, activate the tissue macrophages and recruited monocytes to become more efficient at killing intracellular pathogens.

The Th cells are able to migrate specifically to cutaneous or mucosal sites because of expression of adhesion molecules on the endothelium due to cytokines being made at the site of the DTH reaction. TNFα and IL-1 made by macrophages stimulate local endothelial cells to express adhesion molecules. E-selectin is induced on cutaneous endothelium and is a specific ligand for CLA on the Th cell, thereby enabling the Th cell to bind to the endothelium and enter the DTH site. Similarly MadCAM-1 is induced on mucosal endothelium at sites of inflammation. This binds both LPAM-1 and L-selectin expressed on mucosally stimulated Th cells, enabling them to home back to the mucosal site of DTH.

The selective up-regulation of different adhesion molecules in cutaneous and mucosal sites that enables the effector Th cell to return to the site of antigen is important because many pathogens infect either mucosal or cutaneous sites but not both. Therefore Th cells specific for a mucosal pathogen antigen are required in the mucosa and not the skin and vice versa for skin pathogen-specific Th cells.

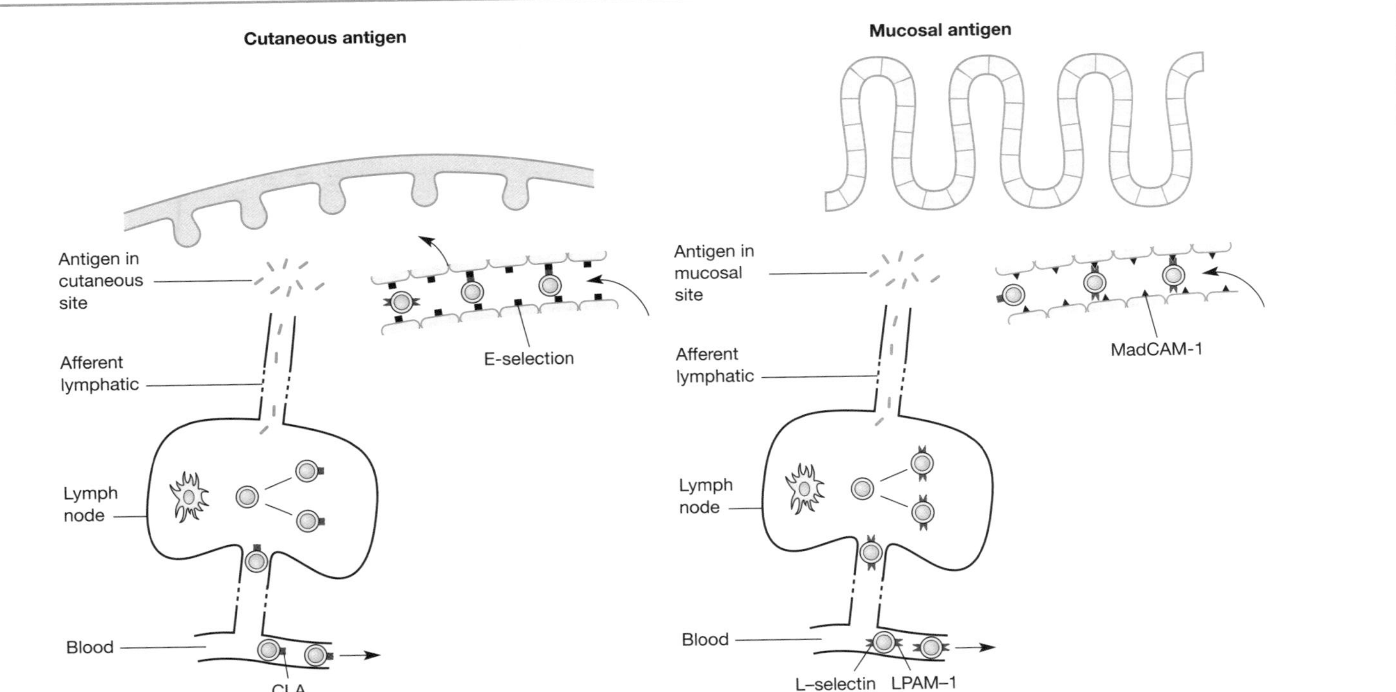

Figure 9.6 Migration of Th. Cutaneous antigen goes to draining cutaneous lymph nodes where CD4 T cells are stimulated. CD4 T cells stimulated in cutaneous lymph nodes express an adhesion molecule called cutaneous lymphocyte-activation antigen (CLA). This binds to E-selectin, which is expressed on cutaneous endothelial cells in response to inflammatory mediators, thereby guiding the Th back to the cutaneous site. Mucosal antigen goes to mucosal lymph nodes and CD4 T cells stimulated here express the adhesion molecules L-selectin and LPAM-1, which bind to MadCAM-1 expressed on mucosal endothelium and guide the Th to the mucosal site.

Recruitment and retention of monocytes

The expression of new adhesion molecules on endothelium at the site of DTH reactions also promotes the recruitment of monocytes from the bloodstream. Important among these adhesion molecules is vascular cell-adhesion molecule-1 (VCAM-1), which is expressed on endothelial cells and binds to the CR3 on monocytes.

Monocytes that are recruited to the site of a DTH reaction are subsequently stimulated by cytokines to differentiate into macrophages, which are then activated. These macrophages are retained at the site of the DTH by a cytokine called macrophage inhibition factor, which is secreted by the effector Th cells.

Activation of macrophages

The macrophages and Th cells that are recruited to a DTH site are able to stimulate each other and provide an amplification loop that increases the magnitude of the DTH response (Figure 9.7). This occurs as follows: the Th cells secrete cytokines that activate the macrophages. The most important of these cytokines is IFN-γ but TNFα and IL-2 can contribute to the macrophage activation in the presence of IFN-γ. The activated macrophages up-regulate class II MHC expression on their cell surface. They are also good antigen-processing cells and can therefore present antigen on their class II MHC to the Th cells. The Th cells are stimulated by recognising antigen/class II MHC on the macrophages to secrete more cytokines that further activate the macrophages, and so the response is amplified. The requirement for the continued stimulation of Th cells to maintain cytokine production also provides a level of control of the response. Once the antigen has been eliminated the macrophages can no longer stimulate the Th cells to produce the cytokines necessary to maintain the response and the response will gradually disappear.

Elimination of pathogens in DTH reactions

Although activated macrophages are better at stimulating Th cells, the main reason for activating the macrophages is to increase their ability to kill phagocytosed pathogens. Activated macrophages have increased levels of nitric oxide and oxygen radicals and higher secretion of proteolytic enzymes. Pathogens that can avoid being killed by less active macrophages become susceptible to the enhanced microbicidal activity of the macrophages and are killed.

9.5 Th2 responses are important against worms

Over one hundred different parasitic worms can infect humans and collectively they can infect many tissues or organs including, blood, brain, eye, intestine, liver, lymphatics, muscle and skin. Worm infections pose special problems for the immune system. Up until the last hundred years or so

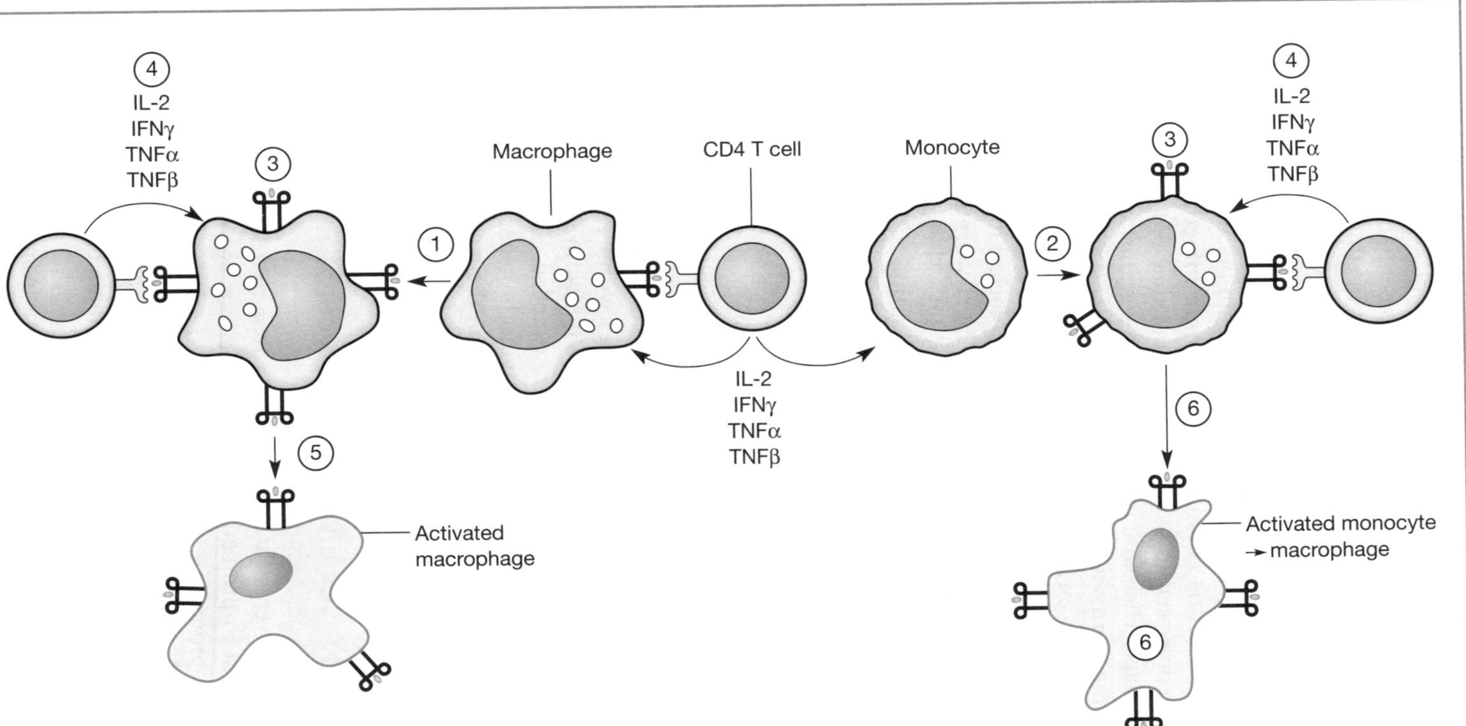

Figure 9.7 Amplification loops occur within DTH sites. Th cells secrete cytokines, which activate macrophages (1) or monocytes (2) to up-regulate class II MHC (3), thereby enabling them to provide more stimulation for the Th to secrete cytokines (4). These in turn activate the macrophages (5) and cause the monocytes to differentiate into activated macrophages (6), which can kill the pathogen.

humans everywhere were infected with worms and worms are still endemic in many parts of the world, particularly the tropics. As you can imagine, worms, being the biggest of pathogens, could potentially stimulate immune responses that cause a lot of damage to the host in efforts to get rid of them. Therefore there is a balance to be struck between controlling worms and limiting the damage caused by immune responses against them and humans and worms have co-evolved to live with each other. This co-existence resulted in a balance so that the immune system did not always get rid of worms and the worms could establish **chronic infections**. However the worms were controlled to some extent so that they did not directly cause too much damage. Therefore immune responses against worms are designed to eliminate the worms where possible without too much damage or to control them in chronic infections. However for most worm infections it is not clear how they are eliminated or controlled and it appears that this occurs through a mixture of immunological and physiological mechanisms.

In immunological terms it is clear that most worms stimulate a Th2 response characterised by the production of IgE. There is, however, very little direct evidence for IgE actually contributing to the elimination or control of worms. Other cell types that have been shown to play a role in some, but not all, worm infections include eosinophils, basophils and mast cells although it is not always clear what they are doing. The activity of these cells is controlled by Th2 cytokines such as IL-4, IL-5, IL-9 and Il-13. *In vitro* eosinophils can kill worms in the presence of worm-specific IgE by antibody-dependent cell-mediated cytotoxicity (see Chapter 8) but it is not known if this occurs *in vivo*.

Recent studies have shown that the immune response may also affect physiological systems in a way that promotes worm expulsion. One way this occurs for gut-dwelling worms such as Whipworm is by increasing epithelial cell turnover so that the worms are expelled. Some Th2 cytokines, such as IL-4, IL-9 and IL-13, can cause muscle hypercontraction in the gut which can lead to expulsion of worms from the gut. Th2 cytokines like IL-13 can also induce excessive mucus production and this can promote the expulsion of worms from the gut and lung.

9.6 Th17 responses involve high levels of inflammation

Th17 are the most recently discovered Th type and less is known about their involvement in protection against infectious agents. They appear to be important in protection against a variety of extracellular bacteria, some viruses and particularly fungal infections. Individuals who cannot make Th17 responses because of mutations in certain genes are particularly prone to staphylococcal skin infections, pneumonia caused by a variety of bacteria and fungi, and mucosal candidiasis.

During Th17 responses the Th17 cells migrate from the lymphoid tissue where they were produced and travel through the bloodstream to the site of infection. At the site of infection they are stimulated by antigen to produce a number of cytokines. Two of these, IL-17 and IL-17F, stimulate keratinocytes or lung epithelial cells to produce chemokines that recruit neutrophils to the site of infection. Another cytokine, IL-22, stimulates the production of antibacterial peptides by keratinocytes. The concerted action of these cell types and factors produced by them hopefully results in elimination of the pathogenic microbes.

9.7 Different effector responses have different costs to the host

The different types of effector responses that the immune system is able to generate are usually effective at protecting us from most pathogens that we encounter. However, the different types of responses can result in more or less damage to the host's own cells and tissues. Antibody responses are generally the least damaging to the host. This is because antibody is targeted directly at the pathogen or its products. Neutralisation of toxins by antibody carries essentially no risk of damage to the host. Similarly, complement-mediated lysis of a pathogen or phagocytosis and killing of a pathogen by a phagocyte will usually not cause damage to the host.

Cytotoxic T cell responses are directed at infected host cells and therefore must involve a degree of damage to the host. Cytotoxic T cells seem to provide the best protection against chronic non-lytic viral infections where killing of virally infected host cells limits the replication of the virus and on balance can be beneficial. Acute lytic viruses are better dealt with by antibody because they replicate rapidly and kill the host cell. There is not much point in a cytotoxic T cell killing the host cell if the virus will do so shortly anyway. In some cases the cost of killing virally infected host cells may be too high. This is presumably why neuronal cells do not express class I MHC. Neuronal cells cannot be replaced, or only very slowly, and on balance less damage may be caused to the host by allowing a non-lytic virus to replicate in neurons rather than killing them.

DTH responses and Th17 responses carry the greatest potential risk of damage to the host. Although in many cases these responses eliminate the pathogen with little damage, if the response is too great the reactive chemicals and proteolytic enzymes produced in the responses can damage host cells and tissue. In severe cases, DTH responses do not eliminate the infection and the DTH response becomes chronic, leading to the formation of **granulomas.** Granulomatous lesions involve chronic stimulation of macrophages with the deposition of fibrin and formation of giant epitheloid cells (Figure 9.8). The excessive production of degradative enzymes by macrophages leads to extensive tissue damage and fibrin can 'wall off the area', leading to a lack of oxygen and tissue necrosis. However, the fibrin walls may limit the spread of microbes from the site.

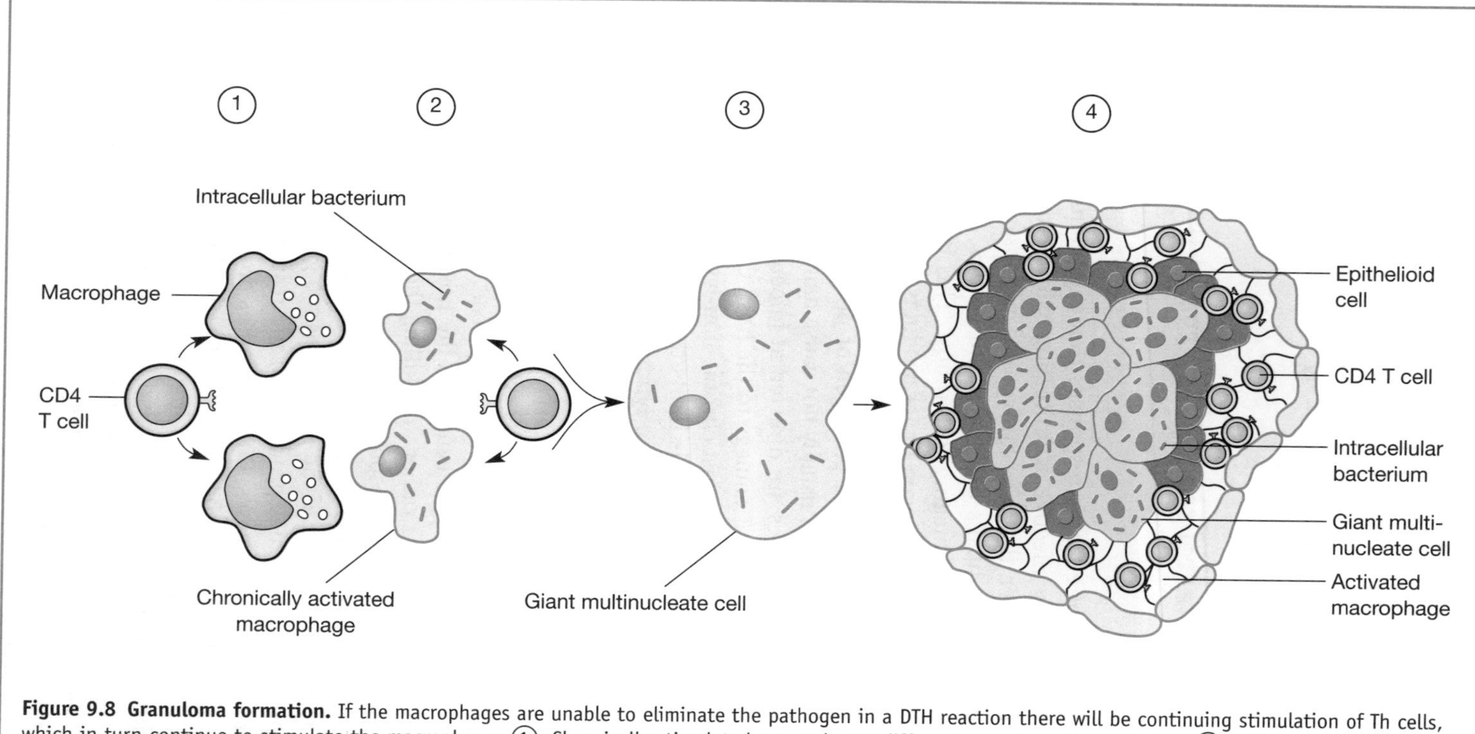

Figure 9.8 Granuloma formation. If the macrophages are unable to eliminate the pathogen in a DTH reaction there will be continuing stimulation of Th cells, which in turn continue to stimulate the macrophages ①. Chronically stimulated macrophages differentiate into epithelioid cells ②, which can fuse to form giant multinucleated cells ③. The mixture of Th cells, macrophages, epithelioid cells and giant multinucleated cells forms a granuloma ④. Lytic agents and enzymes released by the macrophage-derived cells within a granuloma cause extensive tissue damage and necrosis. In some cases the edge of the granuloma is walled off by epithelioid cells and fibrin, leading to further necrosis due to lack of oxygen.

The importance of the different types of immune response to different pathogens is illustrated by the types of infection that are common in people with genetic (congenital) deficiencies of a particular component of the immune system (see Box 9.2).

BOX 9.2: WHAT CONGENITAL IMMUNODEFICIENCY DISEASES TELL US ABOUT THE IMMUNE SYSTEM

Congenital immunodeficiency diseases are caused by mutations in a single gene. These mutations are either inherited or arise spontaneously. These mutations can affect:

- the production of a cell type;
- the production of a molecule;
- the function of a molecule.

Immunodeficiency diseases have been divided into those affecting the specific immune system and those affecting the innate immune system. Diseases affecting the specific immune system are those where the **primary defect** is in the functioning of a B or T cell, and those affecting the innate system are where the primary defect is in a component (cell or molecule) of the innate immune system. However, because of the close interaction between the innate and specific immune systems, diseases that are due to aberrant functioning of one cell type can affect the functioning of other cell types. For example there is a group of immunodeficiency diseases called **common variable hypogammaglobulinaemia.** People with this syndrome show defects in the production of one or more antibody classes. In many individuals with this disease, the primary defect is in the CD4 T cells, but, because of the need for Th cells to make antibody, there is a knock-on effect on antibody production.

Despite some of the complexities in identifying the cell or molecule being affected, immunodeficiency diseases have given us a lot of information on the importance of different components of the immune system in protection against different types of pathogens, particularly at the effector stage. Some examples are given in the table overleaf.

These diseases helped to establish the importance of antibody and complement in extracellular infections and T cells in intracellular infections. Phagocytes are especially important in extracellular bacterial infections but also operate against some intracellular infections through delayed-type hypersensitivity reactions.

Primary cell or system affected	*Types of infection suffered*
Diseases affecting antibody production, e.g. Bruton's agammaglobulinaemia, common variable hypogammaglobulinaemia	Respiratory and GI tract, especially with extracellular pyogenic bacteria such as staphylococci, streptococci, haemophilus
Diseases affecting all T cells, e.g. severe combined immunodeficiency (SCID)	Systemic viral infections, especially affecting respiratory and GI tracts: measles, herpes, cytomegalovirus
	Fungal infections: *Pneumocytis carinii, Candida albicans*
Diseases affecting Th17 cells	Skin and lung infections with bacteria and fungi, especially *Staphylococcus aureus* and *Candida albicans*
Diseases affecting phagocytes: chronic granulomatous disease, Wiscott–Aldridge syndrome	Bacterial infections (especially catalase-positive) affecting GI tract, skin, urinary tract: staphylococci, *E. coli.*
	Fungi: *Candida albicans*
Diseases affecting complement:	
C3 deficiency	Pyogenic bacteria such as staphylococci
C5, 6, 7, 8 or 9 deficiency	Neisserial infections: gonorrhea or meningitis

9.8 Summary

- CD4 T cells can differentiate into at least three functionally different types of helper cell called Th1, Th2 and Th17, which secrete different combinations of cytokines.
- The generation of CD8 cytotoxic T cells and delayed-type hypersensitivity responses are two types of specific immune response that involve T cells but not antibody.
- CD8 cytotoxic T cells kill target cells bearing endogenously derived antigen on their class I MHC molecules.
- CD8 cytotoxic T cells kill by two mechanisms: the granule exocytosis and Fas pathways.
- Granule exocytosis involves the release of granule contents of the cytotoxic T cell. The granule contents include perforin and granzymes, which rapidly induce apoptosis in the target cell.
- If target cells express Fas, Fas-ligand expressed by the cytotoxic T cell can activate Fas death pathways in the target cell, also resulting in apoptosis.
- Delayed-type hypersensitivity (DTH) reactions occur against intracellular pathogens where antibody and CD8 cytotoxic T cells may not be fully effective.

- In DTH reactions, antigens from pathogens residing in a tissue site activate CD4 T cells in the draining lymph node to become Th effector cells. The Th cells travel through the blood and enter the tissue where the pathogen is located.
- At the site of infection Th cells release cytokines, which cause the recruitment of monocytes to the area from the blood and then activate the monocytes and any local tissue macrophages.
- The activated monocytes and macrophages have increased ability to kill phagocytosed pathogens and may kill them with little damage to host tissue.
- Sometimes DTH reactions can cause extensive tissue damage in severe acute reactions or in situations where the reaction becomes chronic, leading to granuloma formation.
- Th1 cells promote the generation of Tcs and DTH reactions and some classes of antibody while Th2 cells promote the production of other antibody classes and recruit eosinophils and mast cells to sites of infection. Th17 cells promote immune responses involving the extensive recruitment of neutrophils.

9.9 Questions

1) The diagram below shows the cytokines involved in differentiation of helper CD4 T cells (A, B, C) and the main cytokine produced by the Th (D, E, F).

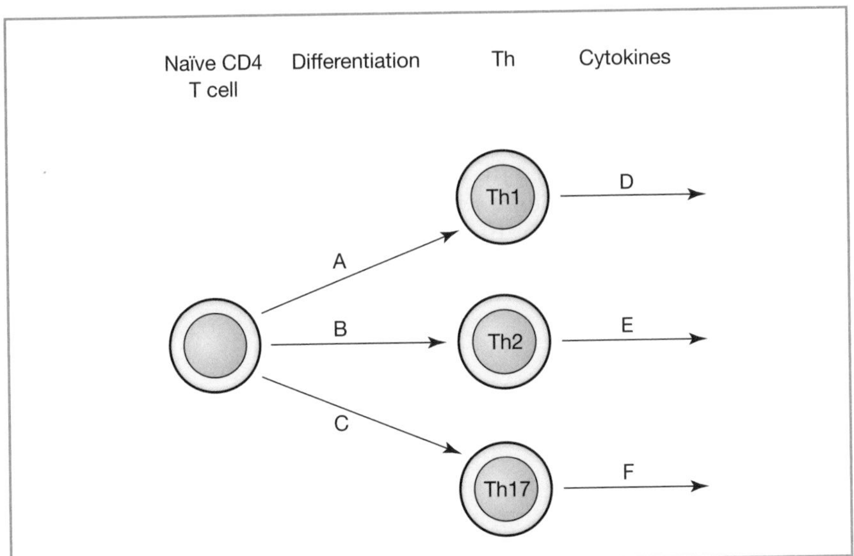

Match the cytokines with the labels:

Labels	**Cytokines**
A	IL-6 + TGFβ
B	IL-4
C	IFNγ
D	IL-12
E	IL-17
F	IL-4

2) Describe the two mechanisms that Tcs use to kill target cells.

3) What are the advantages of killing a host cell that is infected with virus?

4) Which type of immune response would be most effective against:

   (i) A virus
   (ii) An extracellular dwelling bacterium
   (iii) A yeast
   (iv) A parasitic worm?

5) What are the main aims of a DTH reaction?

The answers to these questions can be found on page 338.

9.10 Further reading

1) Wan YY. (2010) Multi-tasking of helper T cells. *Immunology* 130:166–171.

2) Waterhouse NJ, Clark CJP, Sedelies KA, Teng MW, Trapani JA. (2004) Cytotoxic T cells: instigators of dramatic target cell death. *Biochemical Pharmacology* 68:1033–1040.

3) Vukmanovic-Stejic M, Reed JR, Lacy KE, Rustin MHA, Akbar AN. (2006) Mantoux Test as a model for a secondary immune response in humans. *Immunology Letters* 107:93–101.

CHAPTER TEN

Immunological memory and vaccination, the production and use of antibodies

Learning objectives

To understand the basis of immunological memory and how this protects against repeated infection with the same pathogen. To know about current and developing approaches to vaccination. To know how antibodies can be generated and used for clinical and technical purposes.

Key topics

- Immunological memory
  - Memory for antibody production
  - T cell memory
- Vaccines
  - Properties of an ideal vaccine
  - Current vaccines
  - Future vaccines
- Production and use of antibodies
  - Antisera
  - Monoclonal antibodies
- Antibody-based techniques

10.1 Immunological memory – the basis of immunity

Even though they didn't know about bacteria and viruses, people in ancient times had a concept of immunity; they knew there were some diseases (such as smallpox) that if you survived you were protected from getting that again. In fact, during epidemics which ravaged ancient Rome, Athens and other cities, individuals who had survived the disease were used to tend the sick because it was known they were protected.

10.1.1 Protective immunity and memory

An important aspect of the immune system is to provide increased protection against infectious agents that may be encountered more than once. This protection comes in two stages, protective immunity and memory. When you are infected with a microbe you mount an immune response against it. This usually involves the production of antibody and the generation of effector T cells. Essentially all responses will involve CD4 T cells and intracellular pathogens will often stimulate the production of CD8 cytotoxic T cells (Tcs). Hopefully the immune response will eliminate the microbe although there may be illness associated with a first infection because it takes a week or so to generate antibody and effector T cells. In the period immediately following the elimination of the pathogen there will still be enough antibody and effector T cells around so that if you were infected with the same microbe again you would eliminate it without the need to generate a new immune response. This state is known as **protective immunity** and lasts for a certain period depending on the pathogen (see Figure 10.1). After a certain period, the levels of antibody and effector T cells will drop below those needed to deal directly with a new infection and a new immune response will have to be generated to deal with any further infection. However because of the generation of immunological memory during the first immune reponse the new response will be faster and bigger than the initial response to a first exposure to the pathogen (Figure 10.1). This means that upon subsequent exposure to a pathogen you will not get ill or will only get very mild symptoms. This is the basis of immunity; the word derives from the the Latin word 'immunis' which meant exemption from military or other public services or tax payments. The immune response to a first exposure to a pathogen or other antigen is also called the **primary** response and that to the second exposure is called the **secondary** response. Immunity can be naturally acquired following natural infection with a pathogen but it can also be induced by vaccination which aims to induce immunological memory without exposure to pathogenic organisms.

10.1.2 Immunological memory results in faster, bigger and better responses to second encounters with pathogens

Immunological memory can best be illustrated by examining antibody responses after a first and second injection with antigen (Figure 10.2). IgM is made typically within a week after the first immunisation and IgG production follows after IgM. If you look at antibody production following a second exposure to the same antigen the IgM response is essentially the same. However, the IgG response shows dramatic differences from the primary injection. The IgG response is **faster, bigger** and, although not depicted in Figure 10.2, **better.** Another feature of immunological memory is that it is specific for the particular antigen. In the example in Figure 10.2 you only get a bigger response to a second immunisation with the same antigen; if the

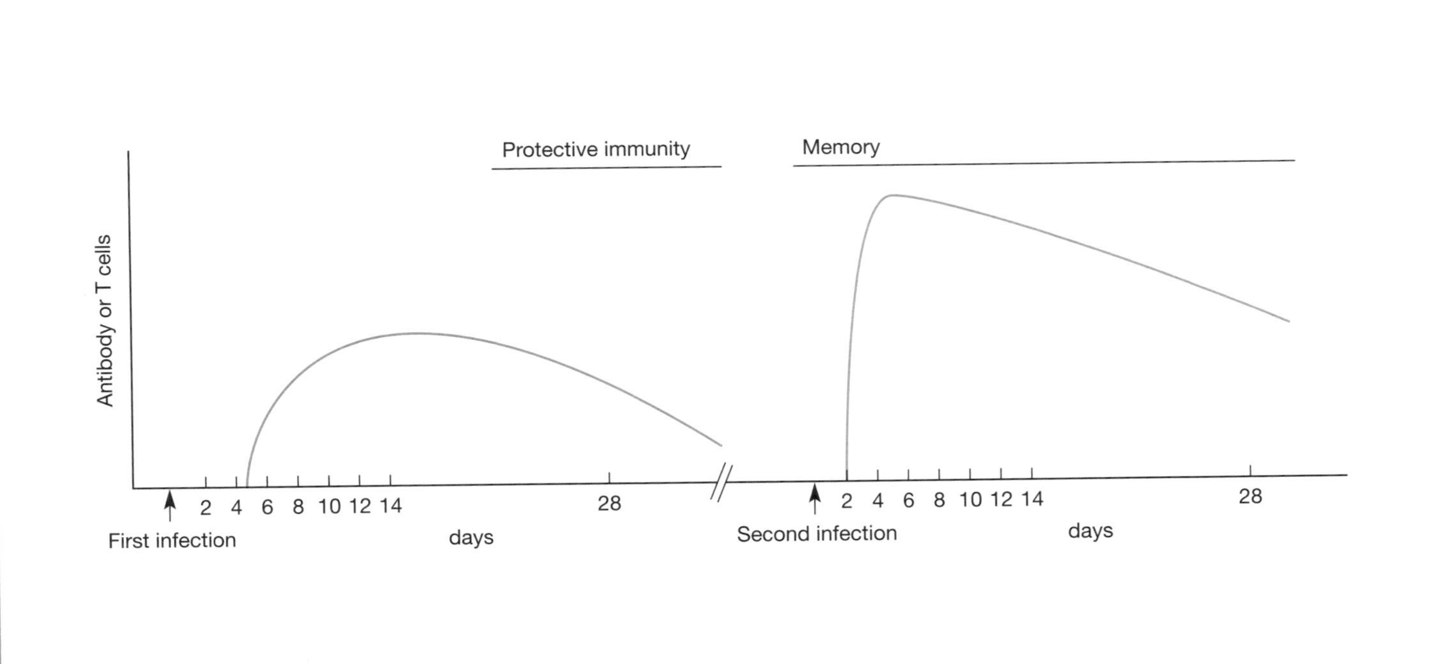

Figure 10.1 Protective immunity and immunological memory. During a first infection antibody and effector T cells are produced. After the infection has been cleared there are enough antibody and T cells to protect against further infection without having to generate new antibody or T cells – this known as protective immunity. If enough time passes without re-infection the antibody and T cell levels decline to such an extent that they are no longer able to clear a second infection. In this case protection is provided by immunological memory. Memory responses are bigger, faster and better than the first immune response to infection.

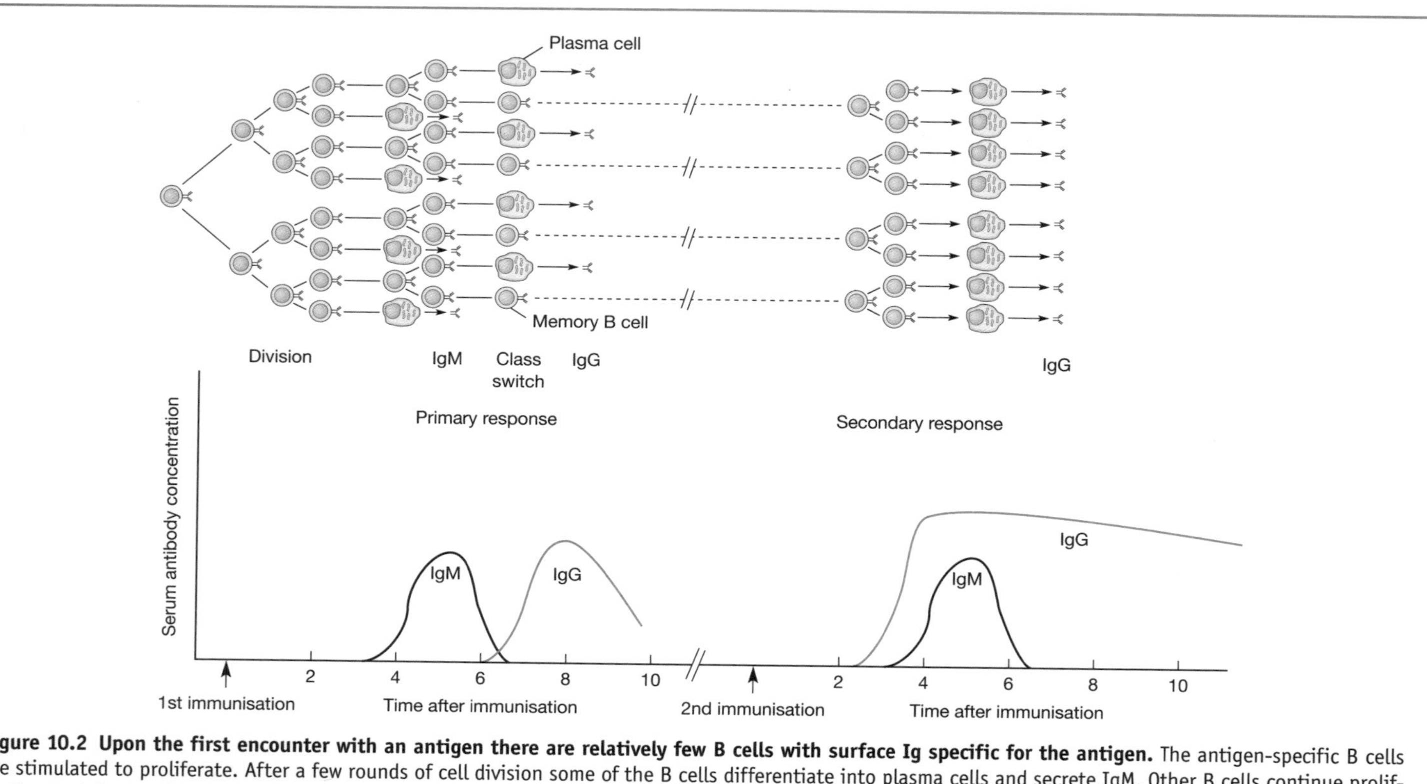

Figure 10.2 Upon the first encounter with an antigen there are relatively few B cells with surface Ig specific for the antigen. The antigen-specific B cells are stimulated to proliferate. After a few rounds of cell division some of the B cells differentiate into plasma cells and secrete IgM. Other B cells continue proliferating and then undergo class switch to IgG. Some of these B cells differentiate into plasma cells and secrete IgG, while other B cells become memory cells. If the same antigen is encountered again there are more memory B cells specific for the antigen and these quickly proliferate and differentiate into plasma cells, giving a quicker and bigger secondary IgG response.

second immunisation was with a different antigen from the first you would get a primary response to the second antigen. This is why if you get infected with one strain of flu, virus you are only protected against that strain but not other strains. Immunological memory is a feature of all specific immune responses and involves B cells, CD4 T cells and CD8 T cells.

10.1.3 B cell memory gives you faster, bigger and better antibody responses

B cell memory is the best understood feature of immunological memory. Memory B cells are generated only in germinal centres and their production is dependent on CD4 T cells, as described in Section 7.4. Memory B cells are B cells that have undergone proliferation, class switch and affinity maturation (Figure 10.2). Memory B cells still have to proliferate and differentiate into plasma cells before they secrete antibody and this process requires help from CD4 T cells. Memory B cells circulate through the lymphatic and blood system as described in Chapter 6 and the way in which they interact with CD4 T cells in lymphoid tissue is thought to be the same as described in Section 7.3. However, there are some differences in the stimulation of memory B cells; they can respond to lower doses of antigen and soluble antigen is good at stimulating memory B cells while antigen needs to be presented on the surface of a macrophage to stimulate naïve B cells. Because antigen-specific B cells have undergone proliferation before becoming memory B cells, there will be more antigen-specific memory B cells than there were B cells before the first encounter with antigen. This will contribute to a bigger and faster antibody response. Because memory B cells have already undergone antibody class switch, and therefore have IgG, IgA or IgE on their surface, they do not have to undergo class switch when restimulated by antigen, thereby speeding up the secondary response. Finally, the fact that memory B cells have undergone affinity maturation means that the secondary response will result in more rapid production of high-affinity antibody than in the primary response. High-affinity antibody is biologically more effective than lower affinity and the successful vaccines so far are those that stimulate production of large amounts of high-affinity antibody and memory B cells.

10.1.4 CD4 T cell memory co-ordinates secondary immune responses

During a primary immune response CD4 T cells and, if they are involved, CD8 T cells proliferate to generate large numbers of effector cells. This proliferation is especially dramatic for CD8 T cells and may result in a 50 000-fold increase in the number of cells specific for the pathogen. The effector T cells are Th cells in the case of CD4 T cells and Tcs in the case of CD8 T cells. The effector cells then carry out their functions. For Th this may be helping B cells, helping in the generation of Tcs or activating cells of the innate immune system. Tcs will be involved in killing infected cells and also secreting cytokines. When (hopefully) the infection has been

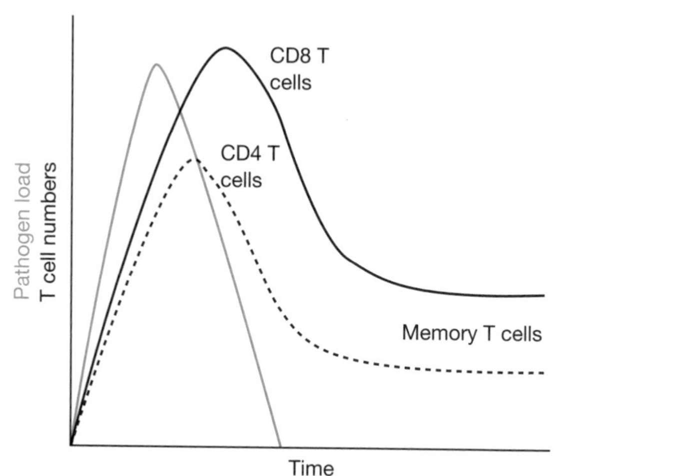

Figure 10.3 Memory T cells are generated during a primary response. Upon infection there is a large increase in the number of T cells specific for antigens on the pathogen. Most of the cells die off but a proportion become memory T cells. Because there are a lot more memory T cells than there were T cells before the infection the memory T cell response is bigger and faster than the first response.

cleared about 90% of the newly generated cells die and the remaining 10% differentiate into memory T cells (Figure 10.3). The generation of memory CD8 T cells is dependent on help from CD4 T cells.

We now look at the situation on secondary infection with a particular pathogen. You will have memory CD4 T cells, memory CD8 T cells and memory B cells. There will be many more of these than you had before infection and the different memory cells can all contribute to a bigger and faster secondary response. The memory CD4 T cells have the ability to rapidly respond to antigen and acquire effector function; they can do this in hours rather than the days it takes naïve T cells to differentiate into effector cells. They can then help the memory B cells and CD8 T cells to become plasma cells and Tcs and there is also evidence that memory CD4 T cells are more efficient at helping B cells and CD8 T cells to become plasma cells and Tcs respectively. The fact that memory B cells and CD8 T cells can also differentiate into plasma cells and Tcs more rapidly than naïve B cells and CD8 T cells also contributes to the faster, bigger secondary response.

10.2 Vaccines induce immunity without causing disease

The understanding that the immune system could protect against further infection with a pathogen provides the basis for vaccination. Vaccination is the process of trying to induce immunity without going through the disease process, and some forms of vaccination have been practised since ancient times (see Box 10.1). It is now appreciated that immunity is brought about by immunological memory.

10.2.1 Properties of an ideal vaccine

Vaccination is a way of manipulating the immune system to provide protection from disease caused by a pathogen without subjecting the person or animal to the disease. To be effective, a vaccine should stimulate the right type of immune response in the right anatomical site. Depending on the pathogen, antibody, CD8 T cells, delayed-type hypersensitivity (DTH), Th17 responses or a combination of these may be appropriate. For many pathogens the induction of secretory IgA at the appropriate mucosal site may protect from infection. There are also other properties that an ideal vaccine should have:

- **Safety.** There should be no side-effects from the vaccine and no risk of procedural errors in vaccine manufacture exposing individuals to pathogens or their toxic products.
- **Price.** Cheapness is obviously desirable and almost essential in many parts of the world with high rates of endemic disease and few economic resources.
- **Stability.** Ideally a vaccine should be able to be stored in high ambient temperatures so that it is available in hot climates with limited refrigeration facilities.
- **Ease of administration.** Children, the main target of vaccination, do not like having needles stuck in them and therefore oral vaccines are the most suitable form for administration and also can cut out the cost of materials for injection.

Vaccines have become safer. Variolation, used to induce protection against smallpox (see Box 10.1), carried a 1% death rate, which would now be considered unacceptable. The first true vaccine, using cowpox to vaccinate against smallpox (see Box 10.1), was much safer. Many of the vaccines that have been developed and are still in use today are quite safe and effective, despite being developed when relatively little was understood about the immune system. Vaccines are now available for a wide variety of pathogens (see Table 10.1).

However, there are still problems. Some vaccines carry a small risk of side-effects. Although the actual risk is small there can be a knock-on effect, which is the loss in a population of a phenomenon called **herd immunity.** Herd immunity occurs when the proportion of people in a population that is immune to a pathogen is so high that the pathogen cannot find enough susceptible hosts to infect and cannot survive in the population. Therefore even the minority of individuals who are not immune to the pathogen do not get infected. However, if people stop being vaccinated, or having their children vaccinated, the proportion of susceptible people increases to a level where the pathogen can become re-established in the population and herd immunity is lost.

BOX 10.1: ANCIENT PRACTICES

Many ancient peoples practised a form of vaccination even though they did not know about the existence of infectious microorganisms. They were aware that certain diseases were contagious; that is, they could be transmitted between people and/or animals, and that people who survived their first bout with the disease were protected from getting the disease again. The disease that has probably received the most attention over the millennia is smallpox. This is probably due to the magnitude of the threat posed by the disease and because even the ancient approaches to protection were partially effective. Until its eradication in 1979, as a result of a global vaccination programme organised through the World Health Organization, smallpox had been a major scourge of humankind. It is hard to imagine today the scale of the problem but in medieval times up to 60% of the population were infected with smallpox, of whom some 15% died. This means that about 10% of the population died of smallpox. Those who survived, especially adults, were often severely disfigured and/or blinded. Ancient Iranians, Chinese and Indians practised a technique known as variolation, which involved deliberately introducing material from a smallpox lesion into a scratch on the individual being treated, who was usually a child. Even though the process induced the actual disease, it was known that smallpox caused a less severe illness in children and those that survived variolation would be protected as adults. A variation among the dwellers in Baluchistan was to encourage children with wounds to touch the cowpox skin lesions of infected cows. Camelpox was claimed to be as effective. The idea that cowpox could provide protection against smallpox thus preceded the vaccination strategy of Jenner by centuries.

Another disease that attracted considerable attention was rabies. Romans in the fifth century advocated protecting cattle from rabies by making infected cows swallow the boiled liver of a rabid dog; quite how you would do this was not described in so much detail. Valli in the seventeenth century claimed that incubating saliva obtained from a rabid dog (obtaining this seems a somewhat risky process) with the gastric juices of a frog provided material that protected against rabies. Both of these treatments could be considered types of attenuation in which the virus or toxin was rendered harmless but provided immunity.

Other 'vaccination' procedures may have also inadvertently resulted in attenuation of pathogenic organisms or toxins. Inhabitants of ancient Iran treated (vaccinated?) goats against pneumonia by taking the lungs from infected animals and grinding them with garlic and vinegar. A needle and thread was then incubated in the material before being passed through the ear of the goat. A variation on this, practised in parts of Africa, was to ferment pieces of infected lung in a mixture of bran, millet and a plant extract used for tanning skin, to produce a potion for protection against bovine pneumonia. It was observed that although the treatment sometimes caused fatal pneumonia, where it was effective, protection was lifelong.

Edward Jenner

Edward Jenner is credited as being the pioneer of modern vaccination. Variolation had been introduced to England in the early 1700s by Lady Elizabeth Montague who had observed the practice in Turkey. Edward Jenner was a country doctor with a practice in Gloucestershire. He had become aware of the local folklore that milkmaids had clear skin, which was not blemished by the ravages of smallpox. He developed the theory that exposure to cowpox protected the milkmaids from smallpox, although the existence of infectious microorganisms was then unknown. To test his theory he performed in 1798 what would today be regarded as an extremely unethical experiment. He took material from a cowpox lesion and introduced it into a scratch on the arm of a young boy. No illness followed. To test whether the exposure to cowpox material protected against smallpox, Jenner deliberately infected the boy with material from a smallpox lesion. Fortunately the experiment was successful and the boy was completely protected. Although it seems highly unethical in today's context to expose someone deliberately to a potentially deadly infection, it should be remembered that variolation, which involved deliberately infecting children with smallpox, was still being practised, and was considered by many to be worth the risk. The term 'vaccination' comes from vaccinia, the cowpox virus.

Table 10.1 Some of the current available vaccines

Organism	*Type of vaccine*
Viruses	
Measles	Attenuated virus
Mumps	Attenuated virus
Rubella (German measles)	Attenuated virus
Polio	Attenuated, killed virus
Chickenpox	Attenuated virus
Rotavirus (infant diarrhoea)	Recombinant vaccine
Influenza	Inactivated virus
Hepatitis (A and B)	Recombinant vaccine
Bacteria	
Diphtheria	Toxoid
Pertussis (whooping cough)	Killed bacteria, subunit vaccine
Meningitis – *Neisseria meningitidis*	Capsular polysaccharide
– *Haemophilus influenzae*	Polysaccharide-conjugate
Tetanus	Toxoid
Streptococcus pneumoniae	Capsular polysaccharide
Cholera	Killed bacteria

A second issue with vaccines is that there are still many diseases for which no effective vaccine exists. These diseases, including malaria, tuberculosis and parasitic worms, kill in excess of ten million people annually and some estimates suggest 30% of infant mortality could be prevented by successful vaccination. There is still no vaccine for AIDS.

10.2.2 Vaccines – the past and the present

The situation with cowpox and smallpox is unusual. Other pathogens do not have naturally occurring harmless equivalents in other species that provide a ready-made source of vaccine. Therefore other approaches were needed to develop vaccines, many of which are still in use today:

- **Killed or inactivated pathogen.** This approach, used primarily for viruses, involves taking the virus and treating it with heat or chemicals so that it is no longer infectious. These vaccines stimulate good antibody responses but, because the virus is no longer infectious, do not stimulate CD8 cytotoxic T cell responses. This approach has been used for influenza, rabies, polio (Salk vaccine) and some bacteria including pertussis, the cause of whooping cough. One problem is the possibility of some pathogen particles surviving the killing or inactivation process; this happened with the polio vaccine in the 1950s and the vaccine caused many cases of polio.
- **Attenuated pathogen.** Attenuated pathogens are still viable and cause infection but do not cause disease. Attenuation is usually achieved by growing the organism in cells of another species so that the pathogen becomes adapted to cells of the other species and grows poorly in human cells. Because they can infect cells they stimulate CD8 cytotoxic T cells as well as antibody. These vaccines have been developed for polio (Sabin vaccine), measles, mumps and tuberculosis, among others. A problem is the possibility of reversion to full pathogenicity, either in the vaccine stock or in the vaccinated individual.
- **Subunit vaccines.** In some situations an antibody response against a particular component of the pathogen is sufficient to provide immunity. Many bacteria produce a polysaccharide coat that prevents phagocytosis in the absence of antibody. Vaccination with the polysaccharide induces antibody, which is enough to provide immunity. This approach has been used against *Haemophilus influenzae*, which causes 'flu, and *Neisseria meningitidis*, a cause of potentially fatal meningitis. One problem with these vaccines is that the polysaccharide antigen does not stimulate Th cells and therefore only IgM is produced. To overcome this, the polysaccharide can be conjugated to a protein such as tetanus toxoid. The tetanus toxoid stimulates Th cells, which can help the B cells specific for the polysaccharide to switch to other antibody classes and make a bigger higher affinity antibody response (Figure 10.4).
- Subunit vaccines can also be proteins as in the case of the vaccine for hepatitis B, where immunisation with the major surface antigen of the virus, called the HbsAg, induces the production of protective antibodies.

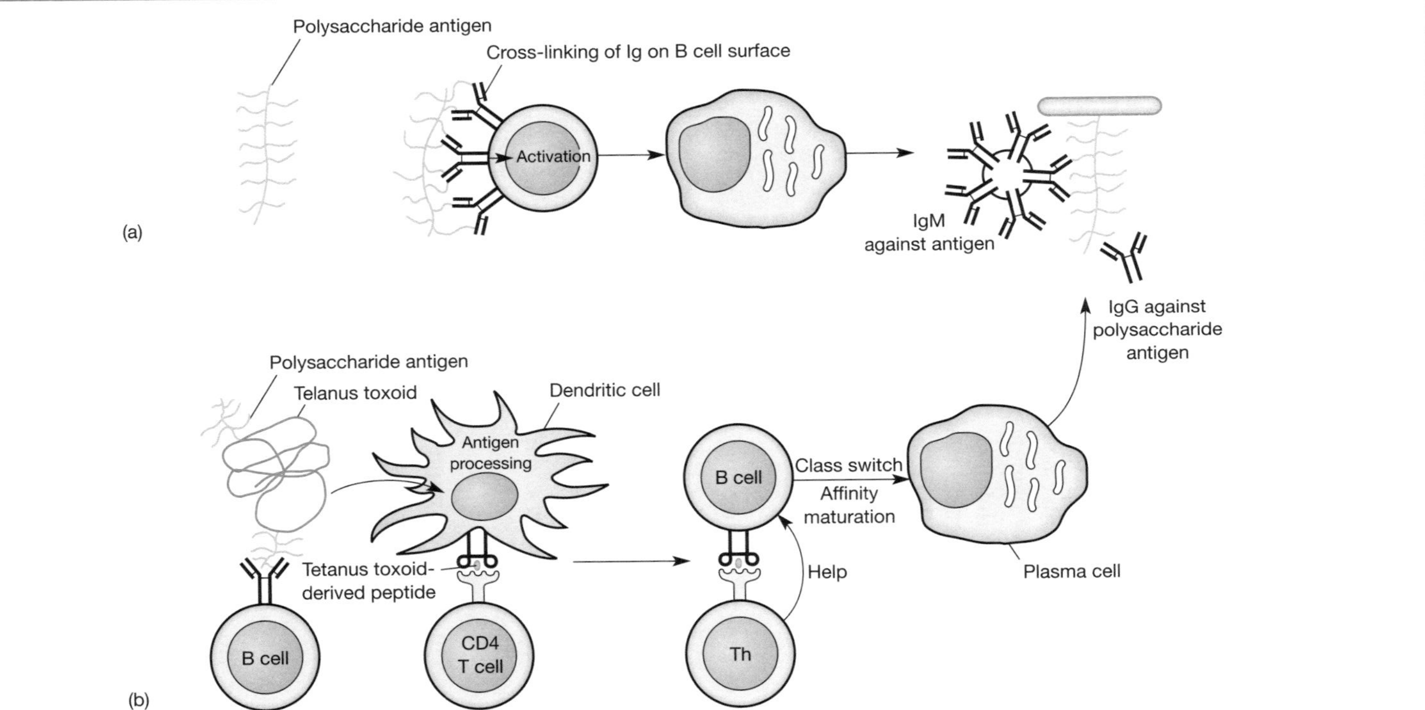

Figure 10.4 Subunit vaccines. (a) Bacterial polysaccharides contain repeating epitopes, which can cause extensive cross-linking of Ig on B cells when they are bound by the Ig specific for the epitope. This is a strong enough stimulus to activate B cells to become antibody-secreting plasma cells in the absence of T cell help. However, because CD4 T cells are not stimulated there is no class switch or affinity maturation and only low-affinity IgM is produced. Because IgM is very efficient at fixing complement this can provide protection against infection by the bacteria. (b) The polysaccharide can be conjugated to a protein such as diphtheria toxoid, to which most people are immunised. The toxoid has T cell epitopes that stimulate toxoid-specific Th, which can help the polysaccharide-specific B cells undergo class switch and affinity maturation, resulting in plasma cells secreting high-affinity IgG specific for the polysaccharide.

- **Toxoids.** Where pathogens cause disease almost solely through the production of toxins it is possible to vaccinate just against the toxin. This prevents disease upon infection and the immune system can then mount a response against the pathogen and eliminate it. To prevent the toxic effects of the toxin upon vaccination, the toxin is treated chemically so that it loses toxicity but retains antigenicity. Examples of toxoid vaccines are those for tetanus and diphtheria.

Nearly all of the vaccines above were developed when much less was known about the immune system than now. It turns out that the most successful vaccines are ones that result in the production of high amounts of high-affinity neutralising antibodies (see Section 8.3). The pathogens for which no effective vaccines exist may be ones where neutralising antibody is not protective and different type of immune responses are required. While our vastly increased knowledge of the immune system theoretically makes rationale vaccine development more feasible, repeated failures to generate vaccines, for instance against malaria and HIV, have highlighted the complexity of developing vaccines that protect in other ways than by neutralising antibody. There are however several new approaches being used for vaccine design.

10.2.3 Vaccines – the present and the future

The development of genetic engineering holds great promise for the future of vaccinations. Already some of the subunit and toxoid vaccines that used to have to be purified from bulk cultures of the relevant organisms can be produced using recombinant DNA technology so that the relevant genes are introduced into bacteria or yeast that produce large quantities of the protein. However, other applications of DNA technology raise the prospect of developing vaccines that are safer, cheaper and more stable than many of the present vaccines. These aspects are all important, especially for the use of vaccines in countries where refrigerated storage conditions are limited. The new styles of vaccine being developed for trial include the following:

- **Recombinant vector vaccines.** These are vaccines in which the genes encoding important antigens for a pathogen are introduced into the genome of attenuated viruses or bacteria (Figure 10.5). An example would be to introduce the gene for a pathogenic antigen into the genome of the vaccinia virus, which previously was used for vaccination against smallpox. The vaccinia virus, in addition to directing the expression of its own antigens, would also cause expression of the pathogen's antigen and stimulate immunity against smallpox (which is not necessary because natural infection with smallpox should not occur) and the pathogen. Because the vaccinia virus can infect cells it should stimulate good antibody responses and CD8 cytotoxic T cell responses against the pathogen's antigen.
- **DNA vaccines.** A somewhat surprising observation led to a new approach for vaccines. When muscle cells were exposed to DNA, it was found that they could take up the DNA and express the proteins coded for by the DNA. This happens *in vivo* and provides a mechanism for inducing the

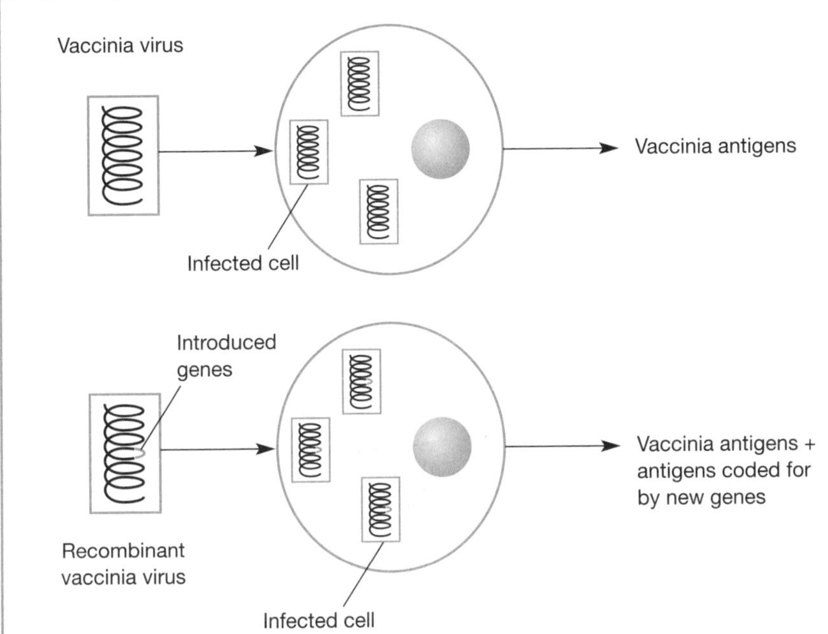

Figure 10.5 Recombinant vector vaccines. Vaccinia virus is an attenuated virus previously used as a vaccine for smallpox. Because it is a live virus it is infectious and stimulates antibody production, Th cells and cytotoxic CD8 T cells. By genetically modifying vaccinia virus it is possible to introduce genes coding for antigens of other pathogens. The virus directs the production of these antigens when it infects host cells and can therefore stimulate protective immune responses against the other pathogen.

production of a wide variety of proteins including pathogenic antigens. In practice the DNA vaccine is in the form of a plasmid (Figure 10.6). This means that promoters can be introduced into the plasmid which cause high production of the pathogen's protein. Experimental (DNA) plasmid vaccines induce both antibody and CD8 cytotoxic T cell-mediated immunity and have been shown to provide immunity against infection with the pathogen from which the antigen DNA was derived. Clinical trials are under way with several DNA vaccines. One major advantage of DNA vaccines is that DNA is very stable and does not require refrigerated storage.

- **Peptide vaccines.** The type of vaccine that has the potential to be the safest, cheapest and easiest to store is the peptide vaccine. This consists of a synthetic peptide that contains a CD4 T cell epitope and, depending on the pathogen, a B cell epitope and/or CD8 T cell epitope (Figure 10.7). They can be produced chemically and therefore in bulk, cheaply and, because they do not involve DNA or the inactivation of toxins, there should be no risk of accidental exposure to toxins or virulent organisms. The main drawbacks to peptide vaccines have been identifying appropriate epitopes and making the peptides immunogenic. Their development has therefore been slower than originally hoped.

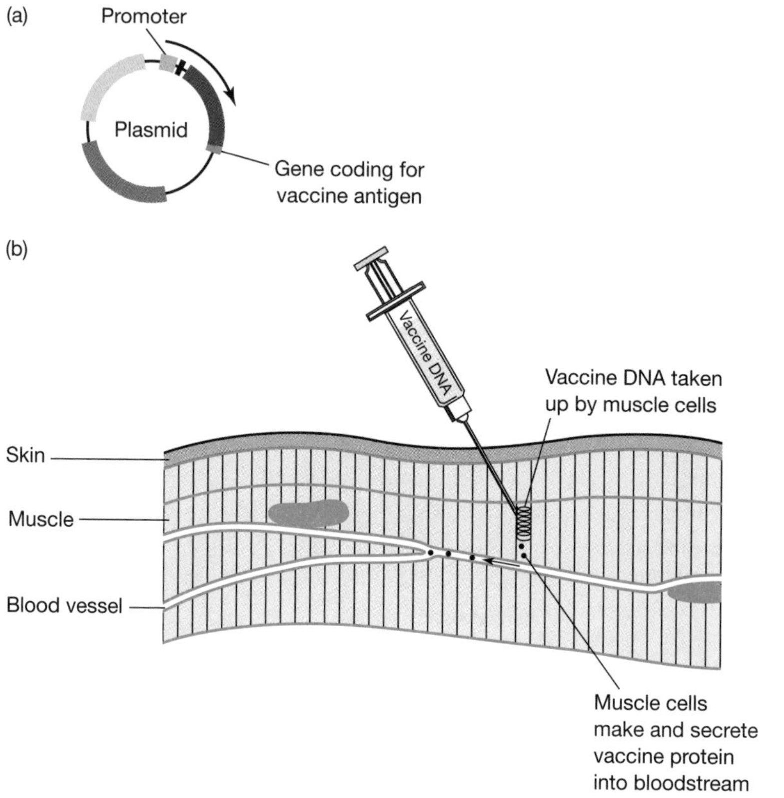

(a) Promoter
Plasmid
Gene coding for vaccine antigen

(b)
Vaccine DNA
Vaccine DNA taken up by muscle cells
Skin
Muscle
Blood vessel
Muscle cells make and secrete vaccine protein into bloodstream

Figure 10.6 DNA vaccines. Plasmid DNA engineered to include genes for antigens of pathogens (a) is taken up by muscle cells after intramuscular injection. The muscle cells express the antigen (b), which can stimulate protective immune responses.

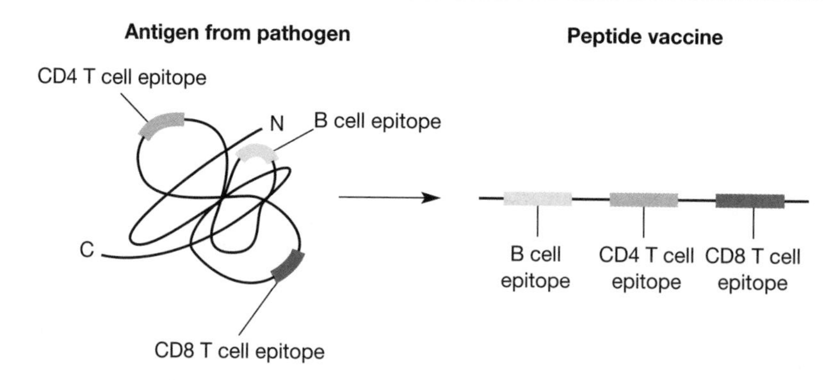

Figure 10.7 Peptide vaccines. In this hypothetical example, a B cell epitope, a CD4 T cell epitope and a CD8 T cell epitope have been identified in a protein made by a pathogen. Knowing the amino acid sequences of these epitopes, it is possible to synthesise a peptide containing all three epitopes and this peptide can be used as a vaccine.

10.2.4 Making vaccines more effective

As mentioned above two of the most important features about a vaccine are that it stimulates a good immune response and it stimulates the right sort of immune response. In addition to the new-style vaccines described above considerable advances are being made in making vaccines more effective in the size and type of immune response they stimulate. Two weapons used in stimulating immune responses to antigens, including vaccines, are **adjuvants** and **cytokines**.

Adjuvants. From the late 1800s most immunology was devoted to the development of vaccines. Successful vaccines had been developed against smallpox and rabies and there was a search for vaccines against the childhood infections that caused significant mortality; these included measles, mumps, diphtheria and whooping cough. Ways were being sought to increase the immune response against the vaccine and in the 1920s it was shown that if diphtheria toxoid vaccine was mixed with substances such as tapioca, starch oil or even breadcrumbs, it stimulated a much better antibody response than toxoid vaccine on its own. These agents, which increased the immune response to antigens were called **adjuvants**. It is perhaps not surprising that breadcrumbs never made a great impact as an adjuvant and up until recently the only adjuvant licensed for use in humans was alum, which is a mixture of aluminium salts in which the antigen is emulsified. Alum was developed in the 1920s and has been successful in vaccines against diphtheria, tetanus, hepatitis A and B and *H. influenzae*. One problem with alum is that it is very good at stimulating antibody responses but does not stimulate other types of immune responses such as CD8 cytotoxic T cells. Recently other adjuvants have been licensed in the EU. These are emulsions of oil and water in which the vaccine is incorporated.

For a long time it was not completely understood how vaccines worked. One property of adjuvants was to form a depot of antigen which was released slowly over time. However this alone was not enough to explain the adjuvant effect. It is now becoming clearer that effective adjuvants have two other important properties; they promote uptake of antigen by antigen-presenting cells, especially dendritic cells, and they also cause activation of dendritic cells. A good way to activate DCs is by stimulating their TLRs. One TLR ligand, monophosphoryl lipid A, has been incorporated into the alum adjuvant for the new vaccine for cervical cancer. Other TLR ligands such as CpG, polyI-C (which mimics dsRNA), flagellin and LPS are being tested for their ability to increase the potency of adjuvants.

Cytokines. Given their role in regulation of immune responses, cytokines were an obvious choice for use in trying to boost and control immune responses, including in vaccination. Cytokines can theoretically boost immune responses by activating antigen-presenting cells or increasing the proliferation and/or survival of B and T cells. They could also be used to control the types of responses stimulated by vaccines by pushing responses towards Th1, Th2 or Th17 responses or by promoting the production of different classes of antibody or the production of Tcs. Although they have been shown experimentally to increase the efficacy of vaccines, cytokines have not been licensed for use in vaccines against infectious agents although they are being tested clinically in cancer vaccine trials (see Chapter 15).

Overall, vaccination has been a tremendously successful form of manipulating the immune system by stimulating specific immune responses against antigens on pathogens and has led to a significant reduction in mortality and morbidity. But vaccines are not the only aspect of immunology that is being used to benefit mankind, and indeed animalkind. Antibodies are being used as tools in the clinic, in the laboratory and for assaying a huge number of substances.

10.3 Antibodies can be produced and used in many ways in treatments and in tests

Vaccination was the first way in which manipulation of the immune system was used to benefit mankind. But there are other ways in which the immune system can utilised and one of these is to stimulate antibody production so that the antibodies can be used for clinical or laboratory use. It had been known since the 1880s that if you injected rabbits with sublethal amounts of bacterial toxins (such as diphtheria or tetanus) they became resistant to normally fatal doses of the toxin. Furthermore this protection could be passed on to other animals by injecting them with serum from the immunised rabbits. At the time the protective factors in the serum were called antitoxins although it is now known that they were neutralising antibodies. The scientists of the day believed that antitoxins could be used in humans as a treatment for diseases like diphtheria which at the time was a common childhood illness with a mortality rate of between 20% and 60%. Because of their larger size, horses were immunised to produce antitoxins for treatment of humans, and children treated with immunised horse serum had a much lower mortality rate. Horse serum was used as treatment for bacterial infections until the late 1930s. However there could be complications such as serum-sickness which was a potentially fatal reaction to repeated injections of horse serum. With the development of antibiotics in the 1940s the use of immune animal sera for bacterial infections declined.

It was shown in the 1890s that a similar approach could be used with snake toxins. Horses could be injected with gradually increasing amounts of toxin so that they developed antitoxins (antibodies) and the serum from these horses could be given to people bitten by a snake. The first successful antitoxin was against cobra venom. A major problem with this approach was you had to know what sort of snake had bitten you but despite this, these so-called antivenoms are still used today. The antivenoms are still produced mostly in horses although sometimes sheep are used because sheep antibodies cause fewer adverse reactions. Another measure to reduce adverse reactions is to use purified antibodies from the serum rather than the whole serum. Today antivenoms are available against toxins from thousand of species of snakes, spiders, scorpions, frogs, poisonous mushrooms and plants.

This procedure of transferring protective antibody is known as passive immunisation because although the antibodies work in the recipient they are not produced by the recipient's immune system. However the ability to immunise animals to produce large amounts of antibody is an example of immunotechnology whereby knowledge of the immune system is used to generate products for use in the clinic or the laboratory. Antibodies could also be used in assays for various substances, such as insulin, and were a useful laboratory tool. However there could be problems with antisera generated in animals because the serum not only contained antibodies against the antigen they were immunised with but also antibodies against antigens of infectious agents that the animal had been exposed to. Therefore sometimes antibodies were difficult to work with and gave false positive results. In the 1970s two scientists, Kohler and Milstein, developed a technique for producing large amounts of identical antibody. These were called **monoclonal antibodies** and their development has revolutionised medicine and science.

10.3.1 Monoclonal antibodies – Ehrlich's magic bullets

Monoclonal antibodies are produced by cells called hybridomas. Hybridomas are made by taking B cells from an immunised animal, usually a mouse, and fusing the cells with immortalised myeloma cells (see Figure 10.8). A myeloma is a plasma cell tumour and because the myeloma cells are transformed they will grow *in vitro* whereas B cells die in tissue culture. The product of the fusion of a B cell with a myeloma cell is called a hybridoma. Hybridomas combine the immortality of a myeloma cell with the antibody-producing capacity of the B cell. After fusion with the B cells the hybridomas can be selected in medium containing hypoxanthine, aminopterin and thymidine (HAT medium). The myeloma cells are deficient in the enzyme hypoxanthine guanine phosphoribosyl transferase (HGPRT) and cannot grow in the presence of aminopterin. The B cells are HGPRT+ but will die after a short time in culture. The hybridomas will be HGPRT+ because of the contribution by the B cell fusion partner and therefore will be the only cells to grow. Cloned hybridoma cells can be grown *in vitro* to produce large numbers of identical cells all producing the same antibody; for this reason the antibody produced by a hybridoma is called monoclonal antibody. It is now possible to produce monoclonal antibody in kilogram amounts using industrial fermenters. This has resulted in over forty monoclonal antibodies being licensed for clinical use for cancer, transplantation and autoimmunity. Furthermore monoclonal antibodies can be used in assays to measure tens of thousands of different compounds. This may be in a clinical setting for measuring levels of hormones or other biological products to aid in diagnosis of disease. Assays can also be used for research purposes or in the technology industries, especially biotechnology.

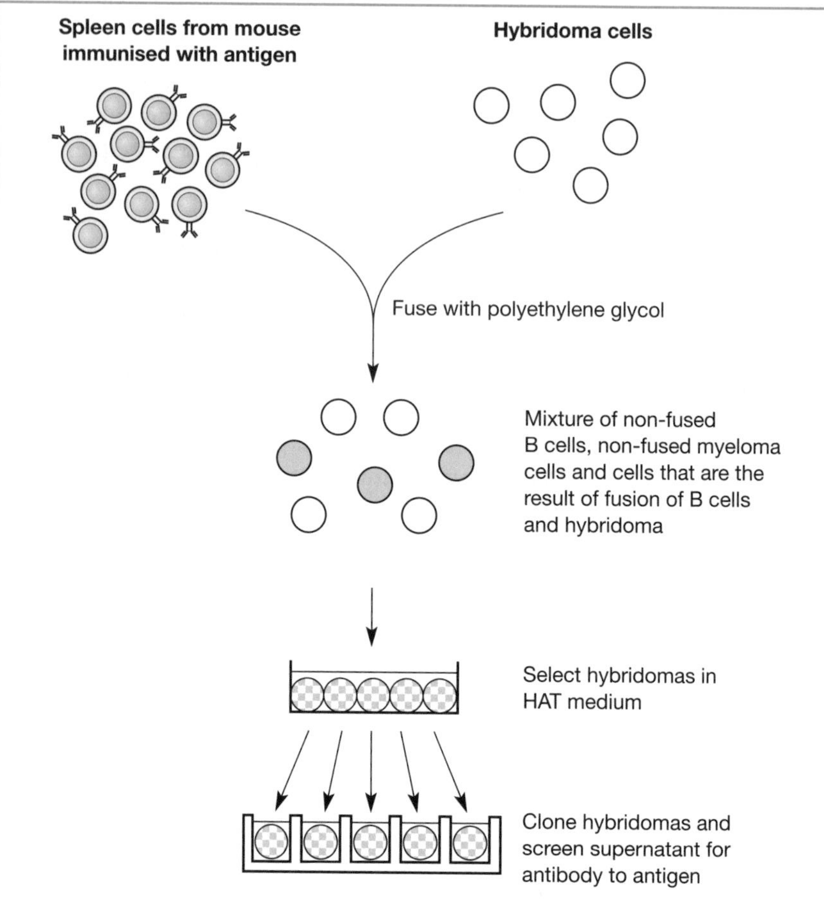

Figure 10.8 Production of monoclonal antibodies. Spleen cells from a mouse immunised with antigen are fused with myeloma cells and the fused hybridoma cells selected in HAT medium. The hybridomas are cloned and the supernatants of the cloned cell cultures are screened for antibody against the antigen.

One of the great pioneers of the early use of antibodies was the great microbiologist, Paul Ehrlich. Over a hundred years ago he suggested that agents could be developed that would kill bacteria without harming humans or animals and because they were specific for the bacteria he called them 'magic bullets'. Although their chemical structure was unknown at the time, antibodies, because of their specificity, were one type of magic bullet and monoclonal antibodies could be regarded as the ultimate magic bullet.

10.3.2 Using antibodies in the laboratory and clinic

Because of their specificity, especially in the case of monoclonal antibodies, antibodies can be used to identify and, in some cases, quantify, a wide variety of products in different settings. Three of the most widely used antibody based techniques are enzyme-linked immunosorbant assay (ELISA), immunohistochemistry and western blotting.

ELISA – an assay for all settings

The ELISA can be used as a highly quantitative assay to measure essentially anything it is possible to make monoclonal antibodies against. ELISAs can be used to measure antibody responses in vaccinated individuals or to test for infection such as in HIV. The antigen to which antibody is being measured is used to coat the wells of plastic plates (Figure 10.9). Serum from the patient is added to the well and if there is antibody in the serum it will bind to the antigen in the well. After washing the well an enzyme-linked monoclonal antibody against human Ig is added to the well followed by a colourless substrate which the enzyme turns to a coloured product. Therefore the amount of colour is proportional to the amount of antibody in the serum sample. The most common type of ELISA is called a sandwich ELISA. It basically involves coating the wells of plastic plates with an antibody against the product being measured, incubating the coated well with the sample and adding a second enzyme-linked antibody against the product. Again a colourless substrate is added to the well which is converted to a coloured product by the enzyme and the amount of substance is proportional to the amount of colour. It is a very sensitive assay and can detect concentrations of products of less than 1 pg/ml (Figure 10.10).

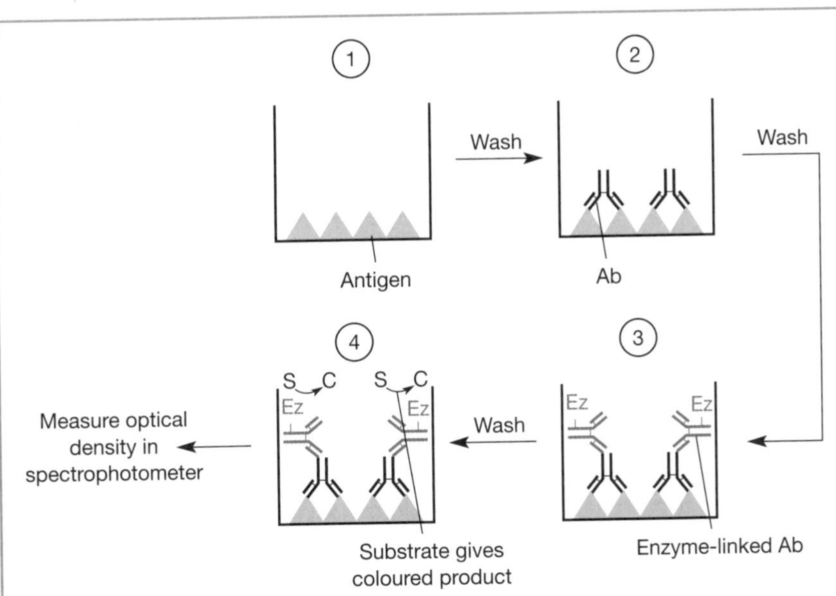

Figure 10.9 Enzyme-linked immunosorbent assay (ELISA). (A) Measuring antibody. Antigen is bound to a plastic well ①. Serum is added to the well and if autoantibody to the antigen is present it will bind to the antigen in the well ②. An enzyme-linked anti-human-IG antibody is added to the well and will bind to any autoantibody bound to the antigen ③. A colourless substrate (S) is added to the well and this will be converted into a coloured product (C) by the enzyme ④. The amount of coloured product produced can be measured using a spectrophotometer. The amount of product is proportional to the amount of enzyme, which in turn is proportional to the amount of autoantibody that bound to the autoantigen.

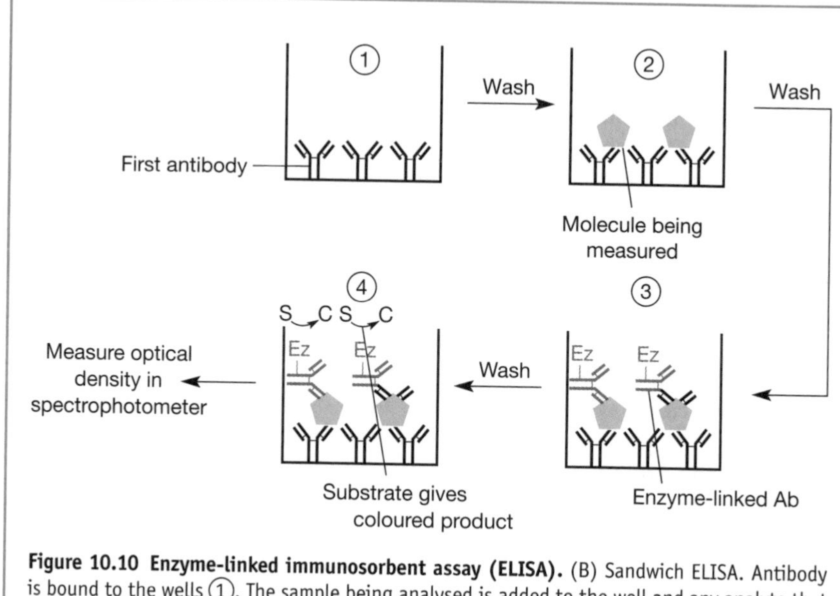

Figure 10.10 Enzyme-linked immunosorbent assay (ELISA). (B) Sandwich ELISA. Antibody is bound to the wells ①. The sample being analysed is added to the well and any analyte that is present will bind to the antigen in the well ②. An second enzyme-linked antibody to the analyte is added to the well and will bind to any analyte bound by the first antibody ③. A colourless substrate (S) is added to the well and this will be converted into a coloured product (C) by the enzyme ④.

ELISAs are used in a wide variety of settings. They are used to measure levels of hormones, enzymes and other products in clinical diagnostic tests. In food and environmental safety they can be used to detect herbicides, pesticides or antibiotics in food. In sport ELISAs are employed to check for doping of athletes or horses. ELISAs are even being used in the art world to analyse the materials used historically and have shown that in medieval times egg yolk and skin were used in the coatings of sculptures.

Immunohistochemistry

Antibodies can be used to visualise the location of specific molecules in tissue sections. Antibodies are added to a section of tissue and bind to their antigen if present in the tissue. Then a second antibody is added that can be visualised. The second antibody can be linked to a fluorescent dye in the technique called immunofluorescence. This dye will shine with a certain colour when looked at with a fluorescence microscope (Plate 13). Different dyes can be used which fluoresce with different colours, enabling a number of different antibodies to be used on the same section. An alternative visualisation process is called immunohistochemistry. Here the second antibody is linked to an enzyme and when it has bound to the first antibody a substrate is added which is converted to a coloured product by the enzyme and can be seen under an ordinary microscope (Plate 13).

Western blotting

Western blotting is used to identify particular products in a complex mixture such as a cell lysate and can be semi-quantitative. First of all the sample is subjected to electrophoresis to separate the different proteins by size, electrical charge or other physical property. The proteins are then transferred to a membrane. The membrane is incubated with enzyme-linked antibody followed by a substrate which will produce a colour where the antibody-enzyme bound. This enables the molecular weight of the protein that the antibody was specific for to be identified (Figure 10.11a). Western blots can therefore identify if individual proteins are present in a sample and can even identify aspects such as the phosphorylation status of a protein. This technique is very good for studying cell signalling and differentiation.

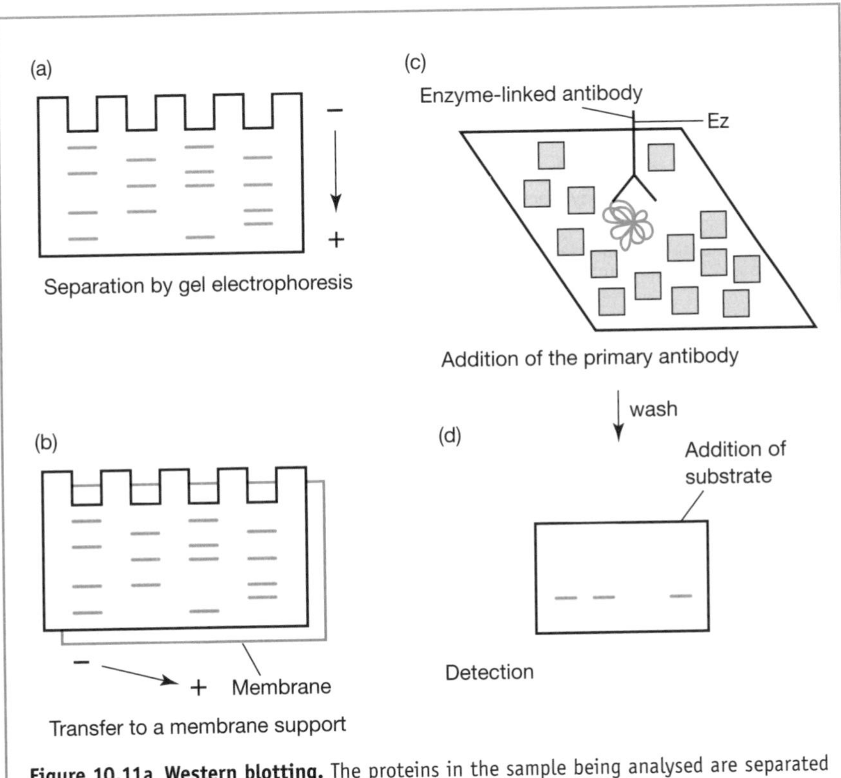

Figure 10.11a Western blotting. The proteins in the sample being analysed are separated by gel electrophoresis (a) and the gel contents transferred to a membrane (b). The membrane is incubated with an enzyme-linked antibody which will bind to proteins, or other molecules, that the antibody is specific for (c). After washing a substrate is added which is turned to a coloured product by the enzyme (d).

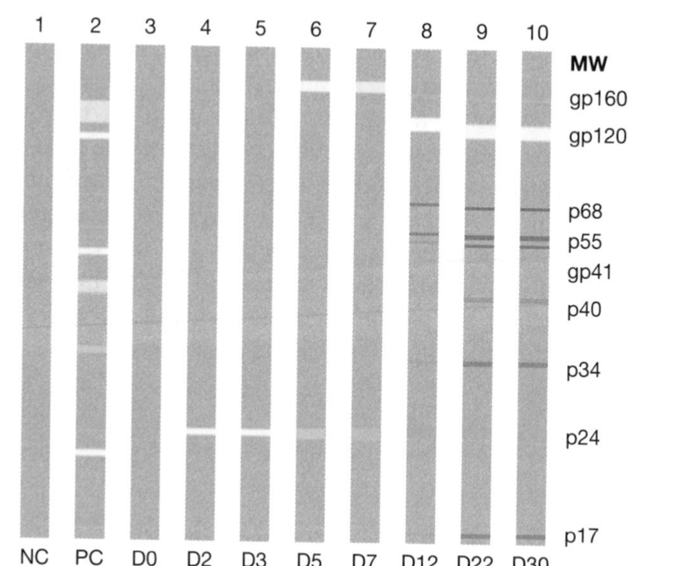

Figure 10.11b Western blotting. Appearance of western blot for diagnosis of HIV infection. HIV lysate was separated by electrophoresis and the membrane incubated with serum followed by enzyme-linked anti-human-Ig and substrate. The negative control had no bands (Lane 1) while the HIV infected serum showed a number of bands indicating the presence of antibodies to different HIV antigens. PC-positive control. D0–D30-day after infection.

10.4 Summary

- Immunological memory is a mechanism for making immune responses to a second, or subsequent, exposure to an antigen faster and bigger.
- Memory for antibody responses is based on memory B cells produced in germinal centres during a primary antibody response. Because memory B cells have already switched antibody class to IgG, IgA or IgE and have undergone affinity maturation, they are able to differentiate rapidly into high-affinity antibody-secreting plasma cells, providing a quicker response. There are also more memory B cells, providing a bigger memory response.
- Memory T cells are also present in higher numbers than prior to initial exposure to antigen and are able to differentiate more rapidly into effector cells, which secrete more cytokines.
- Immunological memory provides the basis for vaccination and there are many different vaccines which are in use or being developed.
- For vaccines to be effective they must stimulate the right type of response and be safe. Ideally they should also be cheap, stable and easy to administer.

- Current vaccines are attenuated or heat-killed versions of the pathogen, subunit vaccines that are antigens from the pathogen, or toxoids that are inactivated toxins. They have provided protection against many illnesses with considerable mortality or morbidity but there are many infections for which no vaccines are available.
- Vaccines under development or being introduced for clinical use are recombinant vector vaccines, DNA vaccines or peptide vaccines. They have the advantages that they may be safer, cheaper and easier to store than many current vaccines.
- Passive immunity involves injecting people with antibodies that were produced by immunising animals. It is used as a treatment for poisonous snake bites.
- Monoclonal antibodies are generated by fusing B cells from immunised animals with immortal myeloma cells to generate hybridomas which secrete identical antibody. Monoclonal antibodies are widely used for various assays and are becoming increasingly used as therapeutic agents.

10.5 Questions

1) Complete the sentences with brief explanatory phrases.

   The secondary antibody response is faster, bigger and better than the primary response.
   It is faster because ____________________.
   It is bigger because ____________________.
   It is better because ____________________.

2) What are the advantages and disadvantages of attenuated and killed vaccines?

3) Explain what is meant by an adjuvant.

4) Draw a diagram to explain how an ELISA may be used to measure the level of insulin in a blood sample.

5) What is the difference between polyclonal antibodies and monoclonal antibody?

The answers to these questions can be found on page 338.

10.6 Further reading

1) McHeyzer-Williams LJ, McHeyzer-Williams MG. (2005) Antigen-specific memory B cell development. *Annual Review of Immunology* 23: 487–513.

2) Kalia V, Sarkar S, Gourley TS, Rouse BT, Ahmed R. (2006) Differentiation of memory B and T cells. *Current Opinion in Immunology* 18:255–264.

3) Seder RA and Ahmed R. (2003) Similarities and differences in CD4+ and CD8+ effector and memory T cell generation. *Nature Immunology* 9:835–842.

4) Beverley PCL. (2002) Vaccination. *British Medical Bulletin* 62:1–230.

5) Guy B. (2007) The perfect mix: recent progress in adjuvant research. *Nature Reviews Microbiology* 5: 505–517.

6) Bierer B. (2005) *Current Protocols in Immunology* Wiley New York.

CHAPTER ELEVEN

Immunological tolerance and regulation – why doesn't the immune system attack ourselves?

Learning objectives

To learn how B and T cells develop from bone marrow precursors. To know about the thymus and its role in T cell development. To understand the mechanisms of preventing the immune system reacting against self-antigens – self-tolerance.

Key topics

- The need for immunological tolerance
- Tolerance in B cells
  - Tolerance in the bone marrow
  - Peripheral tolerance
- Tolerance in T cells
  - Positive and negative selection
  - Induction of Tregs in the thymus
- Peripheral tolerance in T cells
  - Anergy
  - Regulation
  - Ignorance

11.1 Immunological tolerance – what is it and why do we need it?

We have seen how the immune system has evolved many recognition systems that enable foreign pathogens (and other objects) to be recognised

(Figure 11.1). The innate immune system uses a variety of receptors, such as TLRs, mannose-binding proteins and lectins to recognise foreign molecules on pathogens and some of our own molecules released following tissue damage. The specific immune system uses antibody and TCRs to recognise antigenic epitopes. Having these three recognition systems makes it virtually impossible for pathogens to avoid recognition by at least one component of the immune system.

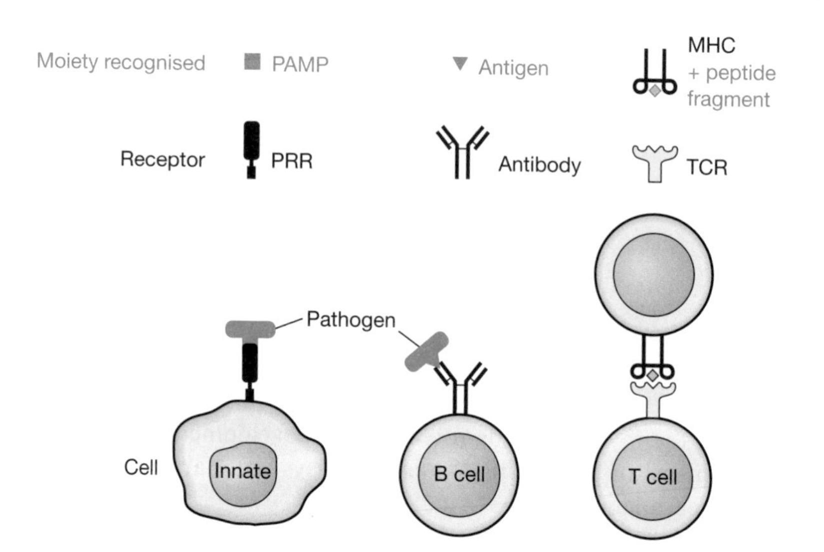

Figure 11.1 Recognition by the immune system. The innate immune system uses pattern recognition receptors (PRRs) to recognise microbial products, pathogen-associated molecular patterns (PAMPS). In the specific immune system B cells use antibody to recognise antigen and T cells use the TCR to recognise antigen-peptides in association with MHC.

The biggest difference between the various receptor systems is the way receptor specificities for their targets are generated. Innate receptors are fixed in the germ-line so that the genes you inherit are expressed in a genetically pre-determined way. By contrast B and T cells rearrange their Ig and TCR genes respectively during development as described in Chapter 5. The mechanisms of gene rearrangement are designed to generate large numbers of different antigen receptors from a limited number of genes. There are two main advantages to this system. One is that huge numbers (estimates in excess of 10^{11}) of different antigen receptors (Ig or TCR) can be generated from a few hundred genes. The other advantage is that the mechanisms introduce variations in nucleotide sequence **at random.** This is important in increasing the number of different receptors but also has another huge advantage in that infectious pathogens cannot evolve to avoid recognition by receptors generated at random. Pathogens have evolved to avoid, or limit, recognition by receptors of the innate immune system. They can do this because the innate receptors are fixed in the germ-line and therefore mutations in pathogens that enable them to avoid recognition by a receptor of the innate immune system give the pathogen a selective advantage. However, because Igs and TCRs generate specificity for antigen at random, it is impossible for pathogens to avoid recognition by at least some lymphocytes. Because of the huge number of different antigen specificities, if a pathogen mutates and expresses a different form of an antigen, the mutated form of the antigen will be recognised by different lymphocytes with Ig or TCR specific for the mutated form of the antigen.

However, there is a big downside to the randomness of receptor generation. This is that it is inevitable that some Igs or TCRs will be generated that are specific for our own proteins or other tissue or cellular components. We are all full of antigens that the immune system can potentially react against. This is shown by transplantation; if a kidney or other organ from one person is transplanted into another, the recipient's immune system recognises the transplant as foreign and tries to destroy it – so-called graft rejection. So, if we can't stop the production of lymphocytes with receptors specific for our own antigens, how do stop these self-reactive lymphocytes from responding to these self-antigens and attacking our own bodies? The answer is that the immune system has evolved a number of different mechanisms to limit the production of self-reactive B and T cells and also to prevent those that are produced from actually generating immune responses against self-antigens. This process is known as immunological tolerance, which is where the immune system does not react against an antigen, and more specifically self-tolerance, where the immune system does not attack our own bodies. It is not surprising that many mechanisms exist for something as important as stopping our immune systems from attacking our own bodies.

Mechanisms of self-tolerance operate for both B cells and T cells. However, because there is interaction between CD4 T cells and B cells or CD8 T cells in the generation of immune responses (see Chapters 7 and 9) it is possible to maintain tolerance as a whole if only one component of a complex interaction system is non-responsive. It is a similar situation to a car engine, if one part is not working the engine will not run even if all the other parts can potentially function properly. Because CD4 T cells play a central role in immune responses it is perhaps not surprising that tolerance in CD4 T cells is most powerful although additional mechanisms operate for B cells and CD8 T cells.

11.2 Self-tolerance in B cells

Two mechanisms exist that act directly on B cells to stop them from reacting against self-antigens: these are clonal deletion and clonal anergy. When B cells develop in the bone marrow they go through a stage known as an immature B cell where they express IgM but not IgD on their surface (see Chapter 5). The IgM on the immature B cells could be specific for foreign antigen or self-antigen. The environment in the bone marrow where B cells develop is usually sterile and devoid of foreign antigens. Therefore the only antigen that developing B cells could encounter is self-antigen. If an immature B cell recognises a self-antigen in the bone marrow through its IgM it can undergo one of two fates (Figure 11.2):

- **Clonal deletion.** In this situation the immature B cell recognising self-antigen is induced to die by apoptosis and obviously will not develop into a mature B cell. For this reason the process is known as clonal deletion.
- **Clonal anergy.** Alternatively the developing B cell may be rendered unresponsive. Even though it goes on to express IgD and can leave the bone marrow it will not be able to respond to their particular antigen if it sees it again. These cells are called anergic.

Functionally speaking, clonal deletion and clonal anergy are the same. With clonal deletion developing self-reactive B cells are eliminated and with clonal anergy the self-reactive B cell is still there but can no longer respond to antigen. In neither situation will self-reactive antibody be produced.

It is not clear why both mechanisms exist but the nature of the antigen seems to affect whether cells are deleted or anergised. Experiments have shown that antigens on the surface of other cells induce clonal deletion but that soluble antigens induce anergy.

11.2.1 The main reason self-specific B cells do not respond is due to lack of T cell help

The mechanisms of clonal deletion and anergy do stop some self-specific B cells from developing or functioning However many self-antigens are not expressed in the bone marrow and therefore mature self-antigen specific B cells do leave the bone marrow and enter the lymphoid system. In fact we are full of self-specific B cells. So why do they not go on to become antibody-producing plasma cells? The answer is a lack of self-antigen specific helper T cells. For B cells to be able to differentiate into plasma cells they require help from CD4 T cells. B cells bind antigen through the Ig on their cell surface and take up the antigen. They then process the antigen and express antigen peptides on class II MHC on the cell surface (Figure 11.3). This peptide/MHC complex is recognised by helper T cells which then provide signals to the B cell that stimulate it to divide and differentiate into a plasma cell. If the B cell is specific for foreign antigen there will be CD4 T cells specific for that foreign antigen. However, if the B cell is specific for self-antigen, even if it can process the antigen and present self-peptide on its class II MHC there will be no self-specific CD4 T cells present and therefore the B cell will not get the signals to become a plasma cell. In fact, in the absence of these T cell signals the B cell will die by apoptosis.

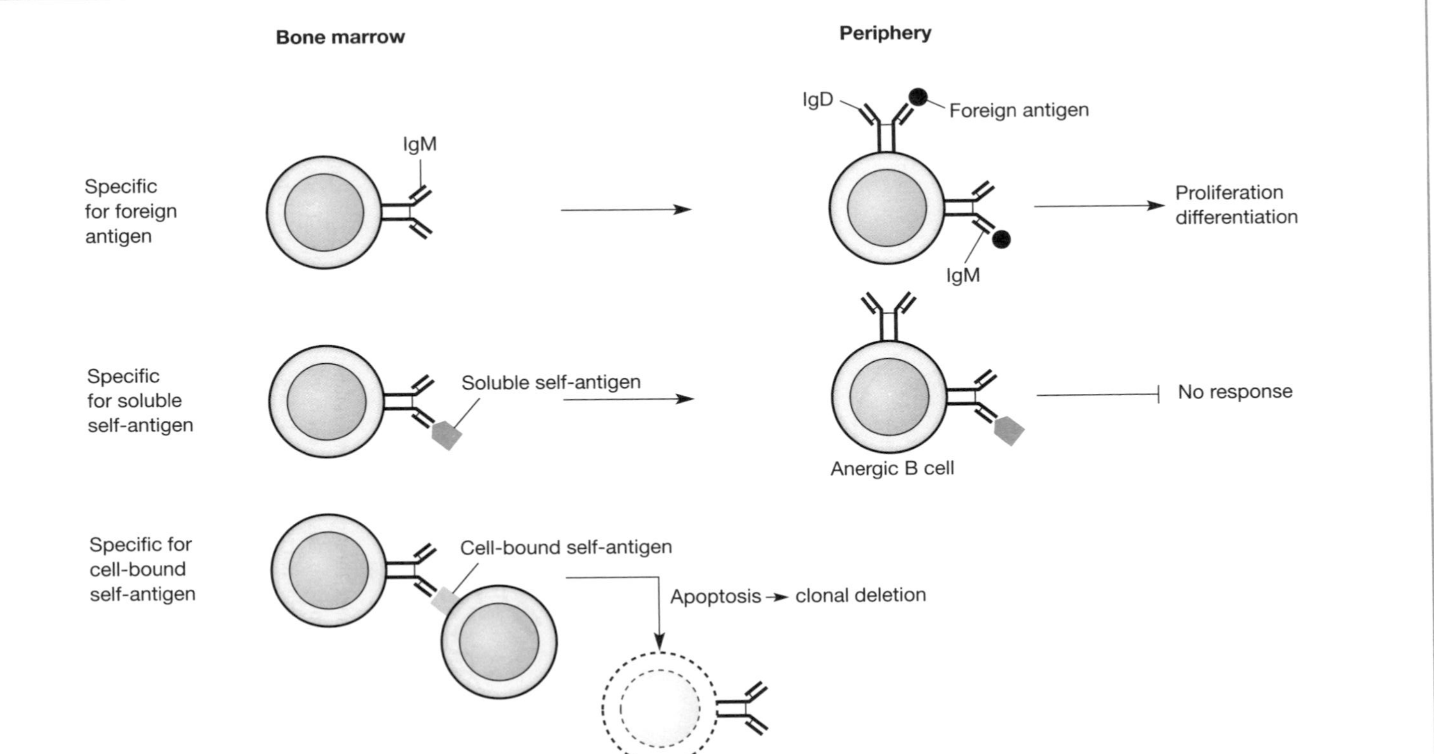

Figure 11.2 Tolerance in developing B cells. Immature B cells expressing IgM, but not IgD, are susceptible to tolerance induction either by being rendered anergic if their membrane IgM binds soluble self-antigen, or by apoptosis if their membrane IgM binds cell-bound self-antigen in the bone marrow.

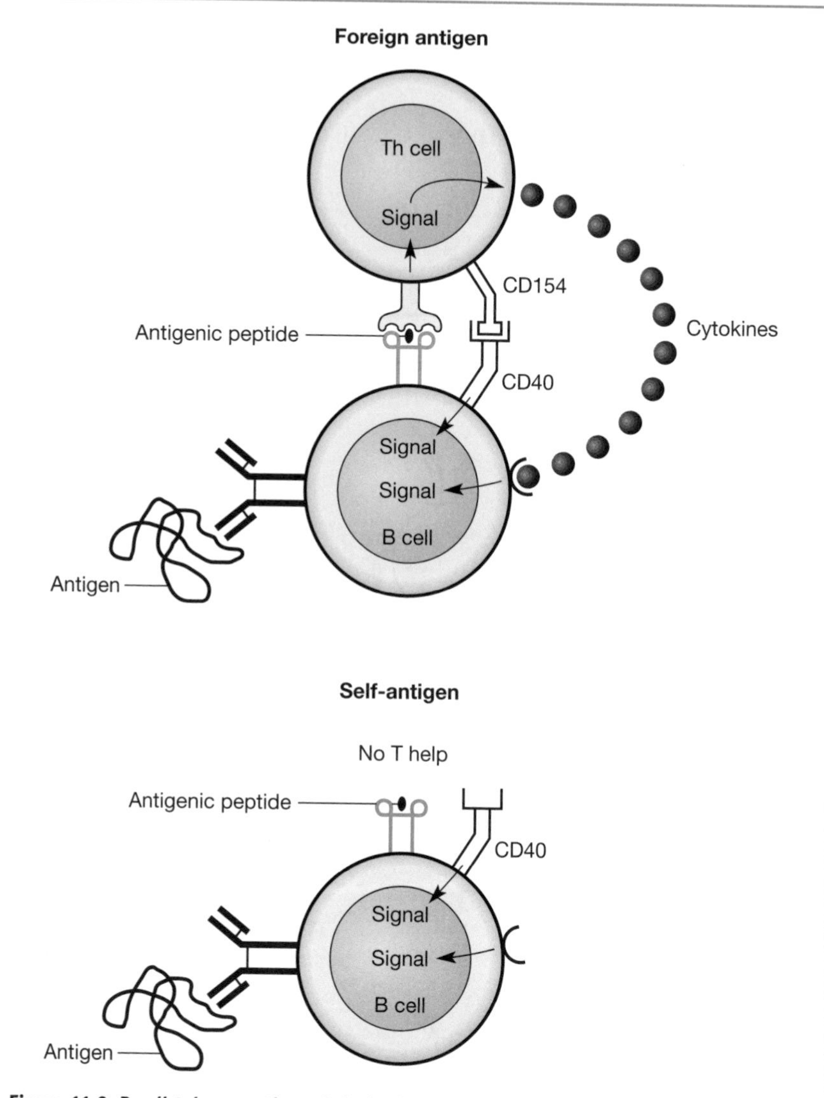

Figure 11.3 B cell tolerance through lack of T cell help. B cells recognise antigen, process it and present peptides on their class II MHC for recognition by helper T cells (Ths). If the B cell is specific for foreign antigen there will be Th that can recognise the foreign peptide and help the B cell become a plasma cell. However if the B cell is specific for self-antigen there will be Th and so the B cell will die by apoptosis.

11.3 Self-tolerance in T lymphocytes – selecting for recognition of self-MHC but not self-antigen

The development of T cells poses some additional problems not seen with B cells. Since B cells recognise free antigen there is a simple distinction to be made with every developing B cell and that is whether the Ig it expresses is specific for self-antigen or foreign antigen.

T cells recognise antigen in an MHC-restricted manner. Some of the TCR binds to MHC and some binds to the peptide (see Section 4.4). Because the rearrangement of TCR genes is random, T cells can be produced with four potential types of specificity in terms of MHC and antigenic peptide. The TCR could be specific for:

1. Foreign antigenic peptide + self-MHC.
2. Self-antigenic peptide + self-MHC.
3. Foreign antigenic peptide + foreign MHC.
4. Self-antigenic peptide + foreign MHC.

Of the four specificity patterns, only cells that are specific for foreign antigenic peptide plus self-MHC (pattern 1) will be useful at recognising antigens derived from pathogens. Cells with specificity for self-antigenic peptide plus self-MHC (pattern 2) are potentially damaging and could cause an immune response against the body's own antigens. Cells that can only recognise antigen in association with foreign MHC (patterns 3 and 4) will be useless in that particular individual because they cannot recognise antigen peptide in association with the individual's own MHC. While these cells would not be harmful to the individual, if they were not got rid of they would clog up the immune system with lots of useless cells. Therefore during T cell development there is the requirement (i) to select T cells that are self-MHC-restricted, a process called positive selection, and (ii) to prevent the production of potentially damaging T cells with specificity for self-antigenic peptide and self-MHC, called negative selection. These are accomplished by special mechanisms that operate during T cell development in the thymus.

11.3.1 Positive and negative selection of thymocytes: thymic education

When thymocytes are developing in the thymus (see Chapter 5) they rearrange their TCR genes and express TCRs on their surface. As they continue their development they will encounter other cell types in the thymus that are expressing MHC and peptides. How the TCR interacts with this MHC/peptide will decide the fate of the developing thymocyte.

Positive selection – getting rid of useless cells

The term 'positive selection' is used to describe the processes by which thymocytes that are specific for self-MHC are distinguished from those that are not specific for self-MHC (Figure 11.4). During thymocyte development, double-positive thymocytes, which have rearranged their TCR genes and express a receptor for antigen, encounter class I or class II MHC on the surface of **thymic cortical epithelial cells**. Thymocytes whose TCR can bind the self-MHC expressed on the epithelial cells receive survival signals and proceed to the next stage of differentiation. Thymocytes whose TCRs cannot bind self-MHC do not receive the survival signals and die by apoptosis. The apoptotic cells are rapidly phagocytosed by thymic macrophages. This process means that only thymocytes whose TCR can bind self-MHC survive, but it does not distinguish between thymocytes that are specific for self-antigenic peptide + self-MHC and thymocytes that are specific for foreign antigenic peptide + self-MHC. Therefore a process is necessary to deal with thymocytes with specificity for self-antigen. Thymocytes with specificity for self-antigen and self-MHC can undergo one of two fates when they reach the thymic medulla. They can be clonally deleted and die in the thymus; this process is called negative selection and can happen to both developing CD4 and CD8 T cells. CD4 T cells can undergo an alternative fate if they encounter self-class II MHC/self-antigen; they can be induced to become regulatory CD4 T cells whose role is to prevent potentially self-reactive lymphocytes from actually responding against self-antigen.

Negative selection – getting rid of self-reactive T cells

The area of the thymus at the cortico-medullary junction contains many dendritic cells expressing both class I and class II MHC, which can present peptides derived from self-antigens to the developing thymocytes (Figure 11.4). Developing thymocytes can encounter the dendritic cells and if they react with self-antigenic peptide plus self-MHC with high enough affinity they are stimulated to undergo apoptosis and are phagocytosed. Thymocytes that do not react with self-antigenic peptide + self-MHC are allowed to finish the maturation process and emerge from the thymus as mature CD4 or CD8 T cells.

Clonal deletion is very important in preventing our immune systems from attacking our own bodies. However, on its own it is not sufficient to prevent all self-reactive T cells from being produced and leaving the thymus as mature CD4 or CD8 T cells. Therefore the immune system has evolved other mechanisms to prevent self-reactive T cells from attacking our bodies. One of these mechanisms is the generation of CD4 T cells called regulatory CD4 T cells.

11.3.2 Some self-reactive CD4 T cells become regulatory cells

Some self-reactive thymocytes do not encounter self-antigen being presented by dendritic cells but instead encounter self-antigen being presented on class II MHC of another cell type in the thymic medulla called the **thymic medullary epithelial cell**, or MTEC for short. When thymocytes recognise self-antigen/MHC on a MTEC they are not stimulated to undergo apoptosis but instead are induced to become **regulatory CD4 T cells**, known also as Tregs.

These self-antigen-specific Tregs are very important in preventing immune attack against our bodies. There are some people who have genetic mutations that mean they cannot make Tregs. These individuals

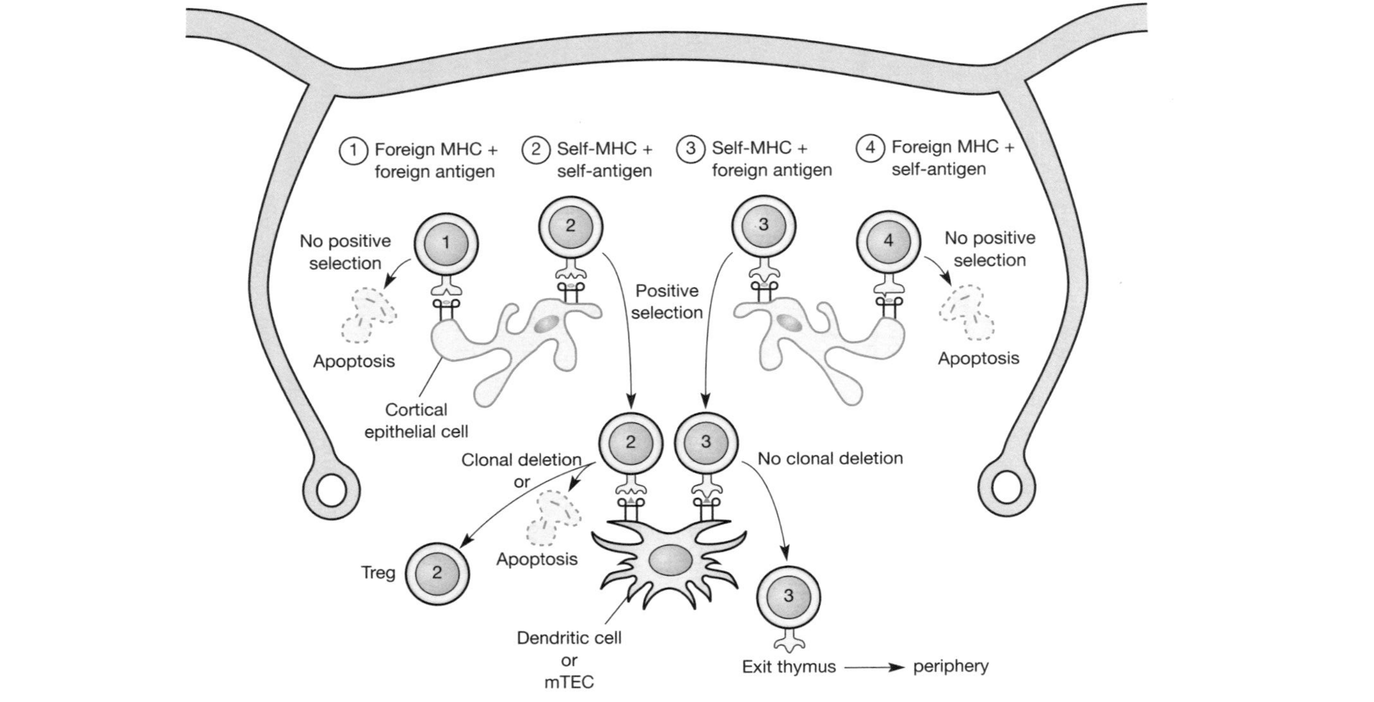

Figure 11.4 Positive and negative selection of thymocytes. *Positive selection:* developing thymocytes that have expressed a TCR encounter cortical epithelial cells expressing class I and class II self-MHC; those thymocytes (cells 2 and 3) whose TCR can bind self-MHC receive a survival signal and survive while those thymocytes whose TCR cannot bind self-MHC (cells 1 and 4) do not receive the survival signal and die by apoptosis. *Negative selection:* thymocytes with specificity for self-MHC encounter dendritic cells around the cortico-medullary junction; these dendritic cells are presenting self-antigenic peptides on their class I and class II MHC. Thymocytes that bind to MHC/self-antigenic peptide with high affinity (cell 2) are stimulated to die by apoptosis; thymocytes that are not specific for self-antigen do not bind self-antigen/MHC (cell 3) and leave the thymus as naïve CD4 or CD8 T cells with specificity for foreign antigen in association with self-MHC.

suffer from a disorder called IPEX (immunodysregulation polyendocrinopathy enteropathy X-linked) where they suffer from autoimmune diseases such as diabetes and thyroiditis plus other immune disorders.

11.4 How do we maintain tolerance to self-antigens not expressed in the thymus?

Although the thymic processes of clonal deletion and induction of Tregs are vital in preventing autoimmune disease they are not in themselves sufficient. This is because not all self-antigens are expressed in the thymus. The self-antigens that thymic dendritic or medullary epithelial cells express on their MHC must either be made in the thymus or travel to the thymus in the bloodstream. In fact, as knowledge of regulation of gene expression increased it was assumed that most self-antigens would not be expressed in the thymus because they are proteins that are made only in specific tissues or organs and are not secreted and therefore do not enter the bloodstream. An example of this would be the thyroid-stimulating hormone receptor, which is only expressed on the surface of thyroid epithelial cells. The self-antigens that are expressed outside the thymus have been called peripheral antigens.

Recent evidence has indicated that thymic medullary epithelial cells may express proteins that would normally be regarded as tissue-specific extra-thymic proteins. One of the first demonstrations of this was that of insulin expression in the thymus; previously it was thought that insulin was only produced by the β cells in the islets of Langerhans of the pancreas. Recently it has been demonstrated that a gene called AIRE (autoimmunity associated regulatory element) is important for thymic expression of tissue-specific genes. The AIRE gene codes for a transcriptional regulatory protein and mice and humans lacking a functional AIRE gene have reduced expression of extra-thymic proteins by their thymic medullary epithelial cells. As a consequence of this they suffer from a number of autoimmune disorders. Therefore there may be more self-proteins expressed in the thymus than was previously thought. However, despite the effects of the AIRE gene there will still be many self-proteins of an individual that are not expressed in the thymus. This means that T cells specific for these self-antigens will be produced in the thymus and enter into the bloodstream and lymphoid system. It is therefore important to have mechanisms that stop these self-reactive T cells from being activated and mounting an immune response against self-antigens. Tolerance induced in T cells that have left the thymus and entered the lymphoid system is known as peripheral tolerance and there are a number of different mechanisms.

11.4.1 Clonal anergy – turning T cells off

It was described in Chapter 7 how CD4 T cells need two signals to be activated by antigen. The first of these signals is through the TCR following recognition of specific antigen in association with MHC. The second signal is through co-stimulatory molecules, especially CD28 on the T cell, which binds to CD80 or CD86 on the antigen-presenting cell. If a CD4 T cell recognises antigen/MHC without receiving the co-stimulus, not only does it fail to be activated but it is made **anergic** (Figure 11.5). This means that the CD4 T cell is rendered incapable of responding to antigen and is a particularly important mechanism for peripheral tolerance in CD4 T cells.

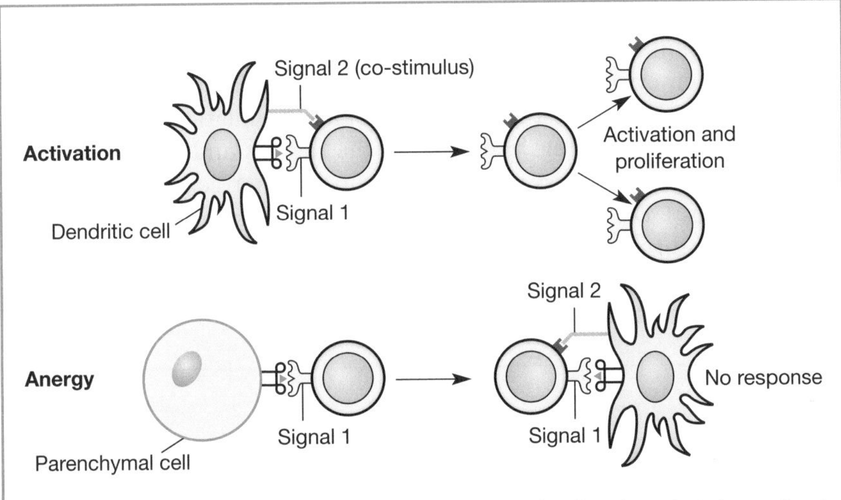

Figure 11.5 Anergy in peripheral T cells. T cells need two signals to be activated, one signal from the TCR-binding antigen/MHC and a co-stimulus provided by the antigen-presenting cell (top). If the T cell receives signal 1 alone it is rendered anergic and cannot respond to a subsequent encounter with the same antigen even if it receives the co-stimulus (bottom).

Most parenchymal cells (i.e. cells, such as hepatocytes in the liver, that make up different tissues) do not express class II MHC. Therefore even if they have peripheral antigen they cannot present them to CD4 T cells so there is no risk of stimulating autoimmune CD4 T cells. Many cell types can be induced to express class II MHC – e.g. by IFN-γ – so that during an immune response to a virus in a tissue the parenchymal cells could be induced to express class II MHC. These cells now have the capacity to present peripheral self-antigen to CD4 T cells. However, these parenchymal cells do not express co-stimulatory molecules and therefore cannot deliver co-stimulatory signals. CD4 T cells that recognise a tissue-specific self-antigen being presented by class II MHC on a parenchymal cell will therefore be anergised, so preventing the activation of the autoreactive cells.

11.4.2 Induction of Tregs outside of the thymus

Tregs that have been produced in the thymus are called natural Tregs or nTregs. However CD4 T-cells that reside in the peripheral lymphoid tissue such as lymph nodes or spleen can be induced to become Tregs when they are stimulated with antigen (Figure 11.6). These are called induced Tregs or iTregs, and iTregs can be induced against self- or foreign antigen. Those induced against self-antigen are important in preventing autoimmunity while those induced against foreign antigen are there to make sure that the immune response against a pathogen does not go 'over the top' and become excessive. Excessive immune responses, even if they are against pathogens, can cause damage; for instance some types of inflammatory bowel disease are caused by excessive immune responses against gut bacteria.

Just as cytokines can influence whether CD4 T cells become Th1, Th2 or Th17 cells, then cytokines and other factors present can cause CD4 T cells to become Tregs. The most important cytokine stimulating iTreg development is TGFβ. Both nTregs produced in the thymus and iTregs produced in peripheral lymphoid tissue can inhibit many other cell types including dendritic cells, CD4 and CD8 T cells, B cells and natural killer cells. They inhibit either through cell contact with the cell being inhibited (although the mechanisms are poorly understood) or through the secretion of inhibitory cytokines such as TGFβ and IL-10 (Figure 11.6).

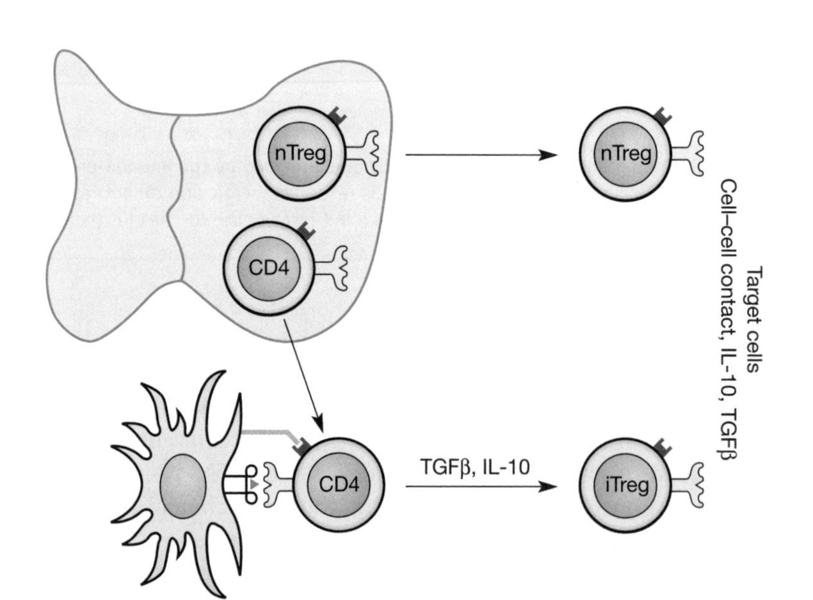

Figure 11.6 Production and action of regulatory CD4 T cells (Tregs) Tregs can be produced in the thymus (nTregs) or generated from peripheral CD4 T cells (iTregs). They can inhibit other cell types through cell–cell contact or by the secretion of soluble factors (cytokines) such as IL-10 and TGF-β.

11.4.3 Clonal ignorance

The term 'clonal ignorance' is used to describe the situation where self-antigen-specific T cells fail to recognise the self-antigen because it is present in too low concentrations or because it is sequestered away from the immune system. An example of the latter is lens protein in the eye. Normally this is not exposed to the immune system but if one eye is damaged (e.g. by a squash ball) in some cases CD4 T cells that are specific for the lens protein and are not tolerant become activated and react against both the damaged eye and the undamaged eye. This condition is called reactive sympathetic opthalmia and can lead to blindness if untreated.

11.5 Summary

- Because of the random nature of Ig and TCR gene rearrangement B and T cells with specificity for self-antigens will be produced during lymphocyte development
- Immature B cells, expressing IgM but not IgD, are stimulated to undergo clonal deletion if their Ig is specific for self-antigen and it binds to self-antigen in the bone marrow.
- Many self-reactive B cells do not see self-antigen in the bone marrow and enter the pool of lymphoid cells. The main mechanism stopping these self-reactive B cells from being activated and becoming plasma cells is a lack of CD4 T cells specific for the self-antigen.
- Thymocytes express TCR that are random in specificity in terms of both antigenic peptide and MHC specificity. There processes which select those T cells that are specific for self-MHC and foreign antigen.
- The first stage is positive selection where only those thymocytes with specificity for self-MHC survive.
- Surviving thymocytes can then undergo negative selection where any cells that recognise self-antigen plus self-MHC die by apoptosis. Other self-reactive CD4 T cells become regulatory CD4 T cells. T cells that recognise self-MHC and do not encounter self-antigen in the thymus leave the thymus as either CD4 or CD8 naïve T cells.
- Because not all self-antigens are expressed in the thymus, mechanisms of peripheral tolerance exist. These include clonal anergy, clonal ignorance and active regulation by Tregs generated from peripheral CD4 T cells following stimulation with antigen/MHC.

11.6 Questions

1) What happens to a developing immature B cell (IgM+IgD–) that recognises a self-antigen in the bone marrow?

2) Describe what is meant by 'positive selection' and 'negative selection' of developing thymocytes. Why are they important?

3) What is meant by the terms 'central' and 'peripheral' tolerance in T cells? Why do you need both?

4) What are the differences between natural regulatory T cells (nTregs) and induced regulatory T cells (iTregs)?

5) What factors determine whether a CD4 T cell is activated or becomes anergic when its TCR recognises ClII/MHC-peptide? How does anergy help in preventing autoimmunity?

The answers to these questions can be found on page 339.

11.7 Further reading

1) Goodnow CC, Sprent J, Fazekas de St Groth B, Vinuesa GG. (2005) Cellular and genetic mechanisms of self-tolerance. *Nature* 435:590–597.

2) Hogquist KA, Baldwin TA, Jameson SC. (2005) Central tolerance: learning self-control in the thymus. *Nature Reviews Immunology* 5:772–782.

3) Sakaguchi S, Miyara M, Costantino CM, Hafler DA. (2010) FOXP3+ regulatory T cells in the human immune system. *Nature Reviews Immunology* 10:490–500.

CHAPTER TWELVE

Autoimmune diseases

Learning objectives

To understand what is meant by an autoimmune disease. To know about the spectrum and nature of autoimmune diseases. To understand how autoimmune responses cause disease. To know that both genetics and the environment contribute to the development of autoimmunity. To understand how tolerance may be lost.

Key topics

- Definition of autoimmune diseases and clinical burden
- Classification of autoimmune diseases
- Immunological features of autoimmune disease
  - Antibody responses
  - T cell responses
- Aetiology of autoimmune diseases
  - Genetic factors
  - Environmental factors
- Loss of tolerance

12.1 Autoimmune diseases occur when our immune systems attack our own bodies

Autoimmune diseases are diseases that involve an immune response against one or more self-antigens. These self-antigens are usually proteins that constitute part of the body; less often they are carbohydrates, lipids or DNA. The self-antigens that the immune system responds to in an autoimmune disease are called **autoantigens** and the immune response against an autoantigen is called an **autoimmune response**.

The first disease to be identified as being autoimmune in origin was Hashimoto's thyroiditis in the mid-1950s. In this disease antibodies are produced against thyroglobulin and other thyroid-associated antigens. Antibodies against self-antigens are called **autoantibodies** and lymphocytes whose antigen receptors are specific for a self-antigen are called **autoreactive cells**. Nearly all autoimmune diseases involve the production of autoreactive CD4 T cells, which are also called **autoreactive Th cells**. Depending on the autoimmune disease there may also be the production of autoreactive B cells and/or autoreactive CD8 T cells. In Hashimoto's thyroiditis there is also an extensive infiltrate of lymphocytes and monocytes into the thyroid (Figure 12.1). The result of anti-thyroid antibody production and the thyroid infiltrate is the destruction of a large proportion of the thyroid glandular tissue, leading to underfunctioning of the thyroid, or hypothyroidism.

Hashimoto's thyroiditis demonstrates another feature of autoimmune diseases – the occurrence of **primary pathology** and **secondary pathology**. Primary pathology is the direct consequence of the autoimmune response; in the case of Hashimoto's thyroiditis this is destruction of thyroid tissue. The secondary pathology is a consequence of the altered tissue function caused by the primary pathology. Again, using Hashimoto's thyroiditis as an example, the secondary pathology due to hypothyroidism differs depending on the severity of the disease and the age at which it occurs. In infants it can cause cretinism, which is the retardation of physical and mental growth. In adults, some of the symptoms are fatigue, intolerance of the cold, dry skin and, in severe cases, mental impairment.

12.2 There are many different autoimmune diseases

The number of different autoimmune diseases is huge and essentially every tissue and organ can be affected by an autoimmune disease. Table 12.1 is a partial list of autoimmune diseases. Some of these diseases, such as type 1 insulin-dependent diabetes mellitus (T1DM), rheumatoid arthritis and multiple sclerosis, are familiar to most people; however, most autoimmune diseases are not that well-known. The full range of autoimmune diseases is probably not yet appreciated and there is speculation that other diseases whose cause is at present unknown, such as some forms of mental illness and alopecia (hair loss), may be due to autoimmunity. Autoimmune diseases are generally chronic in nature and can be quite debilitating; many of them are fatal if not treated.

The number of people affected by all autoimmune diseases is about 5% of the population in industrialised countries, so as a whole autoimmune diseases are quite common. The frequency of individual diseases varies considerably. Rheumatoid arthritis (RA) is the most common autoimmune disease, with an incidence of 1–2% of the population. Other autoimmune diseases are much rarer; pemphigus vulgaris, an autoimmune disease causing severe blistering of the skin, affects fewer than one in 100 000 people. It is not surprising that the more severe diseases with higher fatality rates usually have the lowest incidence. One issue of concern is that the incidence of autoimmune diseases has increased dramatically over the past 50 years or so, especially in industrialised countries.

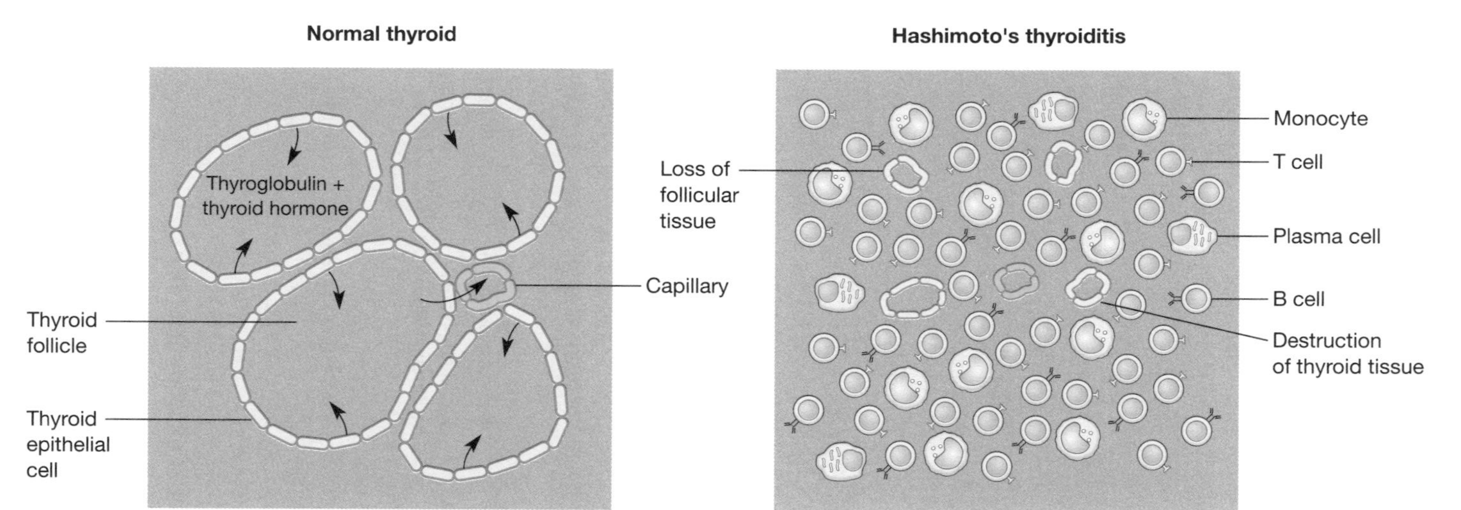

Figure 12.1 Thyroid infiltration and destruction in Hashimoto's thyroiditis. Normal thyroid consists of follicles containing thyroglobulin and thyroid hormone. The follicles are lined by thyroid epithelial cells. In Hashimoto's thyroiditis there is extensive infiltration of T cells, B cells, plasma cells and monocytes into the thyroid tissue, resulting in destruction of thyroid tissue and loss of thyroid hormone production.

Table 12.1 Spectrum of autoimmune diseases

Biological system affected	*Disease*	*Main effects*
Endocrine glands		
Thyroid	Hashimoto's thyroiditis	Thyroid destruction and underfunction
Thyroid	Grave's disease	Thyroid stimulation and overfunction
Islets of Langerhans (pancreas)	Insulin-dependent diabetes mellitus	Destruction of β cells (insulin-producing cells)
Adrenal gland	Addison's disease	Adrenal insufficiency
Haematopoietic system		
Red blood cells	Autoimmune haemolytic anaemia	Anaemia
Platelets	Autoimmune thrombocytopenia	Abnormal bleeding
Intrinsic factor (IF)	Pernicious anaemia	Autoantibody prevents absorption of vitamin B12
Nervous system		
Central nervous system	Multiple sclerosis	Progressive paralysis
Neuromuscular junction	Myasthenia gravis	Progressive muscle weakness
Skin		
Nuclear antigens	Scleroderma	Fibrosis of skin
Epidermal cell junctions	Pemphigus vulgaris	Severe blistering
Joints		
Synovium	Rheumatoid arthritis (RA)	Progressive destruction
Synovium	Systematic lupus erythromatosus (SLE)	Deformity
Kidney		
Basement membrane	Goodpasture's syndrome	Glomerulonephritis
Glomerulus	SLE	Glomerulonephritis

12.2.1 Classification of autoimmune diseases

The wide spectrum of known autoimmune diseases has led to attempts to classify them. One distinction that became apparent between different autoimmune diseases was that some affect only one organ while others are more widespread. Diseases such as T1DM and thyroiditis affect the β cells of the islets of Langerhans of the pancreas and the thyroid specifically; no other organs or tissues are directly affected by the autoimmune response, although other tissues may be affected by the loss of insulin production or thyroid function caused by the autoimmune destruction of the tissue. Other autoimmune diseases affect many tissues, for example systemic lupus erythromatosus (SLE) is an autoimmune disease affecting the skin, kidneys and joints. Autoimmune diseases such as T1DM or thyroiditis, in which one organ is affected, are called **organ-specific.** Diseases such as SLE, where many tissues are affected, are called **systemic** or **non-organ-specific.** (See Box 12.1 for more details on the classification of autoimmune diseases.)

BOX 12.1: CLASSIFICATION OF AUTOIMMUNE DISEASES

The classification of autoimmune diseases as organ-specific or systemic is based on whether one tissue or multiple tissues are affected. This classification is not strictly based on the distribution of the autoantigens that are stimulating the autoimmune response although it was assumed that autoantigens in organ-specific diseases were present only in the tissues affected whereas the autoantigens in systemic autoimmunity were more widespread. In many cases this is true.

The classification of autoimmune diseases as organ-specific and non-organ-specific is based on the distribution of the autoantigens involved in the various autoimmune responses. Autoimmune diseases in which the autoantigen is present only in the tissue affected are called organ-specific diseases. Again T1DM and thyroiditis would be classified as organ-specific. Non-organ-specific autoimmune diseases are those in which the autoantigens are widespread and not confined to the tissue affected. The best characterised non-organ-specific autoimmune disease is SLE in which autoantibodies are produced against antigens such as DNA, histones and phospholipids, which are present in all nucleated cells.

It can be seen that there is considerable overlap between the two classification systems. However, there are problems with both classification systems and although they generally agree as to which diseases are organ-specific and which are systemic or non-organ-specific, there can be some conflicts. These can be illustrated by the autoimmune disease primary biliary cirrhosis (PBC). This is an autoimmune disease in which the bile ducts of the liver are destroyed, leading to loss of liver function. Because only the liver is affected by the autoimmune response PBC could be classified as an organ-specific autoimmune disease. However, the main autoantigens in PBC are mitochondrial antigens, which are present in almost all cells; this would classify the disease as non-organ-specific.

As more autoantigens are identified it is becoming clear that the autoantigens in other organ-specific autoimmune diseases are not always present just in the tissue affected. In T1DM many of the autoantigens are present in tissue outside the pancreas but these are not affected by the autoimmune response. Why the same autoantigen expressed in multiple tissue sites should stimulate autoimmune disease in one site and not others is one of the big puzzles in autoimmunity.

12.3 Immunological features of autoimmune diseases

The types of immune responses in autoimmune disease are no different from the immune responses mounted against infectious pathogens During an autoimmune response the immune system produces effector cells and molecules that attack particular parts of the body. These effectors may include antibody, cytotoxic CD8 T cells and effector CD4 T cells. Therefore, depending on the autoimmune disease, one or more of the serological and histological features associated with an attack by the immune system on a particular tissue or tissues will be seen.

- **Serum autoantibodies**. The serum of many patients with autoimmune disease contains autoantibodies. These can be detected by immunofluorescence or immunocytochemical techniques (see Section 10.3.2). Sections of relevant tissue are incubated with patient's serum followed by fluorescent or enzyme-linked anti-human-Ig antibodies. Fluorescent antibody binding is detected by looking at the slides under a fluorescence microscope. Enzyme-linked antibody binding is visualised by incubation with a substrate producing a coloured product. (See Figure 12.2 and Plate 13.) As more autoantigens are identified it is becoming easier to measure autoantibodies using enzyme-linked immunosorbent assay (ELISA) (see Section 10.3.2).
- **Deposition of antibody and complement in affected tissue**. Another common feature of autoimmune diseases is the deposition of autoantibody in tissue. Autoantibody can bind directly to autoantigens in the tissue or it can be deposited in the form of immune complexes. Both of these situations can lead to complement fixation and activation, resulting in the generation of inflammatory responses. Deposition of antibody and complement can be detected by immunofluorescence or immunocytochemistry (see Figure 12.2 and Plate 13).
- **Infiltration of cells in the affected tissue.** In many autoimmune diseases a cellular infiltrate is seen in the affected organ or tissue (Plate 14). Typically the infiltrate will consist of lymphocytes and monocytes but other immune-related and inflammatory cells can sometimes be seen.

12.3.1 Some autoimmune diseases are caused by autoantibodies

In some autoimmune diseases autoantibody is the only or main autoimmune feature of the disease and the pathology can be completely explained by the actions of the autoantibody. Therefore it is easy to classify these as antibody-mediated autoimmune diseases. Different classes of autoantibody, with different functions, can be produced in autoimmune diseases and cause pathology in a number of ways:

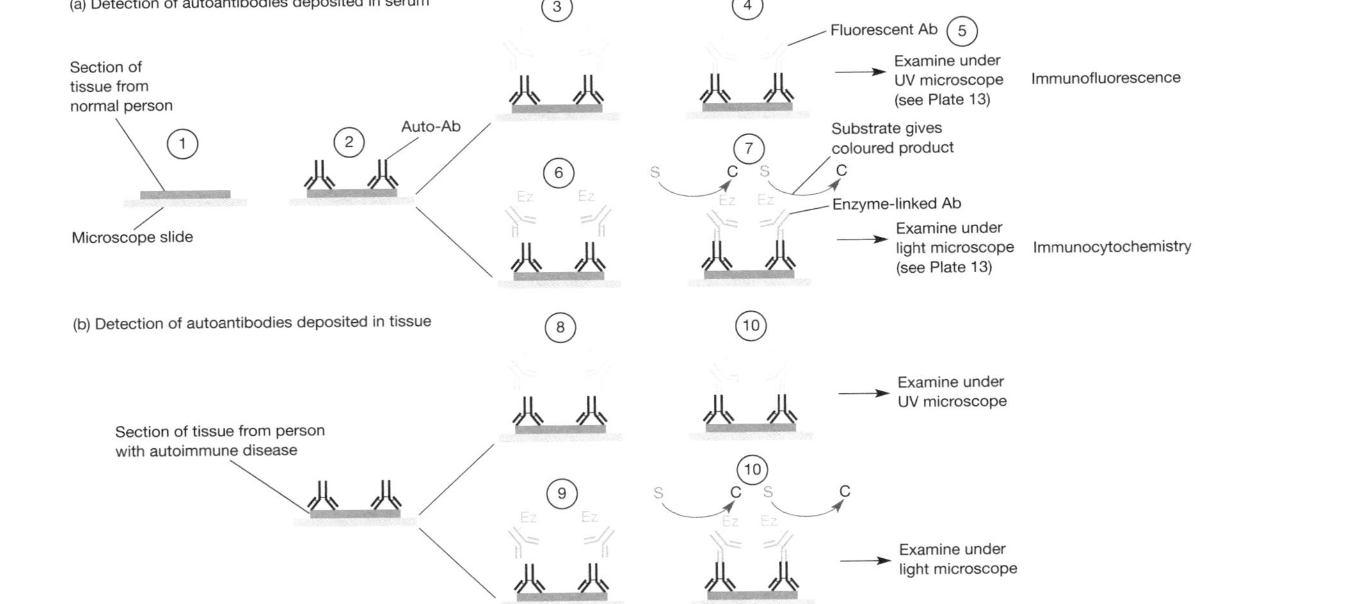

Figure 12.2 Detection of autoantibody by immunocytochemistry. (a) Autoantibody in serum can be detected by adding the serum to a section of normal tissue expressing the autoantigen (1). If autoantibodies are present in the serum they will bind to the autoantigens in the tissue section (2). In the immunofluorescence technique, a fluorescein-conjugated anti-human-Ig antibody is added to the section (3). This will bind to any autoantibody bound to the tissue section (4) and the antibody can be visualised by examining the section under an ultraviolet (UV) microscope (5). An alternative to immunofluorescence is immunocytochemistry. Instead of a fluorescein-conjugated antibody, an enzyme-linked anti-human-Ig antibody is used (6). The binding of the enzyme-linked antibody can be detected by adding a colourless substrate (S), which is turned into a coloured product (C) by the enzyme linked to the antibody (7). (b) Immunofluorescence or immunocytochemistry can also be used to detect autoantibody that has been deposited in a tissue. In this case the fluorescein (8) or enzyme-linked (9) anti-human-Ig antibody is added to a tissue section from the person with the autoimmune disease without prior incubation with the patient's serum. The anti-human-Ig antibody will bind to autoantibody already bound to autoantigens in the tissue bk and the antibody can be visualised under UV microscopy or following the addition of substrate as in (a).

- **Complement-mediated lysis.** Autoantibody binding to red blood cells can result in complement fixation and lysis of the red cells, leading to autoimmune haemolytic anaemia (Figure 12.3).
- **Opsonisation.** In autoimmune thrombocytopenia autoantibody binds to platelets and promotes their opsonisation by phagocytes in the liver and spleen (Figure 12.3). This leads to platelet deficiency (thrombocytopenia) and bleeding problems.
- **Inhibition of receptor function.** The disease myasthenia gravis involves the production of antibodies against the acetyl choline receptor, which is present on muscle fibres of the neuromuscular junction. The autoantibodies bind to the acetyl choline receptors and prevent the binding of acetyl choline released at the nerve endings. This blocks transmission of signals across the neuromuscular junction, leading to muscle weakness (Figure 12.4). The autoantibody can also cause internalisation of the acetyl choline receptors by the muscle cells, further reducing the efficiency of transmission across the neuromuscular junction.
- **Stimulation of receptors.** For reasons that are not understood, in some cases antibodies against receptors can stimulate the receptor rather than block it. This happens in Grave's disease where autoantibodies are produced against the thyroid-stimulating hormone (TSH) receptor present on thyroid epithelial cells (Figure 12.4). Stimulation of these cells results in thyroid overactivity with the symptoms of hyper-thyroidism – nervousness, tiredness, weight loss despite a good appetite and proptosis (bulging of the eyes).
- **Blockage of biological function.** Some autoantibodies block the function of molecules other than receptors. Pernicious anaemia results from autoantibodies against a protein called intrinsic factor. Intrinsic factor is produced in the stomach and binds to vitamin B_{12}, enabling B_{12} to be absorbed from the intestine (Figure 12.5). The autoantibodies against intrinsic factor prevent the binding of B_{12} and hence its absorption, resulting in vitamin B_{12} deficiency. The vitamin deficiency leads to a lack of red blood cells which causes tiredness and neurological changes which result in tingling.
- **Deposition of immune complexes.** In SLE, immune complexes fail to be cleared from the blood (see Section 8.6.4) and are deposited in various sites such as the kidney, skin and joints. The immune complexes then fix complement, which leads to an inflammatory response (see Chapter 2) and damage to the affected tissues. In SLE this results in kidney and joint damage and characteristic skin rashes (Plate 15).

12.3.2 For some autoimmune diseases the cause of damage is unknown

In many autoimmune diseases it is not possible to blame the pathology simply on the action of antibody. In these diseases, which include TIDM, Hashimoto's thyroiditis and rheumatoid arthritis, there is extensive

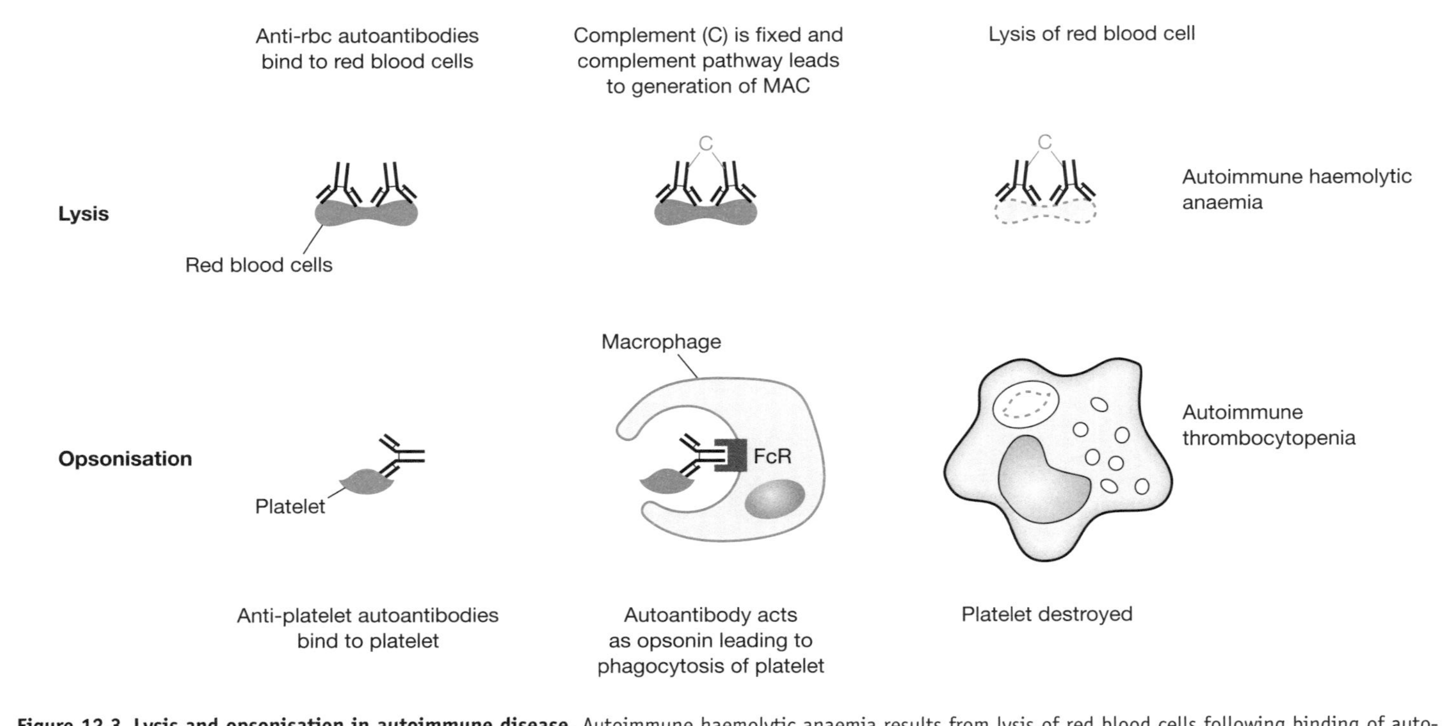

Figure 12.3 Lysis and opsonisation in autoimmune disease. Autoimmune haemolytic anaemia results from lysis of red blood cells following binding of autoantibodies to autoantigens on red blood cells and complement fixation. In autoimmune thrombocytopenia, binding of autoantibody to platelet autoantigens results in opsonisation and phagocytosis of the platelets and their destruction by phagocytes.

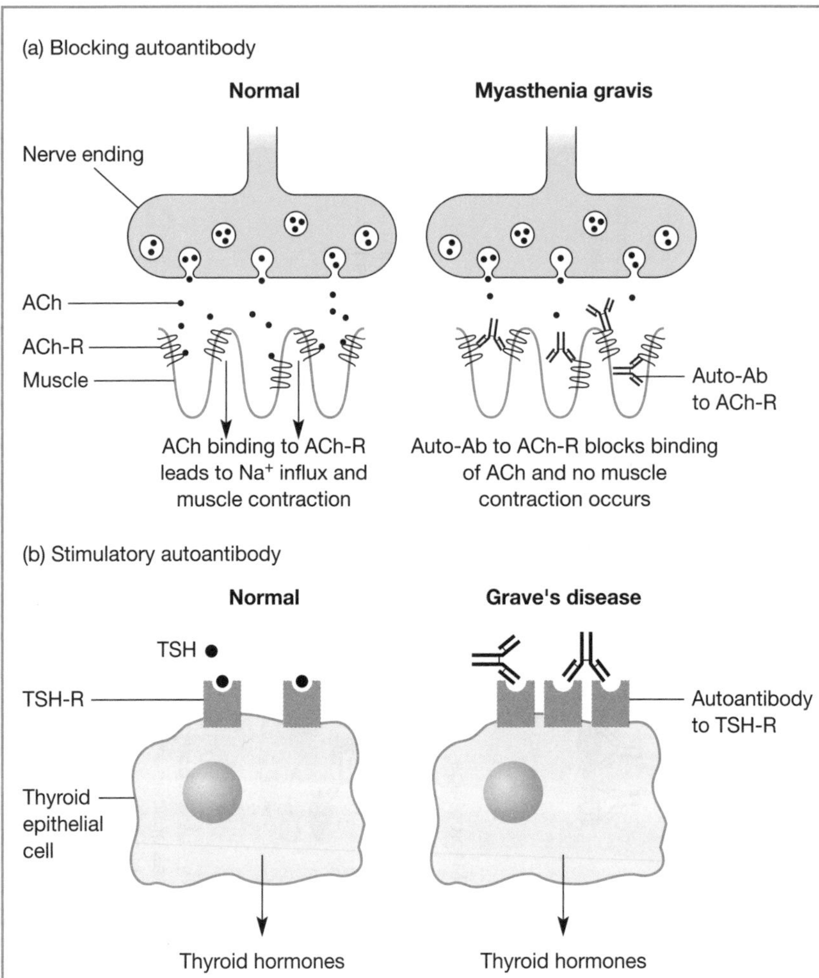

Figure 12.4 Antireceptor autoantibodies can block or stimulate function. In myasthenia gravis, autoantibodies are produced against the nicotinic acetyl-choline receptor present on muscle cells in neuromuscular junctions. The autoantibodies block the binding of acetyl-choline and prevent transmission of signals. The autoantibodies also cause internalisation and degradation of the ACh-R, reducing the number of receptors on the muscle cells. Autoantibodies against the thyroid-stimulating hormone receptor on thyroid epithelial cells stimulate the receptor rather than blocking it. This results in excessive secretion of thyroid hormones by the thyroid cells.

infiltration of immune-associated cells into the affected tissue and the production of autoantibodies (see above). The primary pathology in these diseases is damage to, or destruction of, part of the tissue, but the exact cause of the damage is hard to determine. Often the cellular infiltrate consists of CD4 and CD8 T lymphocytes, B cells, monocytes/macrophages and other inflammatory cells. Therefore the potential exists for any of the immune effector mechanisms described in Chapters 8 and 9 to be responsible for the actual

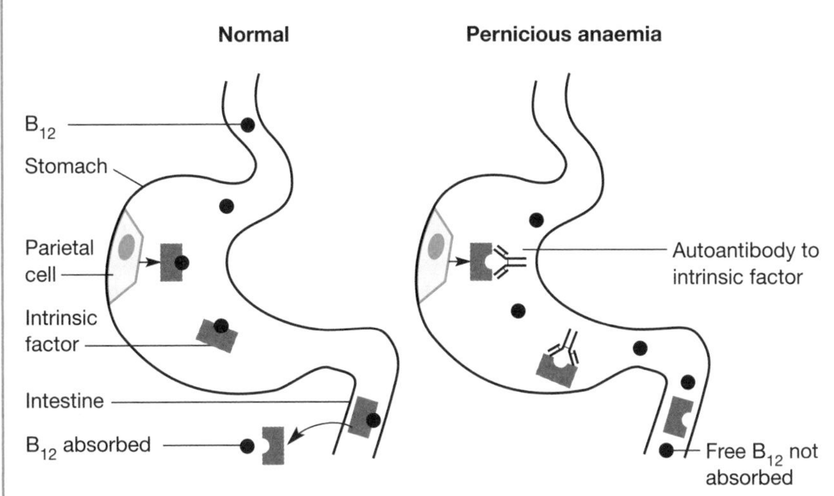

Figure 12.5 Pernicious anaemia. Vitamin B_{12} is normally absorbed in the intestine following binding to intrinsic factor secreted by parietal cells of the stomach. Autoantibodies against intrinsic factor block the binding of vitamin B_{12} and it cannot be absorbed, resulting in vitamin B_{12} deficiency.

tissue damage. Autoantibody bound to cells in the tissue can fix complement, resulting in direct complement-mediated damage. Complement activation can also stimulate an inflammatory response with infiltration into the tissue and activation of neutrophils and monocytes, causing further tissue damage. Antibody can also act as an opsonin, promoting phagocytosis, or contribute to ADCC. CD8 T cells can kill target cells presenting autoantigens on their class I MHC molecules. CD4 T cells, particularly Th1 cells can act as effector cells in promoting a delayed-type hypersensitivity response leading to the recruitment and activation of monocytes that cause tissue damage not involving antibody. More recently Th17 cells have been implicated in some autoimmune diseases although their precise role is not clear.

The difficult task in these autoimmune diseases is sorting out the actual effector mechanism from the potential mechanisms. Unfortunately the presence of a particular cell type in an infiltrate, or the presence of autoantibody, does not necessarily mean it is contributing to the pathology. Most patients with T1DM have autoantibodies against β cell antigens and one question is 'what are these autoantibodies doing?'. One unfortunate child who was born with a genetic immunodeficiency that meant they could not make B cells, and obviously could not produce autoantibodies, went on to develop T1DM in adolescence. This showed that autoantibodies were not essential for T1DM but does not mean they are not doing anything if they are there. For many autoimmune diseases, including T1DM, the nature of the damaging autoimmune effector mechanism is not known. It is also quite possible that for many autoimmune diseases more than one effector mechanism contributes to the tissue damage.

12.4 Both genetic and environmental factors contribute to the development of autoimmune disease

The factors that contribute to the development of autoimmune disease are very complex. It has become clear in recent years that both genetic and environmental factors contribute to the development of autoimmune disease. Autoimmune diseases tend to run in families, so if one family member has an autoimmune disease there is an increased likelihood of another member having the disease. However, because families tend to live together, it is not always clear whether it is the sharing of genes or the environment that increases the incidence of an autoimmune disease in families.

Some of the best evidence for both genetic and environmental contributions to the development of autoimmune disease comes from the study of concordance. In concordance studies a group of individuals with a particular disease are identified. The incidence of the disease in family members who are genetically more or less related to the disease group is compared to the incidence of the disease in unrelated individuals. The incidence of the disease in these different groups is known as 'concordance' and if concordance is higher in genetically more closely related family members this suggests that there is a genetic element to disease susceptibility. This is easiest to understand if you consider a situation where a disease, such as cystic fibrosis, is caused solely by inherited genetic factors. In this situation, if one identical twin gets the disease the other twin will always get the disease because they are genetically identical. The concordance in this situation is 100%. Non-identical twins, or siblings from different pregnancies, share on average 50% of their genes and therefore the concordance for a purely genetically determined disease would be 50% if one gene caused the disease and less if more than one gene was involved. Because twins are usually brought up in the same household and environment it is safe to assume that the difference in concordance in identical and non-identical twins is due to genetic factors.

If a disease is due to a combination of genetic and environmental factors the concordance for identical twins will be less than 100% but it will still be higher than for non-identical twins. This is the case with autoimmune diseases where concordance between identical twins is about 25% and that for non-identical twins is nearer 5%. The exact percentages vary from autoimmune disease to autoimmune disease but identical twins always show higher concordance than non-identical twins. These studies show that susceptibility to autoimmune disease can be inherited and therefore has a genetic element.

Concordance studies also show another very important feature of autoimmune diseases. As mentioned above, if a disease was 100% due to inherited genetic factors the concordance between identical twins would be 100%. The fact that in studies of autoimmune disease the concordance in identical twins is much less than 100% indicates that environmental factors are also contributing to the development of autoimmune diseases. Other evidence also points to environmental factors contributing to the development of autoimmune disease. The incidence of autoimmune disease has risen dramatically in the past 40–50 years, especially in North America, Europe and Australasia. Although some of this increase may be due to better diagnosis it is clear that there is a real increase and that this has happened too quickly to be explained by genetic changes in the population. Furthermore, populations from countries with a low incidence of certain autoimmune diseases, such as Japan and Papua New Guinea, have seen a rapid rise in the incidence of the autoimmune diseases when they have moved to countries with a higher incidence; again this cannot be explained by genetic factors and must be due to environmental effects.

12.4.1 Many genes contribute to the development of autoimmune disease

Genetic studies have shown that autoimmune diseases are multigenic. This means that many genes contribute to susceptibility to the disease. This is different from diseases, such as cystic fibrosis, which are caused by mutations in a single gene.

So what do we mean when we say a gene contributes towards susceptibility to a disease? Many genes are polymorphic; that is they are present in the population in more than one form. Each form of a particular gene is called an allele. Polymorphism can occur in the part of the gene coding for the protein. This polymorphism can lead to different forms of the protein being made and is called **structural polymorphism.** The most prominent form of structural polymorphism is in the MHC genes where many hundreds of different forms of the protein can be made (see Chapter 4). The second type of polymorphism occurs in the part of the gene not coding for the protein. This does not affect the structure of the protein but, if the polymorphism is in the promoter/enhancer part of the gene, it can alter the level of expression of the gene and affect how much of the protein is produced. This is called **non-structural polymorphism.** Polymorphism can therefore affect the biological activity of a protein in two ways: (i) by causing the production of different forms of the protein with different levels of biological activity and (ii) by affecting the amount of protein produced (Figure 12.6). If it can be shown that people with one allele of a polymorphic gene have a different incidence of a disease than people with a different allele of the gene, the gene is clearly affecting susceptibility to the disease and is called a susceptibility gene. This is presumably due to the difference in amount or activity of the gene product.

The degree to which a particular allele of a gene increases susceptibility to a disease can be quantified by calculating what is known as the relative risk, or RR. The RR compares the frequency of a particular allele of a gene in a population with a disease with the frequency of the allele in

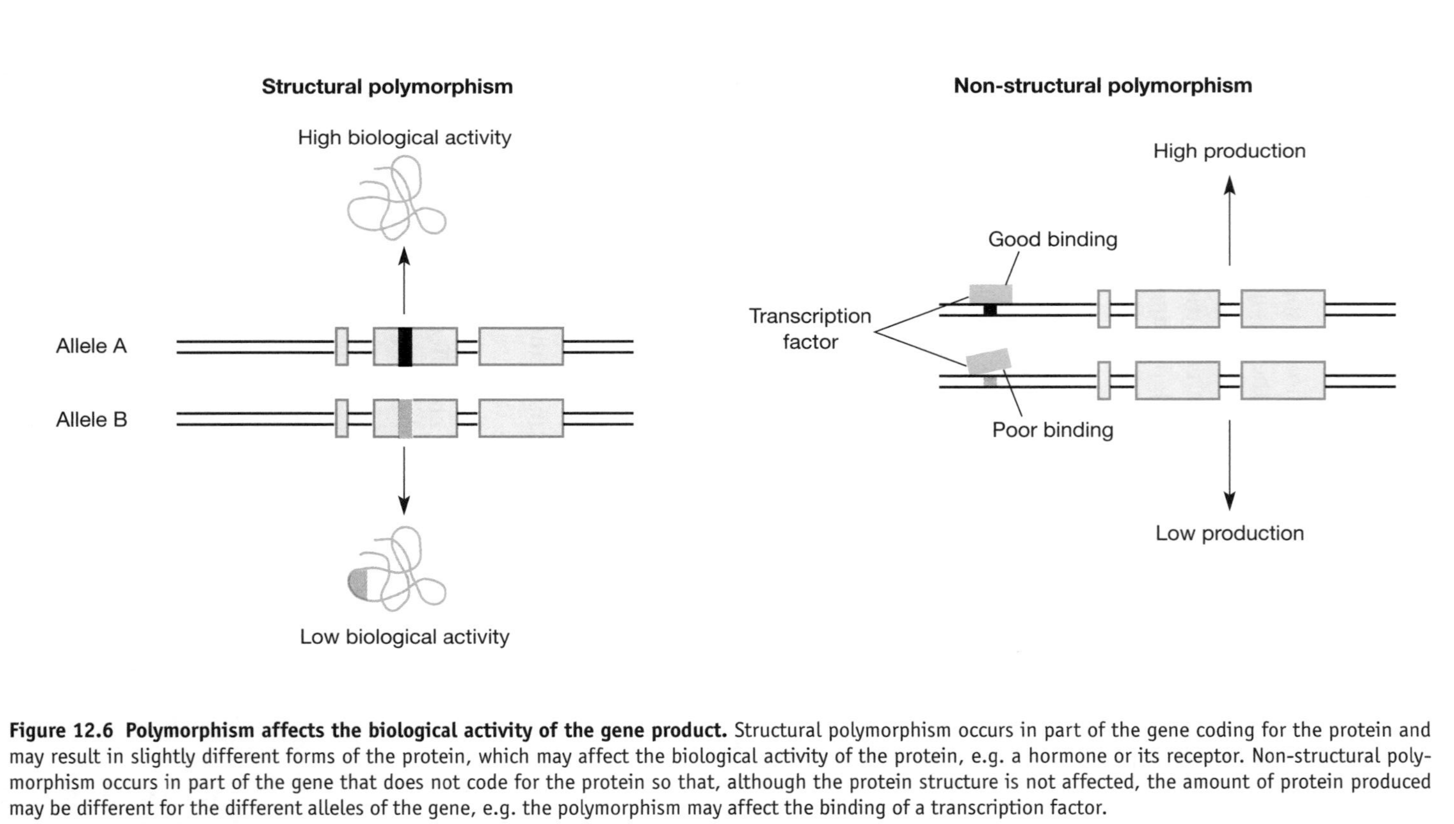

Figure 12.6 Polymorphism affects the biological activity of the gene product. Structural polymorphism occurs in part of the gene coding for the protein and may result in slightly different forms of the protein, which may affect the biological activity of the protein, e.g. a hormone or its receptor. Non-structural polymorphism occurs in part of the gene that does not code for the protein so that, although the protein structure is not affected, the amount of protein produced may be different for the different alleles of the gene, e.g. the polymorphism may affect the binding of a transcription factor.

the population as a whole. Although the actual calculation is more complicated the relative risk can be seen as the following ratio:

$$RR = \frac{\text{Frequency of allele in population with autoimmune disease}}{\text{Frequency of allele in population without autoimmune disease}}$$

Modern advances in DNA technology have made it possible to perform genome-wide screening for susceptibility genes in autoimmune diseases. These screens, combined with animal studies, show that in some autoimmune diseases as many as 50 genes may contribute to susceptibility. These susceptibility genes can be divided into two categories – MHC and non-MHC.

MHC genes

It was realised many years ago that one of the strongest associations between genes and autoimmunity was with the MHC genes, especially class II MHC. This is perhaps not too surprising given that class II MHC proteins present antigen to CD4 T cells. Autoimmune responses are almost always CD4 T cell-dependent in the same way that nearly all specific immune responses against foreign antigens require CD4 T cells for their generation. The relative risk for some HLA alleles with autoimmune disease is shown in Table 12.2. It has been estimated that MHC genes contribute about 50% of the total genetic risk of developing autoimmunity.

Non-MHC genes

Although the most significant susceptibility gene for nearly all autoimmune diseases is MHC, many other genes also contribute to susceptibility to autoimmunity. One set of genes that affects susceptibility to autoimmunity is the **sex-related** genes. Many autoimmune diseases are much more frequent in women than men: Hashimoto's thyroiditis is 50 times more frequent in females and SLE occurs 10 times more frequently in women. Although the immunological basis for this is unknown, experimental studies have shown that it is in part due to the effect of sex hormones on the immune system.

Table 12.2 MHC association with autoimmune diseases

Disease	*MHC association*	*Relative risk (RR)*
Goodpasture's syndrome	DR 2	16
Multiple sclerosis	DR 2	5
SLE	DR 3	6
Myasthenia gravis	DR 3	3
RA	DR 4	7
Hashimoto's thyroiditis	DR 5	3
T1DM	DR 3	3
T1DM	DR 4	3
T1DM	DR 3 + 4	14

Other genes are less well characterised. While new DNA technology has made it possible to identify the number of genes and their approximate chromosomal location, identifying the exact susceptibility gene is a time-consuming, arduous and expensive business. Additionally the non-MHC genes individually only affect susceptibility by a small amount so that the susceptibility allele has an RR of less than 2 compared to the other allele. This means that large sample sizes, in the thousands, are needed to detect these small differences. However more genes are being identified all the time and they seem to fall into two main categories. Some of the susceptibility genes, such as CTLA-4 and PTPN22 (a signalling protein in B and T cells) are involved in signalling or regulation of lymphocyte activity, especially CD4 T cells. These genes generally affect susceptibilty to a number of autoimmune diseases. Other non-MHC genes affect susceptibility to only one autoimmune disease and these are often genes coding for important autoantigens of the diseases. For example insulin is an important antigen in T1DM and individuals with T1DM often have autoantibodies and CD4 T cells that are specific for insulin. The insulin gene has a polymorphism in the non-coding part of the gene and different alleles of the insulin gene are associated with different susceptibility to T1DM. Similarly, the thyroid-stimulating hormone receptor (TSH-R) gene is a susceptibility gene in autoimmune thyroid disease and TSH-R is an important autoantigen in thyroid disease.

12.4.2 Environment and autoimmune disease

Again, while the evidence for environmental effects contributing to the development of autoimmune disease is substantial, it is much more difficult to identify what the environmental influences are. Part of the problem is the chronic nature of autoimmune diseases. An autoimmune response may be initiated by an environmental event and cause damage to tissue for a number of years before any clinical symptoms develop and the person sees a doctor. Therefore it will not be known exactly when the autoimmune response began and it will be difficult, if not impossible, to associate an environmental event with the triggering of an autoimmune response. Furthermore, people may not remember too clearly events that happened to them years ago. However, some behavioural traits are associated with increased incidence of a particular autoimmune disease and allow some environmental candidates to be identified. The main types of environmental factor associated with autoimmunity include the following:

- **Infectious agents.** The best-known example of infectious agents contributing to autoimmunity is rheumatic fever, where antibodies against streptococcal M-antigen react against heart myosin, joints and kidney, resulting in arthritis and heart disease (see Section 12.5.1).
- **Drugs.** Drugs or their metabolites can bind to self-antigens and make them appear foreign. This can result in the development of autoantibody

against the self-antigen itself (see Section 12.5.1). Examples of this are penicillin metabolites, which bind to red blood cells, and sedormid, which binds to platelets.

- **Toxins and pollutants.** In an autoimmune disease called Goodpasture's syndrome, autoantibodies are produced against type IV collagen, which is present in the basement membranes of the kidney and lung. All individuals with this disease suffer from glomerulonephritis, which leads to impaired kidney function. Cigarette smokers, but not non-smokers, also have a very high incidence of pulmonary haemorrhage, which can be fatal. This is thought to occur because the cigarette smoke damages the lungs, making the basement membrane accessible to the autoantibodies.
- **Food.** There have been a number of reports associating different foods (e.g. cows' milk and diabetes) with autoimmune diseases but none has been substantiated.

Because so few environmental agents have been identified it is difficult to know exactly how they contribute to the development of autoimmune disease. The general assumption is that they contribute to the loss of immunological tolerance that is the central feature of autoimmune diseases.

12.5 How is immunological tolerance lost in autoimmune disease?

Autoimmune disease is caused by either the production of high-affinity autoantibodies or the generation of autoreactive T cells. Everyone has B cells with specificity for some self-antigens but they normally do not produce autoantibody because there are no self-antigen-specific Th cells (see Chapter 11). If autoreactive Th cells are generated because of a loss of T cell tolerance they can help autoreactive B cells to make antibody, and autoimmunity will follow. However, there are ways in which autoreactive B cells can make antibody in the absence of autoreactive Th cells, and these will be described first.

12.5.1 Autoantibody can be produced in the absence of autoreactive Th

There are two ways in which B cells can make antibody to a self-antigen in the absence of Th cells specific for that self-antigen. One, which is quite simple to understand, is to bypass the need for Th cells. The other, which is more complicated, is due to cross-reactive epitopes between self-antigen and antigens on infectious agents. These processes occur as follows.

B cell mitogens stimulate autoantibody production in the absence of helper T cells

Bacterial products, such as lipopolysaccharide (LPS), can directly stimulate B cells to proliferate and differentiate into antibody-producing cells without the need for Th. This stimulation is not antigen-specific and the agents are called **mitogens.** The way in which LPS works is through binding to TLR4 on B cells and activating signalling pathways. Because any B cell can be stimulated, B cells specific for autoantigens as well as B cells specific for foreign antigens will be stimulated (Figure 12.7a). This stimulation is transient and once the bacteria have been eliminated the stimulation of autoreactive B cells and production of autoantibody will stop.

Antigens on pathogens and our own self-antigens can share similar epitopes

Some proteins on infectious agents have epitopes that are very similar to self-epitopes and this can lead to the production of autoantibody in the following way. Let us take a hypothetical self-antigen that has on it a self-B cell epitope and a self-T cell epitope (Figure 12.7b). B cells exist that are specific for the self-B cell epitope. They can bind the antigen through the B cell epitope, process the antigen and present the self-T cell epitope on their class II MHC. However, because there are no Th specific for the self-T cell epitope, the B cell will not receive the help it needs to make antibody (Figure 12.7b).

Now imagine an antigen on a pathogen that has the same B cell epitope as the self-antigen but has foreign T cell epitopes. This antigen can also be bound by self-reactive B cells; these B cells process the antigen and now present *foreign* antigenic peptide on their class II MHC (Figure 12.7b). Because the antigenic peptide is foreign there will be Th cells present that are specific for the foreign epitope and therefore the B cell will receive the T cell help and can be stimulated to produce autoantibody. The best known example of this is streptococcal M-antigen, which has B cell epitopes that are cross-reactive with cardiac myosin. Following infection with group A streptococci some individuals make antibodies against myosin, resulting in damage to the heart and rheumatic fever.

Another scenario is that a drug or environmental chemical can bind to a self-antigen, modifying it and creating a T cell epitope that appears foreign. Again the autoreactive B cell binds to the B cell epitope and presents the foreign-looking T cell epitope to Th cells and autoantibody is produced (Figure 12.7b).

In these situations, where pathogenic antigens or chemicals provide the foreign T cell epitopes, autoantibody production is usually transient. Once the pathogen has been eliminated or the person has ceased contact with the drug or chemical, the foreign T cell epitope is no longer present and the autoreactive B cell no longer gets T cell help. However, in genetically susceptible individuals, environmental triggers can lead to the activation of Th cells that are specific for a self-antigen; therefore the autoimmune disease becomes self-perpetuating even in the absence of the environmental agent. Why these autoreactive CD4 T cells are present and how they become activated is a subject of much investigation. At present the answer is unclear but a number of mechanisms have been suggested.

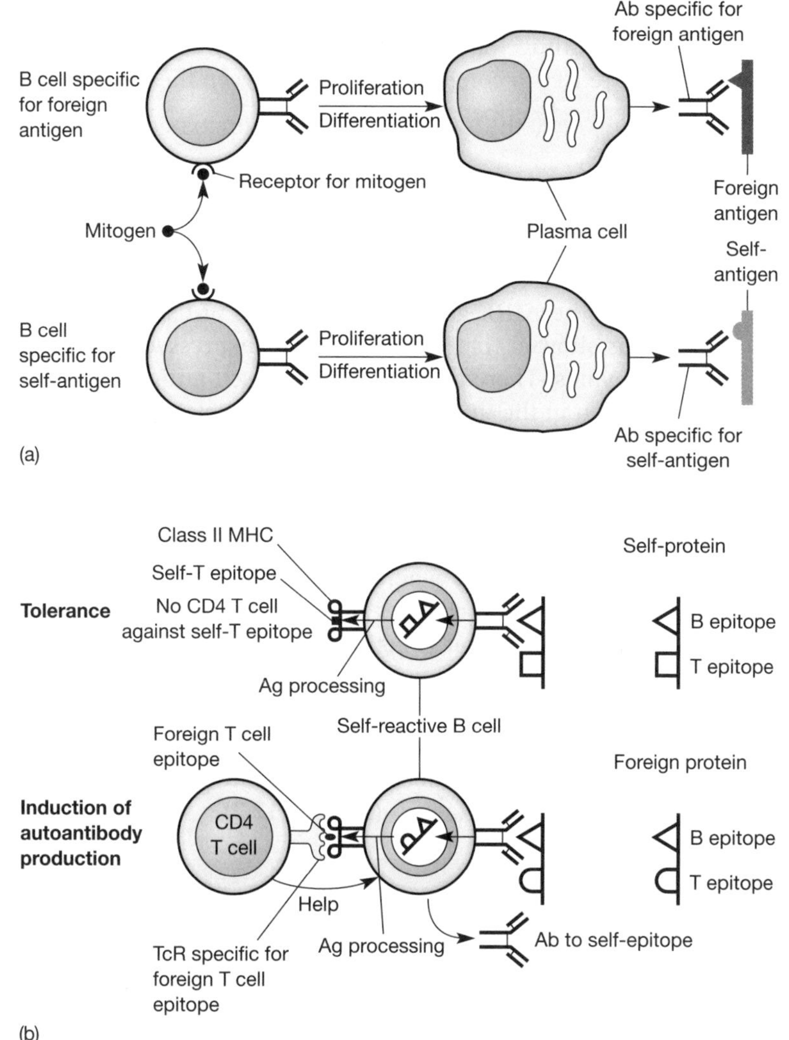

Figure 12.7 Induction of autoantibody production. (a) Autoantibody production induced by B cell mitogens. B cell mitogens stimulate B cells to proliferate and differentiate into plasma cells irrespective of the antigen specificity of the B cell, leading to production of autoantibody as well as antibody to foreign antigens. (b) Induction of autoantibody by cross-reactive epitopes. In the top example a self-protein is depicted with a self-B cell epitope and self-T cell epitope. Autoreactive B cells with specificity for the B cell epitope may bind the self-protein through the B cell epitope, take up the antigen, process it and present the self-T cell epitope on their class II MHC. However, because there are no CD4 T cells with specificity for the self-T cell epitope, the B cell receives no help and is not activated. In the bottom panel a foreign protein (e.g. from a pathogen) shares the same B cell epitope as the self-protein but has a foreign antigen T cell epitope. The same autoreactive B cell binds to the B cell epitope on the foreign protein and presents the foreign T cell epitope on its class II MHC. Because the T cell epitope is foreign there will be CD4 T cells with specificity for the T cell epitope and the B cell will be activated to produce antibody that reacts with the B cell on both the foreign protein and the self-protein.

12.5.2 How can autoreactive Th be generated?

The emergence of pathogenic autoimmune Th cells is probably a multistage process that reflects a complex interplay between the many genes and environmental factors described in Section 12.4.

Genetic factors

The way in which genes could affect the emergence of autoreactive Th could be at several levels. One could be at the level of development influencing whether self-reactive CD4 T cells are clonally deleted in the thymus. Studies on the AIRE gene and expression of tissue-specific antigen in the thymus showed that different people expressed a different set of genes in the thymus. Therefore, for any given self-antigen, people who expressed the antigen in the thymus would clonally delete CD4 T cells that are specific for the self-antigen but people who did not express the antigen in the thymus will not clonally delete self-reactive CD4 T cells. This latter group of people will have self-reactive CD4 T cells which have to be prevented from developing into Th through peripheral tolerance mechanisms such as anergy or Tregs. This is where other genes involved in signalling and regulation can affect the chances of self-reactive CD4 T cells being activated. It is now thought that genetic susceptibility to autoimmunity is due to the inheritance of a specific combination of alleles of many different genes. Most people will have some susceptibility alleles but not enough to be at risk. Those people with the combination of enough susceptibility alleles are at risk of getting an autoimmune disease and whether they get it or not depends on environmental factors (see Box 12.2).

BOX 12.2: VARIATION IN SELF-TOLERANCE

Dispersed throughout the genome are sequences of DNA called minisatellites, which are sequences of 10–100 bps that are repeated in tandem arrays of 0.5–40 kb. Some minisatellites are polymorphic in that the number of repeats varies. These are called **variable number tandem repeats**, or VNTRs. Because they affect the DNA structure, VNTRs near genes may affect expression of the gene product. There is one such VNTR 5′ to the insulin gene and different alleles of this VNTR affect susceptibility to T1DM. The different alleles affect susceptibility to T1DM and also affect levels of expression of insulin protein in the thymus. Alleles which cause high expression of insulin are associated with a lower risk of T1DM while alleles which are associated with lower thymic levels of insulin are associated with a higher risk. These results suggest that high thymic insulin expression results in more efficient clonal deletion of insulin-specific T cells in the thymus and reduced risk of T1DM.

It appears that the variable expression of tissue-specific antigens in the thymus may apply to other self-antigens. Different individuals express some, but not all, tissue-specific proteins in the thymus. The implication of these observations is that people who express a peripheral self-antigen in the thymus will clonally delete thymocytes that are specific for these self-antigens but others will not. This means that for many (all?) peripheral self-antigens you may have the situation that different people are relying on different forms of self-tolerance to the same autoantigen. The different forms of tolerance carry different risks of autoreactive cells being activated *in vivo*. People who clonally delete self-reactive CD4 T cells against a self-antigen clearly carry the smallest risk of generating Th against the antigen. However, people who have autoreactive CD4 T cells in their periphery are at much greater risk of generating an autoimmune response against this autoantigen.

Why should there be such variety in the way people maintain self-tolerance? Surely the safest bet would be to evolve mechanisms to clonally delete all self-reactive CD4 T cells. The best explanation for this variation relates to the balance between the threat posed by pathogenic organisms and the threat posed by autoimmunity. Individuals who clonally delete CD4 T cells specific for a self-antigen may also delete cells specific for a similar antigen on a pathogen. These individuals have a reduced risk of autoimmunity but also a reduced response to the pathogen and therefore an increased risk of infection. Conversely individuals who have 'less vigorous' tolerance to a specific self-antigen have an increased risk of autoimmunity but would give a stronger response to the pathogen and therefore be more likely to survive infection. Within a population there will be a variety of potential responses to each autoantigen, and therefore to different pathogens, so that of the population as a whole a balance is achieved between the two threats.

Environmental activation of autoreactive Th

Although many experimental procedures can lead to the activation of autoreactive Th in animals it is not clear exactly how environmental factors lead to the activation of autoreactive Th in human autoimmune diseases. One aspect that seems to be very important is the involvement of dendritic cells. Autoreactive CD4 T cells, like those against any antigen, need to be initially activated by antigen presented to them on class II MHC expressed by dendritic cells (see Chapter 6). Two main mechanisms have been suggested to explain how environmental agents could promote the activation of autoreactive CD4 T cells by dendritic cells (Figure 12.8):

- **Molecular mimicry.** An environmental agent (such as an infectious organism) will be regarded as foreign by the immune system and dendritic cells will present antigenic peptides derived from the agent on their class II MHC. If one of these antigenic peptides is very similar to a self-antigen peptide, CD4 T cells specific for that peptide will be activated to become Th. These can then react against self-peptides as well as those from the environmental agent and cause autoimmunity.

- **Tissue damage.** Tissue damage, whether caused by a toxic substance or a pathogen, can lead to an inflammatory type of response with the production of cytokines and the release of self-antigens from damaged cells. These two factors can activate dendritic cells to present self-peptides derived from the self-antigens on their class II MHC, which could then activate self-reactive CD4 T cells.

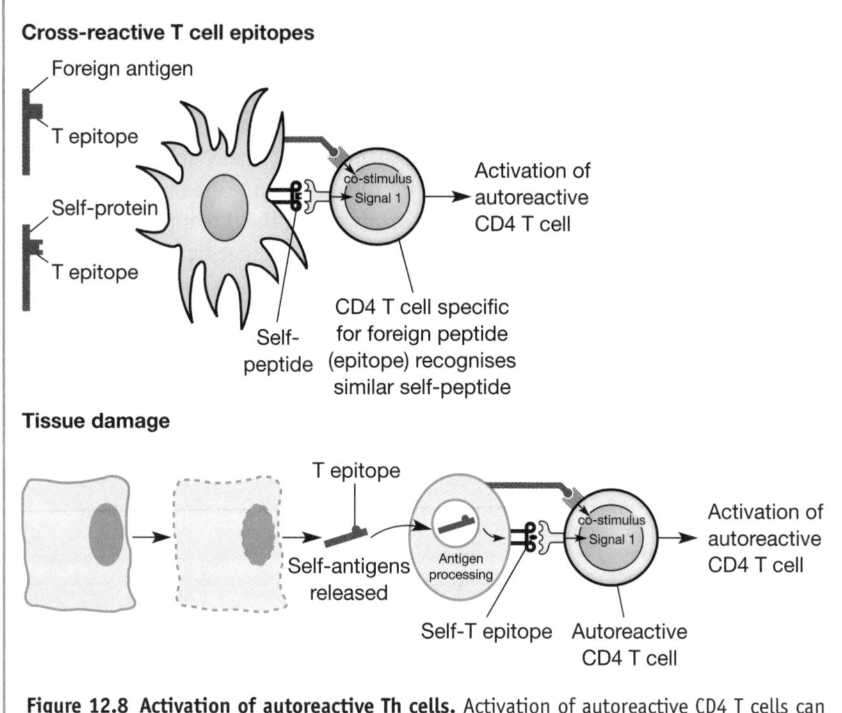

Figure 12.8 Activation of autoreactive Th cells. Activation of autoreactive CD4 T cells can occur due to a pathogen having a very similar epitope to a self so the CD4 T cells activated by the epitope on the pathogen react with the closely related self-epitope (top panel). Alternatively, tissue damage may result in the release of self-antigens, which are taken up and processed by dendritic cells and can stimulate autoreactive CD4 T cells.

12.6 Summary

- Autoimmune diseases constitute a large group of diseases that are characterised by an immune response against one or more components of the body. Overall they affect 5% of the population in industrialised countries.
- In some autoimmune diseases the pathology is solely due to the production of high-affinity autoantibody against self-antigens. In other diseases

there is also an extensive inflammatory infiltrate, consisting of many different cell types in the affected tissue, making it difficult to identify the exact nature of the immune-mediated damage.

- The aetiology of autoimmune diseases is complex, with many genes and environmental factors contributing to susceptibility. One of the major genetic associations is with class II MHC. Many new susceptibility genes have been identified.
- The way in which the environment contributes to the development of autoimmune disease is not well understood but it is thought either to provide autoreactive B cells with a source of help for antibody production or to contribute to the activation of autoreactive CD4 T cells.

12.7 Questions

1) Draw diagrams to illustrate how indirect and direct immunofluorescence can be used to detect autoantibody.
2) How may autoantibodies cause disease?
3) What factors contribute to susceptibility to autoimmune disease?
4) What are the differences between monogenic and polygenic diseases?
5) What is meant by 'molecular mimicry' and how does it contribute to the development of autoimmune disease?

The answers to these questions can be found on page 340.

12.8 Further reading

1) Davidson A, Diamond B. (2001) Autoimmune diseases. *England Journal of Medicine* 345:340–350.
2) Ranjeny T. (2010) The balancing act of autoimmunity: central and peripheral tolerance versus infection control. *International Reviews of Immunology* 29:211–233.
3) Jäger A, Kuchroo VK. (2010) Effector and regulatory T-cell subsets in autoimmunity and tissue inflammation. *Scandinavian Journal of Immunology* 72:173–184.
4) Pearce SHS, Merriman TR. (2006) Genetic progress towards the molecular basis of autoimmunity. *Trends in Molecular Medicine* 12: 90–98.

CHAPTER THIRTEEN

Allergy and other hypersensitivities

Learning objectives

To learn about the different ways in which inappropriate immune responses can cause disease. To understand the immunological, clinical, genetic and environmental aspects of asthma and related allergies. To know about other types of hypersensitivities caused by antibody responses. To be aware of contact sensitivity reactions and their relationship to delayed hypersensitivity.

Key topics

- Type I hypersensitivity (allergy)
  - Immunological basis of type I allergic responses
  - Clinical symptoms of allergy
  - Genetics of allergy
  - Environmental factors
  - Treatment
- Type II hypersensitivity
- Type III hypersensitivity reactions
- Type IV hypersensitivity: delayed-type hypersensitivity and contact sensitivity

13.1 Introduction

Although antibody can provide vital protection from disease, there are situations where excessive production of antibody against harmless antigens actually causes disease. This is known as **hypersensitivity** and there are three types of antibody-mediated hypersensitivity. The most common and well known is type I hypersensitivity or allergy. This is mediated by the production of IgE and is commonly seen as an allergic response. Type II hypersensitivity is caused by cytotoxic antibodies against normal or modified tissue components and type III hypersensitivity is caused by the deposition of antibody–antigen complexes in blood vessels of various tissues. Neither type II nor type III hypersensitivity involves the production of IgE. A fourth type of hypersensitivity is delayed hypersensitivity which does NOT involve antibody.

13.2 Type I hypersensitivity and allergy

The most familiar syndrome associated with allergy is asthma. Approximately one in seven children in Britain suffers from asthma and the incidence has risen dramatically over the past 50 years. Until the twentieth century asthma was a rare condition. Like autoimmunity, asthma rates are highest in more industrialised societies.

Asthma is one clinical manifestation of diseases caused by the production of IgE against otherwise innocuous antigens and is an example of an allergic reaction. Antigens that stimulate an allergic response are called **allergens**. Allergens can be derived from several sources:

- **Pollens.** Many pollens cause allergy, especially rhinitis and asthma. Pollens from plants, trees and grasses can all cause allergy, with ragweed being a particular problem in North America.
- **Insects.** Allergens derived from the house-dust mite are the most common cause of asthma. Cockroaches are also a common source of allergen. Bee and wasp stings also stimulate strong allergic reactions, possibly because they contain pharmacologically active products.
- **Animals and birds.** Animal dander (fur, excretory products) and feathers are another major source of allergens. Cats, dogs and rodents are the most common source of animal allergens although this probably reflects the degree of exposure rather than any special feature of these animals.
- **Drugs.** Many drugs cause allergic reactions, with penicillin and sulphonamides being among the most common.
- **Food.** A number of foods cause allergies with sometimes severe effects. Peanuts, shellfish and milk products are the best known causes of food allergies.

Allergic reactions can cause other illnesses apart from asthma: rhinitis (hayfever), eczema (rashes), conjunctivitis and diarrhoea are all examples of allergic reactions. These allergic reactions are localised and usually reflect the site of exposure. Airborne allergens are more likely to cause asthma and hayfever, while food allergens tend to cause gastric symptoms, although this is not always the case, e.g. food allergens can cause skin reactions. In its severest form an allergic reaction can be systemic, leading to anaphylaxis, an extremely dangerous condition that can be fatal if not treated. The present concern about peanut allergy is due to a number of deaths caused by anaphylactic reactions to foods containing peanut products.

13.2.1 Atopy and allergy

The use of the terms '**atopy**' and '**allergy**' can be a source of confusion. Although often used synonymously, the two terms do have slightly different meanings. Allergy is described as an IgE-mediated hypersensitivity to a particular substance that manifests itself clinically in one or more of the forms described above. Atopic individuals include those with allergies but would also include individuals who have a positive skin prick test (see Section 13.4) or specific IgE against an allergen but have no clinical symptoms. In the UK about 30% of the population is atopic and about 10% of the population has allergies with clinical symptoms.

13.2.2 The immunological basis of allergy

The development of allergy occurs in a number of stages, the first of which is **sensitisation**.

Sensitisation occurs on first exposure to allergen

Sensitisation involves the production of IgE in response to exposure to allergen (see Figure 13.1). This IgE production is dependent on CD4 T cells that respond to antigens associated with the allergen and differentiate into Th2 cells. Secretion of IL-4 by the Th2 cells is particularly important in making B cells specific for the allergen switch to IgE. The IgE binds to mast cells, which have receptors called FcεRs on their surface that bind to the Fc portion of IgE. Mast cells are found in most tissues and two types can be identified in humans based on their location and granule contents; these are mucosal mast cells and connective tissue mast cells. IgE can stay bound to mast cells for months and possibly years. The initial exposure to allergen does not usually involve any symptoms.

Mast cell activation occurs upon further encounters with allergen

Upon re-exposure to the allergen, the allergen will bind to the allergen-specific IgE that is bound to the FcεRs on mast cells. This will cause cross-linking of the FcεRs, resulting in a series of intracellular signalling events that lead to mast cell activation. The immediate consequence of mast cell activation is degranulation and the release of preformed mediators from the mast cell granules. This is followed by the synthesis and release of newly formed mediators (Figure 13.1). A list of preformed and newly formed mediators, together with their main functions, is given in Table 13.1. These mediators cause vasodilation, increased vascular permeability, smooth muscle contraction and mucus secretion, as well as having other effects. The symptoms caused by these events depend on the anatomical site and are described below. These events generally occur within minutes of exposure to the allergen.

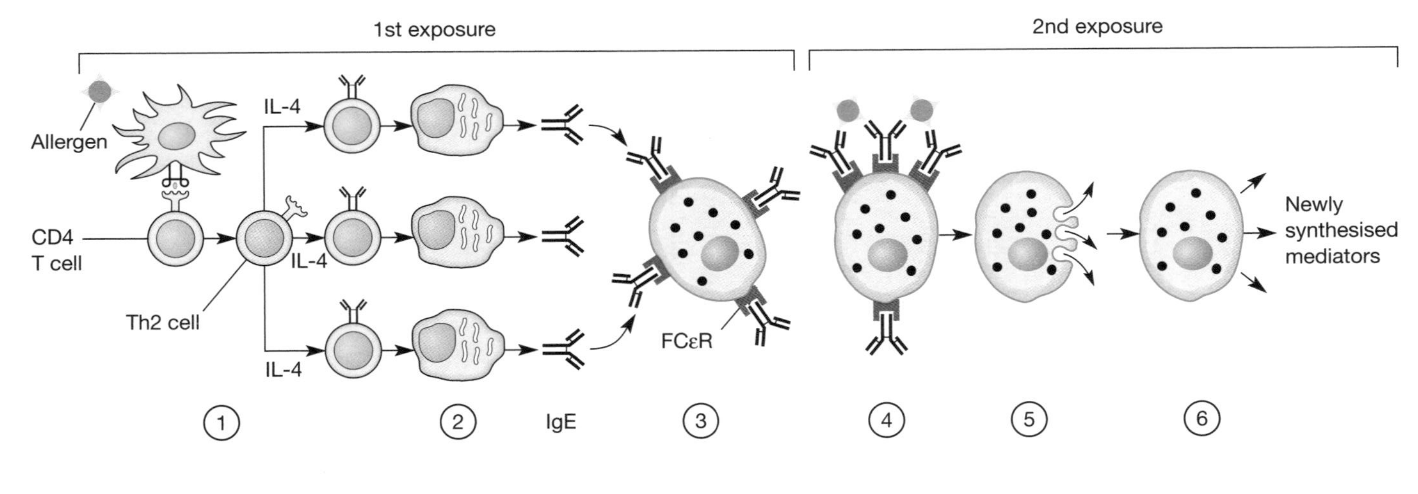

Figure 13.1 The allergic response. Upon first exposure to antigen, CD4 T cells specific for the antigen are stimulated to become Th2 cells (1), which help B cells differentiate into plasma cells secreting allergen-specific IgE (2). The IgE binds to FcεRs in mast cells and basophils (3). A second or subsequent exposure to the allergen results in the allergen binding to the IgE on the mast cells, which causes cross-linking of the FcεRs, resulting in mast cell activation (4). The immediate response of mast cell activation is degranulation of the mast cells with the release of the granule contents (5); these are preformed mediators such as histamine and heparin. The activated mast cells also begin to synthesise new mediators (6), especially leukotrienes, prostaglandins and cytokines (TNFα).

Table 13.1 Inflammatory mediators released by mast cells

Mediator	*Main biological effects*
Preformed (released upon degranulation)	
Histamine	Vasodilation, increased vascular permeability
Enzymes (proteases)	Degradation of blood vessel endothelial membrane
Eosinophil chemotactic factor	Attract eosinophils from blood
Neutrophil chemotactic factor	Attract neutrophils from blood
Tumour necrosis factor-α	Promote inflammation
Newly synthesised after mast cell activation	
Prostaglandins (PGs)	Vasodilation, platelet aggregation
Leukotrienes (LTs)	Increased vascular permeability, mucus secretion
Platelet activating factor	Aggregation and degranulation of platelets
Tumour necrosis factor-α	Promote inflammation

Allergic responses have early and late phase reactions

Activated mast cells secrete TNFα, which activates the endothelium at the site causing expression of adhesion molecules that promote the migration of leukocytes from the blood (see Chapter 2). Chemotactic factors, including PGD2, LTB4, IL-8, MIP1α and eotaxin, are also produced. The net effect of these factors is the later recruitment of eosinophils, basophils, neutrophils and T cells to the site. Activation of these cells produces further inflammation, which is seen as a late phase reaction (Figure 13.2).

13.3 Allergies result in a variety of clinical symptoms

Allergic responses can be localised to the site of allergen or spread throughout the body and become systemic. The clinical symptoms of localised allergic reactions depend on where the mast cells are localised and can result in asthma, hayfever, diarrhoea/vomiting or skin reactions (eczema and hives). A systemic allergic reaction is called anaphylaxis.

13.3.1 Local allergic reactions

Asthma

The main symptom of asthma is shortness of breath. This is caused by narrowing of the large and small airways, a condition called bronchospasm. The narrowing of the airways can be due to three factors (Figure 13.3 and Plate 16). Histamine, LTC4 and PGD2 released by mast cells cause smooth

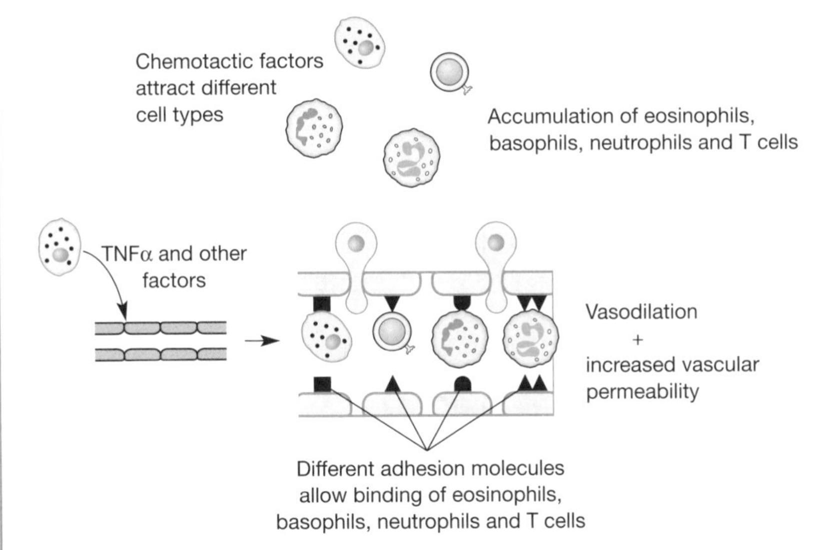

Figure 13.2 Late phase reaction. This occurs a few hours after exposure to the allergen. TNFα secreted by mast cells activates local endothelial cells, which up-regulate expression of adhesion molecules that promote the migration of eosinophils, basophils, neutrophils and T cells to the site. These cells are activated to give a late phase inflammatory response.

muscle contraction and therefore bronchoconstriction. Histamine and LTC4 also increase mucus secretion into the airways, which further narrows the available passage for air flow. The late phase reaction results in an inflammatory exudate consisting of eosinophils, neutrophils, basophils, macrophages and platelets, which causes further blockage of the airways.

Prolonged exposure to the allergen can lead to a condition known as bronchial hyper-reactivity. In this condition the bronchi are exquisitely sensitive to histamine and other mediators so that bronchospasm and asthmatic attacks can be triggered by stimuli such as exercise, cold air or passive cigarette smoke without exposure to the allergen. If a person can avoid the allergen for a period of time this hyper-reactivity may disappear.

Rhinitis (hayfever)

Hayfever is characterised by a blocked and runny nose (rhinorrhoea), coughing, sneezing and itchy eyes and is the result of activation of mast cells in the nose and conjunctivae of the eye. Because the allergen is often pollen, hayfever can be seasonal and occur only during the time of year when a particular pollen is present. The mast cell mediators cause localised vasodilation, increased vascular permeability and mucus secretion. This may be accompanied by the entry of mucosal mast cells into the nasal mucosa, where normally only connective tissue mast cells are found. The presence of mucosal mast cells can complicate the chemotherapy of rhinitis since mucosal mast cells are not sensitive to sodium cromoglycate, a commonly used anti-allergic drug (see Section 13.7.2).

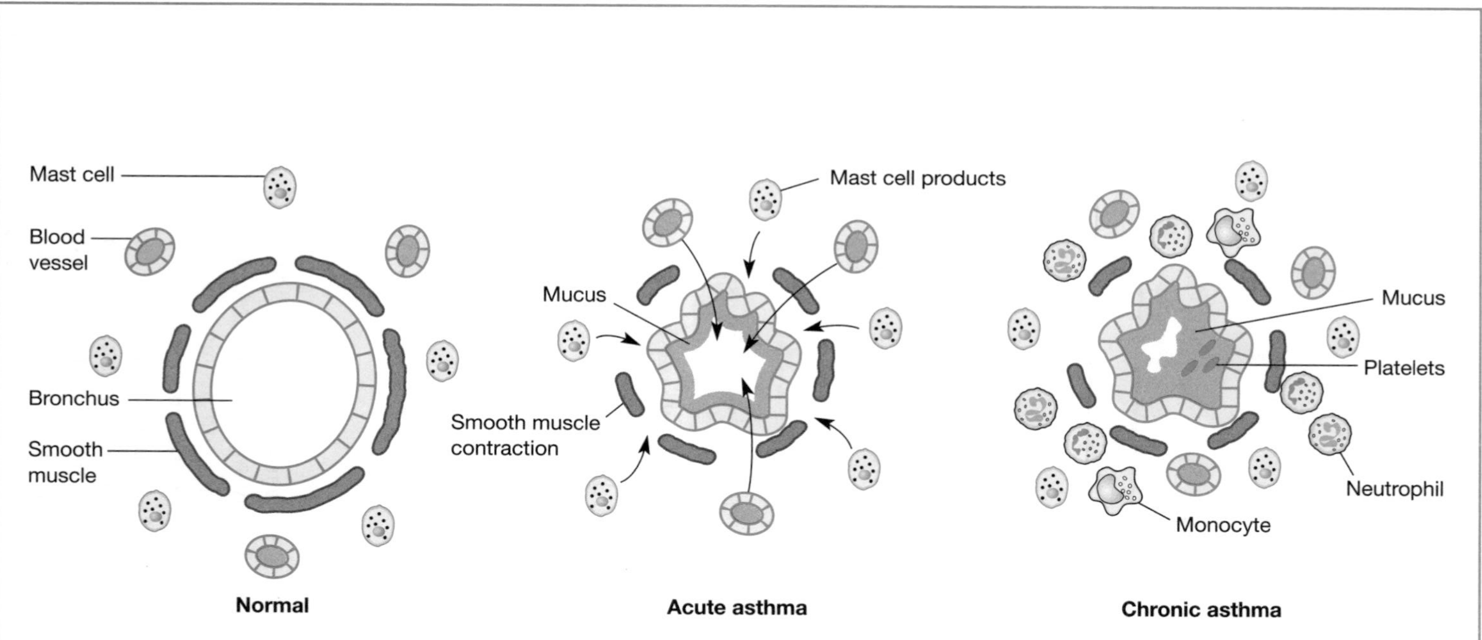

Figure 13.3 Asthma. Acute asthmatic attacks are due to the release of histamine, LTC4 and PGD2, which cause smooth muscle contraction of the lower airways (bronchi). Histamine and LTC4 also stimulate mucus secretion, which contributes to narrowing of the airways. The late phase asthmatic reaction is due to the influx of other inflammatory cells that stimulate further mucus secretion, epithelial cell damage and the activation of platelets leading to microthrombi in the airways.

Food allergies

Activation of mast cells in the gut causes smooth muscle contraction, vasodilation, increased secretion of fluids and reduced peristalsis, resulting in vomiting and/or diarrhoea. The inflammation of the mucosa can also allow food allergens to enter the bloodstream, where they can activate mast cells at other sites, causing asthma or skin allergies, particularly eczema.

Skin allergies

Activation of mast cells in the skin results in vasodilation and increased vascular permeability. These cause increased blood flow and oedema, which cause itchy red swellings (called hives or urticaria) that, if chronic, lead to a severe itchy skin rash (eczema).

Different types of allergies seem to show different ages of onset with food allergies appearing in the youngest followed by dermatitis, rhinitis and asthma – this has been termed 'the atopic march'.

13.3.2 Anaphylaxis: the most dangerous form of allergy

This is the result of a systemic allergic reaction and can be life-threatening. Depending on its severity it can cause symptoms similar to those seen in localised allergic reactions or it can be a life-threatening condition known as anaphylactic shock. The mildest manifestation of anaphylaxis is hives (urticaria). More severe forms result in swelling of the lips, tongue and larynx, nausea, vomiting and asthma. Anaphylactic shock causes severe swelling of the larynx (laryngeal oedema), severe asthma and/or severe hypotension due to massive loss of fluid from the blood to the tissues, which can lead to circulatory collapse. Death can result from respiratory or circulatory failure.

13.4 Testing for allergy

The most common test for allergy is the skin prick test. In this test a few microlitres of allergen are introduced into the epidermis using a lancet. If a person is sensitised to the allergen they will give a **wheal and flare** reaction (Plate 16). The basis of the wheal and flare reaction is the same as that for eczema or hives. The allergen cross-links the FcεRs on skin mast cells causing mast cell activation. The mast cell mediators cause vasodilation, which increases blood flow to the area causing redness or a flare. The mediators also cause increased vascular permeability, which makes the blood vessels leaky. The leakage of blood fluid into the tissue results in oedema and swelling, the wheal. The mast cell mediators also cause itching. The flare is seen within seconds or minutes and the wheal occurs shortly after. The immediate wheal and flare reaction usually disappears in less than an hour after application of the allergen but can be followed by the late phase reaction (Section 13.2.2). This reaction typically occurs four to six hours later and is seen as a lump, which may be accompanied by pain (Plate 16). The main histological feature is the accumulation of eosinophils, basophils and neutrophils at the site.

Skin tests usually give good agreement with allergic status. However, some people with clinical allergy to known allergens fail to give a positive skin test to the allergen and others give a positive skin test to allergens but display no clinical signs of allergy to the same allergen.

Total and antigen-specific IgE can be measured by ELISA (Section 10.3.2). Antigen-specific IgE is now considered to be a more reliable predictor of atopy than total IgE.

13.5 Both genetics and the environment contribute to allergy

Allergy, like autoimmunity, has dramatically increased in incidence in the past 50 years. Particularly alarming has been the rise in the incidence, severity and mortality due to asthma, especially in children. Again, like autoimmunity, both genetic and environmental factors contribute to susceptibility to allergy. Family studies show that although the overall rate of atopy is 15%, the incidence rises to 30% in children with one atopic parent and 50% where both parents are atopic, suggesting a genetic component to susceptibility. Concordance in identical twins is about 60%, indicating that environmental factors are also involved.

13.5.1 Genetics of allergy

Another feature that allergy shares with autoimmunity is that it is multigenic, i.e. many genes determine susceptibility to developing allergy. Genes affect many aspects of the allergic response, some of which are immunological and some non-immunological. Not surprisingly for an immunologically based syndrome, both MHC genes and non-MHC genes are involved.

MHC genes

MHC genes do not appear to affect the overall risk of developing allergies but they may influence what you become allergic to. For instance allergy to rye grass is associated with HLA-DR3 and that to birch pollen with HLA-DR5.

Non-MHC genes

There have been major advances recently in identifying allergy susceptibility genes. Those identified so far fall into three categories: genes affecting the immune response, epithelial barriers and tissue remodelling.

- **Immune response.** Polymorphism in the α chain of the FceR1 receptor for IgE affects total levels of IgE and therefore susceptibility to allergy (Figure 13.4). The cytokine interleukin 33 and its receptor IL1RL1 are involved in the maturation, survival and activation of eosinophils. Both genes are polymorphic and different alleles of both affect susceptibility to asthma. There is a cluster of genes on chromosome 5q which contains genes for the cytokines IL-3, IL-4, IL-5, IL-9, IL-13 and granulocyte-monocyte colony stimulating factor and although the exact gene(s) have not been identified, different alleles of the cluster are associated with risk of allergy.
- **Epithelial barrier function.** Exposure to allergens is via mucosal surfaces of the gut or respiratory system. For both initial sensitisation and manifestation of clinical symptoms allergens must cross the epithelial barriers of these sites. It is not surprising therefore that different alleles of genes coding for proteins with barrier function affect susceptibility to allergy. These genes are expressed in epithelial cells and include flagellin which is involved in keratinocyte differentiation and protocadherin-1 which is an adhesion molecule in airway epithelium.
- **Tissue integrity.** With chronic or acute allergic disease there is considerable inflammation and tissue damage. How well an individual can respond to this damage can determine the severity of the clinical symptoms and, again, polymorphism of genes involved in these processes affects allergy. Some examples include ADAM-33, a proteinase involved in tissue remodelling, and collagen XXIX, a novel collagen expressed in skin, lung and gut mucosa.

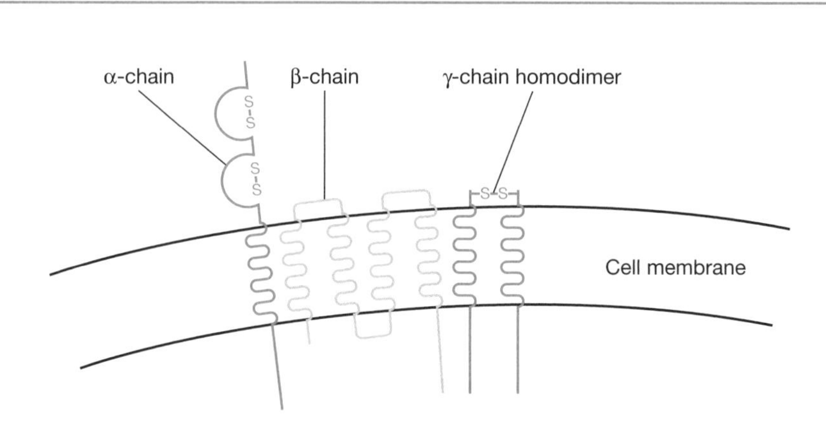

Figure 13.4 The high-affinity receptor for IgE (FcεR). The α-chain of the FcεR binds to the Fc portion of IgE. The disulphide-linked homodimer of γ-chains is responsible for initiating intracellular signalling following cross-linking of FcεRs by allergen and is linked to the α-chain by the β-chain, which spans the cell membrane four times.

13.5.2 Environmental factors in allergy

Genetic studies, and the sudden dramatic increase in the incidence of allergy, demonstrate that environmental factors contribute to the development of allergy. Environmental factors could affect both the incidence and the manifestations of allergy. However, although environmental factors are important, it is not clear how they have contributed to the sudden alarming increase in allergy incidence.

Three main environmental factors have been suggested as contributing to the increase in allergy. One that is obvious but often overlooked is the levels of allergens that people are exposed to. Many studies have shown a positive relationship between the levels of allergens in the home and incidence of allergy. Modern lifestyles may have increased exposure to allergens; this may be especially true for the house-dust mite, a major cause of allergies. Changes in housing, with central heating, fitted carpets and better sealing of houses, have provided an ideal breeding ground for house-dust mites. A second possibility is increasing pollution, although clear-cut associations have not been shown between any single pollutant and allergy. Finally it has been suggested that the modern lifestyle in developed countries, with good sanitation, vaccination and the use of antibiotics, has reduced our exposure to microbes and this has contributed to the increase in allergy and, incidentally, autoimmunity. This has been referred to as the 'hygiene hypothesis'. This is discussed in more detail in Box 13.1.

13.6 Why have IgE in the first place?

With all the problems associated with IgE production to innocuous substances, it would seem that IgE causes more problems than it solves. However, IgE would presumably not have evolved if it was not of some benefit, and certainly not if it was more harmful than beneficial. The situation where IgE production is seen most prominently, apart from allergy, is in parasitic worm infections. However, it has to be said that the evidence that IgE is beneficial in worm expulsion is not extensive (see Chapter 9). Therefore IgE may have evolved to deal with parasitic worm infections, which affect billions of people and animals worldwide. In evolutionary terms the immune system has not learnt to deal with the rapid change in lifestyle experienced in the past 100 years or so, so that in industrial societies, where parasitic infections in humans are less prevalent, the balance of IgE production has swung away from benefit to harm (see Box 13.1).

13.7 Treatment of allergy

Treatment of allergy can be divided into three main approaches: allergen avoidance, pharmacological treatment and immunological intervention.

BOX 13.1: THE HYGIENE HYPOTHESIS

There has been a direct inverse correlation between the incidence of infectious diseases and allergy and autoimmunity over the last fifty years in many countries. Studies in the 1990s seemed to show that children living in very clean environments had a higher incidence of hayfever than those in 'dirtier' environments. This led to the 'hygiene hypothesis' to explain this large increase in the incidence of allergy. In its original form the hygiene hypothesis was explained on the basis of Th1/Th2 responses and was later revised on the basis of Tregs and another form suggests it is due to lack of worm infections.

Modern day living has switched us from 'Th1ness' to 'Th2ness'

Until forty years or so ago most children were exposed to many bacterial and viral diseases such as mumps, chickenpox, measles and rubella (German measles) and many were exposed to polio or smallpox. These are intracellular infections that would be expected to stimulate Th1 responses. Exposure of children to allergens in this Th1 environment may result in a Th1 response to the allergen and therefore not favour the Th2 response that promotes IgE production. Nowadays, not only are children vaccinated against these diseases but up until recently the only adjuvant used in human vaccination was alum, which promotes a Th2 response. Therefore children are exposed to allergen in the relative absence of Th1 responses and in the presence of increased Th2 responses, which could promote the production of IgE in response to the allergen.

We no longer programme our Tregs properly

The switch from Th1 programming to Th2 programming cannot explain the rise in incidence that has also been seen in organ-specific autoimmune diseases, such as T1DM and multiple sclerosis, which are associated with Th1 responses. Therefore the hygiene hypothesis has been modified and suggests the lack of exposure to 'germs' in infancy and childhood leads to altered programming of the immune system and especially a lack of induction of Tregs.

A lack of worms is the reason for defective Treg induction

Up until one hundred years ago all humans would have had various worm infections. Unlike most bacterial and viral infections which last a few days or weeks, worms are able to establish chronic infections that persist for months or years. To do this they must somehow subvert the immune system and it has been shown that worms can induce Tregs to non-worm antigens as well as worm antigens although how they do this is not known. Therefore the lack of worm infection in developed countries leads to lack of proper regulation of immune responses and increased allergy and autoimmunity.

13.7.1 Allergen avoidance

A number of studies have shown that if individuals can avoid contact with allergens their clinical symptoms improve and can disappear. However, it is not always possible to identify the allergen and, if identified, to avoid it totally. Allergen avoidance does not eliminate the underlying state of allergy, so that exposure to the allergen, even after many years of avoidance, can lead to reappearance of the symptoms.

13.7.2 Pharmacological intervention

Pharmacological treatment of allergy is targeted at preventing mast cell degranulation or inhibiting the inflammatory response associated with allergic reactions. These agents treat the symptoms of allergy but do not change the underlying immune response and therefore do not provide a cure.

The most common treatments for allergy are antihistamines, which block the actions of histamine, and inhibitors of mast cell degranulation. Drugs such as sodium cromoglycate and nedocromil inhibit mast cell degranulation but it is not clear whether they work on all mast cell types.

More severe allergies are treated with steroids. These may be given locally, i.e. applied to the skin or inhaled, or systemically. They act primarily by inhibiting the production of cytokines or other inflammmatory mediators by various cell types, including eosinophils and monocytes. They may also inhibit the migration of inflammatory cells into the site of the allergic reaction.

Finally adrenaline is given to treat anaphylactic shock. This binds to β-adrenergic receptors on smooth muscle cells and cause smooth muscle relaxation, leading to a reversal of the vasodilation and increased vascular permeability seen during anaphylaxis. It cannot be used as a long-term treatment for allergy.

13.7.3 Immunological intervention

In contrast to pharmacological approaches, immunological intervention attempts to modify the immune response against the allergen so that the production or effects of IgE are negated. A number of different approaches are now being used or developed, some of which are described below.

Desensitisation involves giving small amounts of allergen to the patient and gradually increasing the dose. This can result in a reduction of allergic symptoms. The immunological basis is not clear but may involve the production of IgG against the allergen which binds to the allergen and stops the allergen binding to IgE on mast cells, thereby preventing mast cell activation (Figure 13.5). It can also result in the induction of Tregs which suppress the allergic response. Although it can be effective, the procedure has the major drawback that giving too much allergen may trigger a potentially fatal anaphylactic attack.

Another approach is to give anti-IgE antibodies to block IgE from binding to the FcεR on mast cells and activating them. For reasons that are not understood, it was found that the anti-IgE antibodies inhibited the production of IgE as well as inhibiting its activity; this may be due to anti-IgE binding to IgE on B cells and inhibiting them or causing their opsonisation. This approach has been successful at reducing allergic responses to the house-dust mite in clinical trials.

Another approach to allergy immunotherapy is to target cytokines or their receptors. Antibodies against IL-5, which stimulates eosinophils, and IL-13 which is involved in the Th2 response, have been shown to reduce symptoms in some patients in clinical trials.

13.8 Type II hypersensitivity

Type II hypersensitivity involves the production of IgG or IgM antibodies that react with antigens on cells or tissues. There are three major syndromes associated with type II hypersensitivity: these are blood transfusion

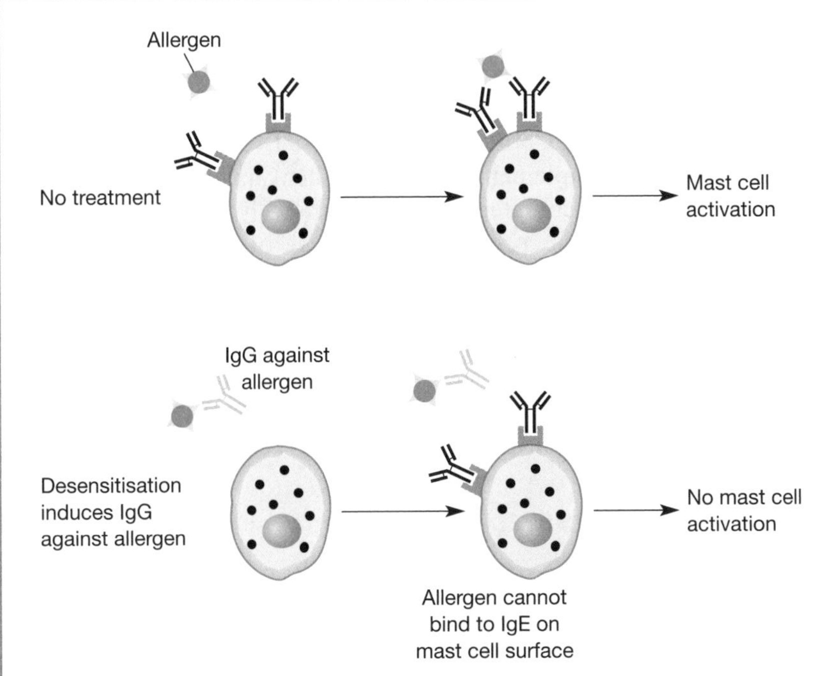

Figure 13.5 Desensitisation in allergy. Small amounts of allergen, insufficient to trigger an allergic response, are given repeatedly to an allergic individual to stimulate production of IgG specific for the allergen. If the person is accidentally exposed to a large amount of allergen (enough normally to produce an allergic response) the IgG binds to the allergen and prevents the allergen binding to the allergen-specific IgE on the mast cells, thus preventing mast cell activation and an allergic attack.

reactions, haemolytic disease of the newborn and drug-induced hypersensitivity. Some autoimmune diseases involve the production of autoantibodies against cells or tissues, causing a type II hypersensitivity-like reaction (see Chapter 12).

13.8.1 Blood transfusion reactions

It was appreciated early in the twentieth century that it was not possible to transfuse blood at random between individuals. There are over twenty antigens on erythrocytes, and they vary between individuals. The most important and well-known are those that make up the ABO blood group system. Differences in glycosylation of red cell glycoproteins give rise to three variants of the glycoprotein called A, B or O. The groups are co-expressed so that individuals can be O, A, B or AB (Figure 13.6). These carbohydrate groups are similar to those seen on bacteria and therefore even if someone has not been exposed to a foreign blood group through blood transfusion, they may develop antibodies against the bacterial antigens that also recognise foreign blood groups. Individuals with blood group O make antibodies to groups A and B; individuals with blood group A make antibodies to B; and those with blood group B make antibodies to A. These antibodies are usually of the IgM class. AB individuals do not make antibodies to A or B and can receive blood of any type.

If someone receives an incompatible blood group to which they have antibodies, the antibodies bind to the transfused red blood cells. IgM is very good at fixing complement and complement fixation results in activation of the classical complement pathway and lysis of the transfused red blood cells. This can result in the release of massive amounts of

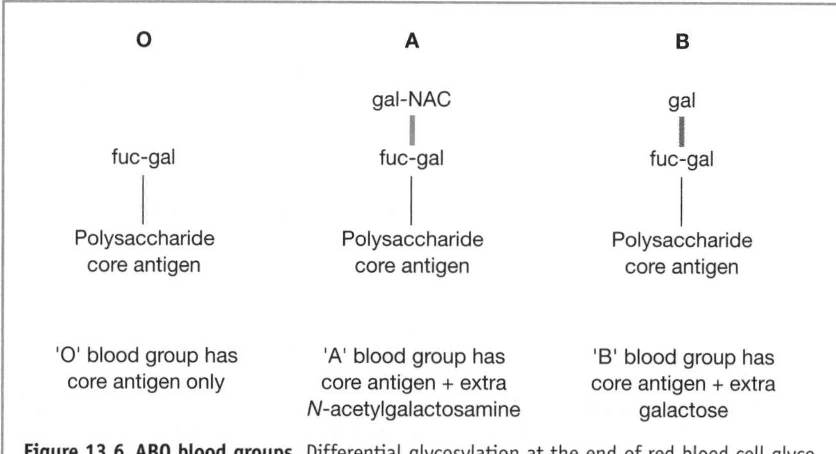

Figure 13.6 ABO blood groups. Differential glycosylation at the end of red blood cell glycoproteins result in blood group O, which contains the core glycoprotein only, blood group A, which has a terminal *N*-acetylgalactosamine, or blood group B, which has a terminal galactose.

haemoglobin, some of which may be metabolised to bilirubin, which is toxic at high levels. Clinical symptoms include fever, chills, nausea and vomiting, and chest and lower back pains.

Blood transfusion reactions can also be caused by antibodies to minor blood group antigens. However, these antibodies do not usually arise through cross-reactivity with microbial antigens but instead are the result of repeated transfusions.

13.8.2 Haemolytic disease of the newborn

Another antigen on red blood cells is called the rhesus (Rh) antigen. About 70% of people are Rh+ but the rest lack the rhesus antigen and are Rh–. During childbirth foetal blood can enter the mother's bloodstream. If an Rh– mother has an Rh+ baby she may be immunised against the Rh+ antigen during childbirth and produce IgG antibodies to Rh (Figure 13.7). If the mother has another Rh+ baby, the anti-Rh IgG can cross the placenta and bind to the foetus's red blood cells. This can result in opsonisation of the red blood cells and their phagocytosis and destruction in the liver or spleen. This results in an enlarged spleen and liver and toxicity due to bilirubin. In severe cases the condition can be fatal.

Sensitisation of the mother can be prevented by giving the mother a preparation of antibodies against the Rh antigen shortly before birth. These antibodies bind to the foetal red blood cells and cause their immediate removal and destruction before they have time to sensitise the mother.

13.8.3 Drug-induced hypersensitivities

Some drugs or their metabolites, can bind to red blood cells or platelets. Although the drugs are too small themselves to stimulate an immune response, by binding to self-proteins the drug–self-protein complex can create a new antigen that appears foreign to the immune system. In a small proportion of people these new antigens stimulate antibody production against the drug. The antibodies bind to the drug and cause either complement-mediated lysis or opsonisation through Fc receptors and C3 receptors on phagocytes (Figure 13.8). Some drugs, such as penicillin, bind to red blood cells resulting in anaemia; quinidine can bind to platelets, resulting in thrombocytopenia. Usually, when the drug is withdrawn the symptoms resolve as non-drug-bound blood cells replace the drug-bound ones.

13.9 Type III hypersensitivity

Type III hypersensitivity reactions are caused by immune complexes of antigen and antibody. They are seen in a number of situations and can be local or systemic (Figure 13.9).

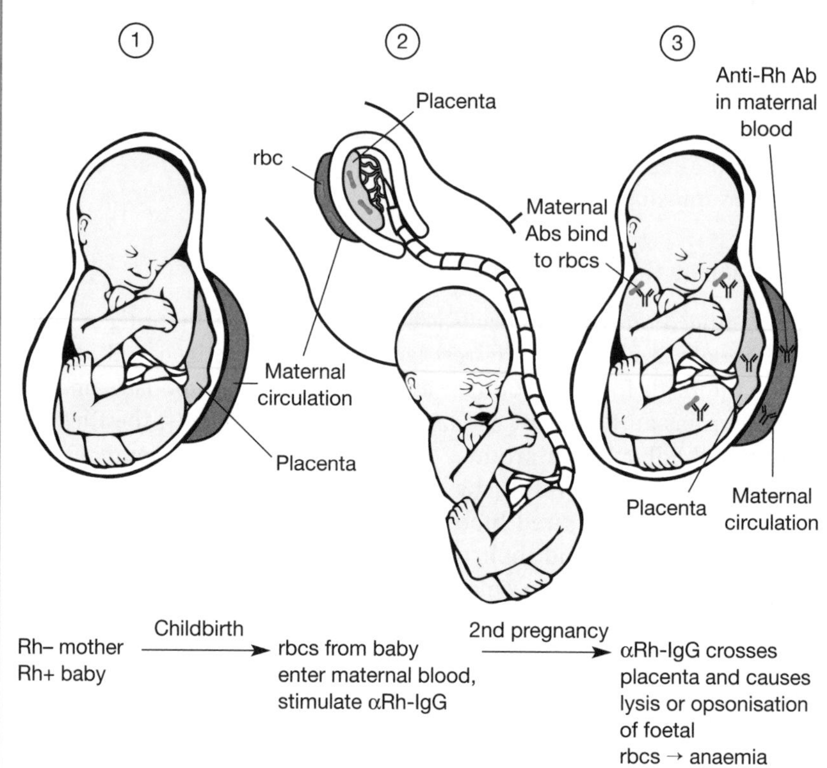

Figure 13.7 Haemolytic disease of the newborn. If a mother who is rhesus-negative (Rh−) has a rhesus-positive baby (Rh+), there is the possibility that at birth some of the baby's red blood cells (rbcs) will enter the mother's circulation (2) and stimulate the production of antibodies (Abs) to the Rh antigen. If the mother has another Rh+ baby, the maternal IgG antibodies against Rh will cross the placenta and enter the baby's blood circulation where the antibodies will bind to the Rh antigen on the baby's rbcs, resulting in complement activation and lysis of the rbcs, causing anaemia.

Systemic immune complex disease

The historical example of systemic immune complex disease is serum sickness. Animal serum (usually horse), containing antitoxin antibodies, was given as a treatment for diphtheria or tetanus to neutralise the toxins secreted by the bacteria (see Chapter 10). Because the serum came from another species, the serum proteins stimulated a strong antibody response. If the person was given a second dose of serum, the antibodies bound to the serum proteins forming immune complexes. These complexes are deposited in blood vessel walls, especially those of the kidney, skin and joints, where they fix complement, leading to an inflammatory response. The resulting symptoms include fever, rashes, arthritis and kidney malfunction. Although serum sickness can now be avoided, systemic immune complex disease can occur in some chronic infectious diseases, such as leprosy, malaria, hepatitis and streptococcal infection.

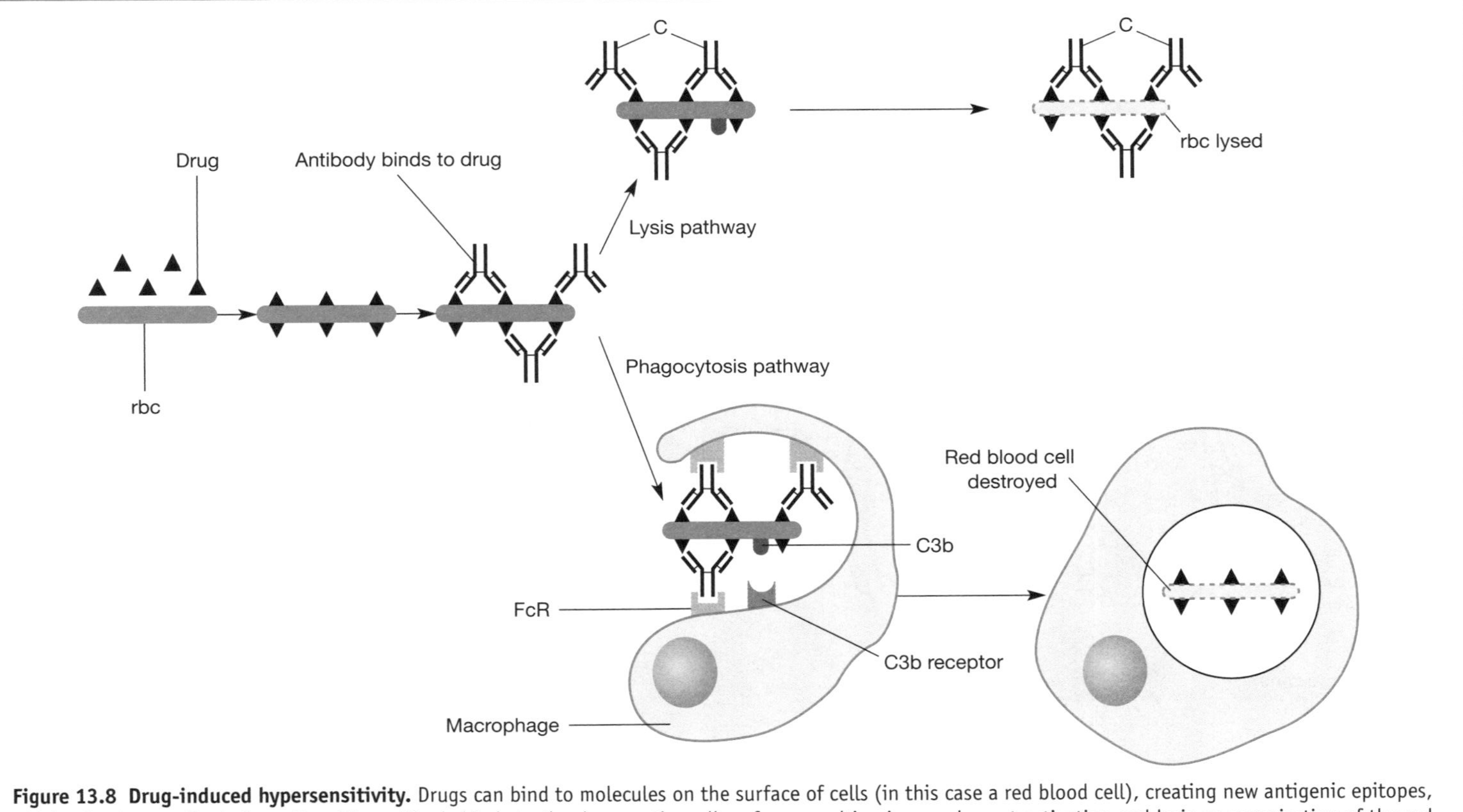

Figure 13.8 Drug-induced hypersensitivity. Drugs can bind to molecules on the surface of cells (in this case a red blood cell), creating new antigenic epitopes, which stimulate antibody formation. The antibody binds to the drug on the cell surface, resulting in complement activation and lysis or opsonisation of the red blood cell.

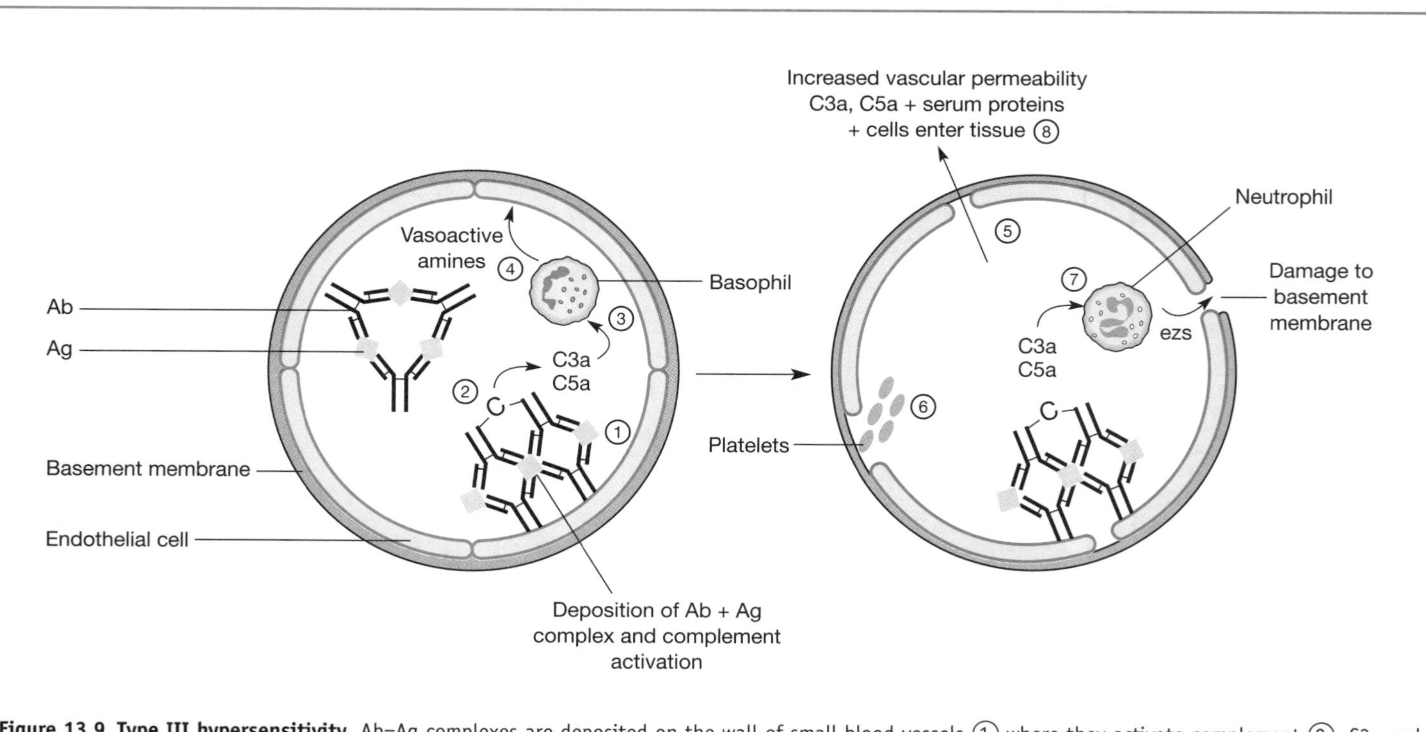

Figure 13.9 Type III hypersensitivity. Ab–Ag complexes are deposited on the wall of small blood vessels ① where they activate complement ②. C3a and C5a activate basophils ③ which release vasoactive amines (e.g. histamine) ④. The vasoactive amines cause vasodilation and increased vascular permeability ⑤, which exposes the basement membrane, allowing platelet binding and activation and the formation of microthrombi ⑥. The C3a and C5a activate neutrophils ⑦, which secrete degradative enzymes causing damage to the basement membrane. If the response persists, the inflammatory mediators enter the tissue where they activate mast cells, initiating an inflammatory response ⑧.

Localised immune complex disease

Type III hypersensitivity can also cause a localised disease. This is most commonly seen after repeated inhalation of antigen. The two best-known examples are farmer's lung, caused by the inhalation of mould spores on hay, and pigeon fancier's disease, caused by the inhalation of pigeon antigens in dried faecal particles.

13.10 Differences between type II and type III hypersensitivity

There is often confusion about the differences between type II and type III hypersensitivity, which both involve the same Ig classes (usually IgG) and complement activation leading to inflammation. The crucial difference between type II and type III hypersensitivity is the nature of the antigen. In type II hypersensitivity the antigen is expressed on the target tissue that is affected by the disease, e.g. in haemolytic disease of the newborn the rhesus antigen is on the red blood cells that are removed or destroyed by the antibody. By contrast, in type III hypersensitivity the antigen is not associated with the target tissue. Type III hypersensitivity is the result of the formation of immune complexes of antibody and antigen. It is not the nature of the antigen that determines where systemic type III hypersensitivity reactions occur. The sites of type III hypersensitivity reactions are determined by where the immune complexes are deposited, and this is usually in small blood vessels such as those seen in the joint, skin and kidneys.

13.11 Delayed hypersensitivity and contact hypersensitivity

Delayed hypersensitivity was initially described as an allergic type response following a skin test. However as opposed to the immediate skin reaction seen in type I allergic responses, in a delayed-type hypersensitivity response the swelling and redness took 24–72 hours to appear, therefore it was called delayed or type 4 hypersensitivity. As described in Section 9.4, delayed-type hypersensitivity does not involve antibody. Delayed-type hypersensitivity also differs from the other hypersensitivities in that it is not necessarily a response against harmless material but is seen in many infections and contributes to protection against pathogens.

One type of delayed hypersensitivity response that is against harmless materials is **contact hypersensitivity.** These reactions occur in response to encounter of small chemicals with the skin. Following a repeated exposure to the chemical there is an eczematous reaction involving swelling, redness and severe itching, which occurs 24–72 hours after contact with the agent in a sensitive individual and is known clinically as **contact dermatitis**. One of the best-known causes of contact hypersensitivity reactions, especially in the USA, is poison ivy. The leaves of the poison ivy plant contain pentadecacatecol, a small chemical that stimulates a very strong contact hypersensitivity reaction. Other causes of contact hypersensitivity reactions are some metals, such as nickel, gold and chromium, used in jewellery, and chemicals found in rubber. The actual chemicals that stimulate contact hypersensitivity reactions are small, often with a molecular mass of less than 100 Da. These chemicals are too small to stimulate an immune response in their free form and are called **haptens.** These haptens have the ability to chemically attach themselves to proteins of the host where they modify the proteins, creating new antigenic epitopes that stimulate the contact hypersensitivity response.

13.12 Summary

- There are three types of antibody-mediated hypersensitivity: type I, type II and type III.
- Type I hypersensitivity involves the production of IgE against usually innocuous antigens such as pollen, house-dust mite faeces or food.
- The IgE binds to Fc receptors for IgE (FcεRs) on the surface of mast cells. Upon re-exposure to the allergen, allergen cross-links the FcεRs, leading to mast cell degranulation.
- Depending on the site of mast cell degranulation, allergy can be manifested by asthma, hayfever, vomiting and diarrhoea, hives, eczema or anaphylactic shock.
- Both genetic and environmental factors contribute to the development of allergy. The incidence has increased dramatically in the past fifty years.
- Treatment of allergy involves allergy avoidance, antihistamines, inhibitors of mast cell degranulation, steroids and immunological intervention, such as desensitisation or administration of anti-IgE antibodies.
- Type II hypersensitivities are caused by IgG or IgM antibodies reacting against antigens on cells or tissues, causing opsonisation and complement fixation.
- There are three major types of type II hypersensitivity reactions: blood transfusion reactions, haemolytic disease of the newborn and drug-induced hypersensitivities.
- Type III hypersensitivity reactions are caused by the formation of immune complexes, which are deposited and initiate inflammatory reactions. Antigens may be benign, such as foreign serum proteins (causing serum sickness) and pigeon faecal antigens (causing pigeon fancier's lung), or because of chronic infection with pathogens, such as malaria, leprosy or hepatitis.
- A fourth type of hypersensitivity reaction is delayed hypersensitivity, which does not involve antibody but involves CD4 and CD8 T cells and monocyte/macrophages.

13.13 Questions

1) In the diagram below, which items from the following list do the letters refer to? (Letters may be used more than once or not at all.) (i) Mast cell, (ii) IgE, (iii) FceR, (iv) Mast cell granule, (v) Allergen.

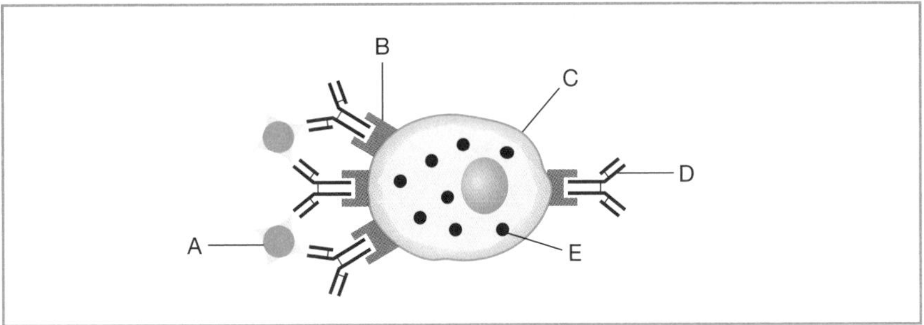

2) Which of the following are COMMON sources of allergens?

    (i) Peanuts
    (ii) Animal fur
    (iii) Paper
    (iv) House-dust mite faeces
    (v) Plastic
    (vi) Cockroaches
    (vii) Shellfish
    (viii) Ryegrass
    (ix) Fibreglass
    (x) Penicillin

3) Fred gets hayfever which is worst in June. What does this suggest about the likely cause and how could you try to identify what Fred was allergic to?

4) What immunological approaches are being used/tested to treat allergy?

5) What are the main features that distinguish type II from type III hypersensitivities?

The answers to these questions can be found on page 341.

13.14 Further reading

1) Rindsjöa E, Scheynius A. (2010) Mechanisms of IgE-mediated allergy. *Experimental Cell Research* 316:1384–1389.

2) Grammatikos AP. (2008) The genetic and environmental basis of atopic diseases. *Annals of Medicine* 40:482–495.

3) Mohapatra SS, Qazi M, Hellermann G. (2010) Immunotherapy for allergies and asthma: present and future. *Current Opinion in Pharmacology* 10:276–288.

CHAPTER FOURTEEN

AIDS

Learning objectives

To know about the history, incidence and clinical progression of AIDS. To understand the biology of the AIDS virus and how it causes immunosuppression. To learn about the immune response to the AIDS virus and be aware of forms of chemotherapy and issues of vaccine development in AIDS.

Key topics

- History of AIDS
- The human immunodeficiency virus (HIV)
  - Structure
  - Replication cycle
- Clinical course of HIV infection
- Immunology of HIV infection
  - Effects of HIV on the immune system
  - Immune response to HIV
- Chemotherapy of AIDS
- HIV vaccines

14.1 History and incidence of AIDS

The story of AIDS began in 1981 when a cluster of unusual diseases was observed in certain groups of people. The two main diseases were pneumonia caused by a yeast, *Pneumocystis carinii*, and an unusual tumour called Kaposi's sarcoma. These diseases were seen initially in homosexual men but later the same symptoms appeared in intravenous drug users and haemophiliacs who were injecting blood-clotting factors. This pattern of disease occurrence suggested that a transmittable agent was responsible for the diseases. The unusual aspect of these diseases was that they are usually only seen in immunosuppressed people and not in people with a fully functioning immune system. The observation that individuals with these diseases had low numbers of CD4 T cells was consistent with immunosuppression and in 1982 the term **acquired immunodeficiency syndrome,** or AIDS, was used by the Centers for Disease Control in Atlanta, USA, to describe the disease.

In 1983 the virus that causes AIDS was isolated from the lymph node of an infected person and was called the **human immunodeficiency virus,** or HIV for short. A second strain of HIV was identified in 1986; this was called HIV-2 and the first strain was renamed HIV-1. HIV-1 and HIV-2 differ in their virulence and geographical location. HIV-2 is less virulent than HIV-1 and is found primarily in western Africa. Genetic studies have shown that both HIV-1 and HIV-2 are natural viruses of primates that have jumped species to infect humans. HIV-1 came from chimpanzees and HIV-2 from the sooty mangabey. Both of these animals are killed for food and it is assumed that it was during this process that the virus initially infected humans. Evidence suggests that the initial transmission from chimpanzee to man that led to the current AIDS epidemic occurred in Cameroon before 1933. HIV-1 and HIV-2 do not cause immunosuppression in chimpanzees or sooty mangabeys; only when the virus crossed into humans did it cause the profound immunosuppression seen in AIDS.

14.1.1 Current incidence of HIV infection and AIDS

It is estimated that, in 2009, 33 million people worldwide were infected with HIV and 2 million died of AIDS. During the whole course of the epidemic 25 million people have died. The highest rate of HIV infection is seen in sub-Saharan Africa, where an estimated 20–40% of young adults are infected. As a result of this, the life expectancy in some sub-Saharan African countries has almost halved; it is now in the 30s–40s instead of approaching 70, which would have been the estimated life expectancy if the AIDS pandemic had not occurred. The pattern of spread in Africa appears to be primarily by heterosexual contact and has a similar incidence in men and women. In central Europe heterosexual spread is numerically the most common but homosexual spread carries the highest risk and it is the most common form of spread in Western Europe, the USA and Oceania.

14.2 The human immunodeficiency virus

14.2.1 There are many different strains and variants of HIV

Some of the terminology used to describe HIV can be confusing. The two strains of HIV are called HIV-1 and HIV-2. There are also many different subtypes of HIV-1 so vaccines will be needed to protect against all subtypes. Additionally, HIV has a very high mutation rate, giving rise to different forms of the virus known as variants. These variants are important because, as described below, they differ in which cell types they can infect.

Although there are different strains, subtypes and variants of HIV, they are very similar in structure and replication and therefore will be described together and referred to collectively as HIV.

14.2.2 HIV is a fairly simple retrovirus

HIV is a retrovirus (Figure 14.1) and contains RNA as its genetic material. The HIV genome contains two molecules of single-stranded RNA, each bound by a molecule of reverse transcriptase. Within the genome are also a p10 protease and a p32 integrase. The genome is surrounded by a nucleocapsid consisting of an inner layer of protein called p24 and an outer layer of protein called p17. The outer portion of the virus consists of a lipid envelope derived from the host cell membrane containing two viral proteins gp120 and gp41, which collectively are called viral **envelope proteins**.

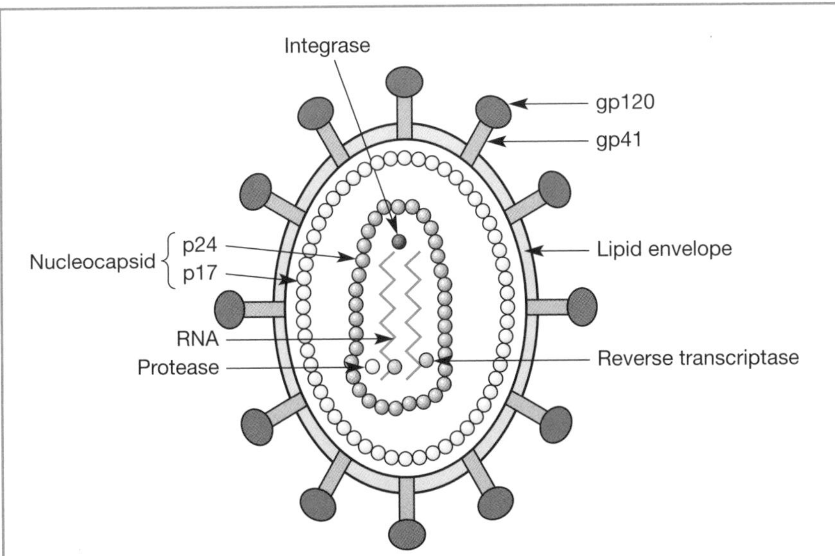

Figure 14.1 Diagram of an HIV particle. HIV is a retrovirus and has two single-stranded RNA molecules as its genetic material. The RNA is associated with reverse transcriptase, integrase and polymerase enzymes, which are necessary for viral DNA and RNA synthesis. Surrounding this is the nucleocapsid, which consists of an inner layer of p24 protein and an outer layer of p17 protein. The outer portion of the virus consists of a lipid layer derived from the host cell into which is inserted the viral gp41 envelope protein. Each gp41 protein molecule is associated with a molecule of the gp120 envelope protein.

14.2.3 The life cycle of HIV

Like any other virus, HIV must infect a host cell before replicating with the viral progeny, leaving the cell to infect others.

HIV infects CD4+ cells

HIV infects cells that express CD4 on their surface. Although CD4 T cells express the most CD4, monocytes and dendritic cells also express low levels of CD4 and are therefore susceptible to infection by HIV. There are two stages to HIV infection: attachment to the host cell, and fusion with the cell membrane to allow the virus to enter the cell (Figure 14.2). The initial binding of HIV to the host cell involves the gp120 protein on the surface of the HIV particle binding to CD4 on the host cell surface. The binding of gp120 to CD4 causes the gp120 to undergo a conformational

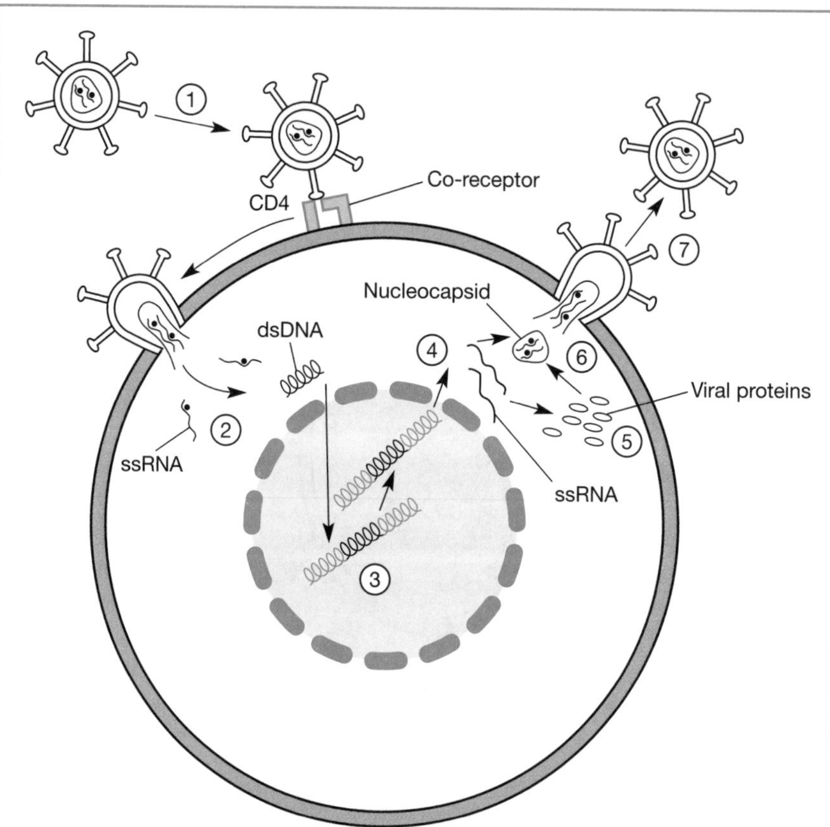

Figure 14.2 Life cycle of HIV. ① The virus binds to CD4 +ve cells through binding of gp120 to CD4 and interactions between the virus and chemokine co-receptors. ② The nucleocapsid enters the cell, where it unfolds, releasing viral RNA, which is reverse-transcribed to double-stranded DNA. ③ The viral DNA integrates in the host genome, where it lies dormant as a provirus. ④ Following cell activation, viral DNA directs the transcription of viral RNA. ⑤ Viral proteins are translated from the RNA. ⑥ The viral proteins and single-stranded viral RNA assemble to form new viral particles. ⑦ The virus buds from the cell, picking up some of the cell membrane, and the complete viral particles can infect other cells.

change so it can bind to a co-receptor. The co-receptor differs between variants of HIV. In some variants the gp120 binds to the chemokine receptor CCR-5, which is on the surface of activated CD4 T cells, monocytes and dendritic cells. These variants can therefore infect all of these cell types and have been called M-tropic. Other HIV variants bind to another chemokine receptor called CXCR-4, which is present on CD4 T cells but not on monocytes or dendritic cells. These variants can only infect T cells and are called T-tropic. The CCR-5 gene is polymorphic and has a null allele which results in a non-functional form of the CCR-5 protein. Approximately 1% of Caucasians are homozygous for the null allele and therefore express no functional CCR-5 protein. These individuals are resistant to infection with HIV. Since their expression of CXCR-4 is normal, this suggests that initial infection is by M-tropic variants using the CCR-5R. Indeed isolates of HIV in recently infected individuals are predominantly M-tropic. There is now evidence that the first cells to be infected are CCR5 expressing CD4 T cells. These can pass the virus on to dendritic cells which can pass the virus on to other CD4 T cells thus initiating a vicious circle. As the virus mutates within infected individuals T-tropic variants emerge which can infect resting CD4 T cells and this is often a sign of rapid progression to AIDS.

Following binding of Gp120 to CD4 and the relevant chemokine receptor, gp41 mediates fusion between the viral envelope and host cell membrane, allowing the viral nucleocapsid to enter the cell.

HIV has a very fast replication rate

Once inside the cell the nucleocapsid of the virus is removed and the reverse transcriptase copies the RNA into double-stranded DNA. The DNA integrates into the host cell DNA, where it is known as a provirus. This stage of viral infection is known as the **latent** stage and the virus can lie dormant for a long time.

When the infected cell is activated, viral RNA is transcribed from the proviral DNA and viral proteins are translated using host cell protein synthesis machinery. The viral proteins and RNA assemble into particles that bud from the cell, taking some of the host cell membrane to form the viral envelope. These particles can now infect other cells.

14.3 Clinical course of HIV infection

The clinical progression of HIV infection can be divided into three stages: infection, the latent stage and development of AIDS (Figure 14.3).

14.3.1 Infection is often clinically asymptomatic

Most people show no symptoms immediately after infection but about 15% demonstrate symptoms that are reminiscent of influenza: these include fever, malaise, aching muscles, sore throat and swollen lymph nodes. Some

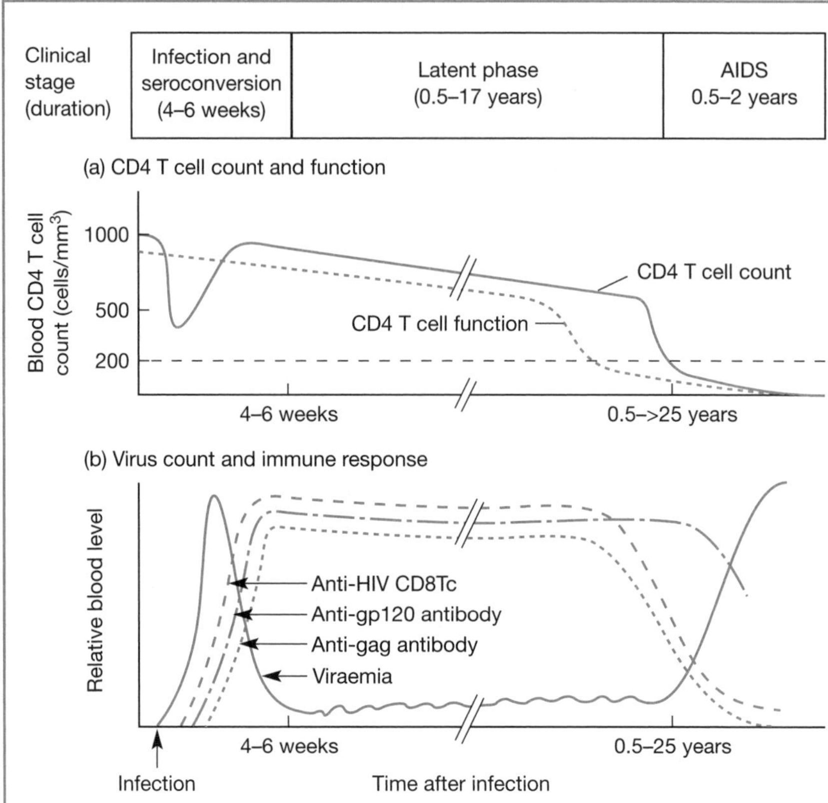

Figure 14.3 The development of AIDS. Following infection the generation of anti-HIV antibody and Tcs clears most of the virus and the disease enters the latent phase. After a variable period of time the loss of CD4 T cell function results in failure to control the virus and the development of AIDS.

people also develop swollen lymph nodes without any other clinical symptoms. Following infection, antibodies to HIV antigens are produced, a process called **seroconversion**. The detection of antibodies to HIV is used to test for infection. The infection phase lasts 4-6 weeks after which HIV infection enters what is called the latent phase.

14.3.2 The latent phase of HIV infection can last many years

This phase of HIV infection is generally asymptomatic although about 33% of infected people will have swollen lymph nodes. The average time from infection to the development of AIDS is about 10 years without treatment but the length of the latency period is extremely variable, lasting less than a year in some people and greater than 25 years in others. Over 90% of HIV+ people will develop AIDS in the absence of treatment.

14.3.3 Development of AIDS

The end of the latency period is accompanied by the emergence of various symptoms that indicate progression to AIDS without treatment. These symptoms include weight loss, night sweats, fever and diarrhoea. There are also infections such as oral candidiasis, herpes simplex virus and shingles caused by minor opportunistic infectious agents.

AIDS is defined clinically by the appearance of major opportunistic infections or by a drop in the CD4 T cell count to below 200 cells/μl of blood. The opportunistic infections are caused by all categories of pathogens as shown in Table 14.1. Without treatment AIDS invariably leads to death. Death is caused by combinations of the infections described in Table 14.1 although some infections are more prominent according to geographical location. In Europe, the USA and Oceania one of the commonest infections is pneumonia caused by *Pneumocystis carinii*. In Africa and Asia, infections such as *Cryptosporidium*, a protozoan that causes severe diarrhoea and weight loss, and *Mycobacterium tuberculosis* are more common.

Table 14.1 Opportunistic infections in AIDS

Pathogen	*Type of organism*	*Disease*
Candida albicans	Fungus	Thrush, disseminated mucocandidiasis
Pneumocystis carinii	Fungus	Pneumonia
Cryptococcus neoformans	Fungus	Meningitis
Herpes simplex	Virus	Pneumonia
Varicella-zoster (chicken pox)	Virus	Shingles, pneumonia
Cytomegalovirus	Virus	Pneumonia
Mycobacterium tuberculosis	Bacterium	Pneumonia
Salmonella spp.	Bacterium	Diarrhoea, septicaemia
Cryptosporidia	Protozoan	Diarrhoea
Toxoplasma gondii	Protozoan	Encephalitis

14.4 Immunological events associated with HIV infection

HIV has a very complicated relationships with the immune system. Because of the nature of the cells it infects and the molecules that it binds to, HIV has a dramatic effect on the immune system, eventually causing profound immunosuppression. Before that stage, and contrary to earlier beliefs, the immune system, far from failing to mount a significant response against the virus, generates a very powerful response against HIV. With almost any other virus this response would be enough to eliminate the pathogen but because of the special features of HIV the virus is able to survive and eventually destroy the immune system.

14.4.1 Changes in the immune system during HIV infection

The hallmark of HIV infection is the gradual decrease in CD4 T cell numbers. However, there are a number of other changes to the immune system that occur following HIV infection.

Mucosal CD4 T cells in are rapidly lost after HIV infection

One of the most dramatic early events following HIV infection is the loss of mucosal CD4 T cells which are deleted by up to 90%. This occurs within a few weeks of infection and is thought to be due to the fact that mucosal CD4 T cells are in an activated state and therefore express CCR5.

Changes in lymph nodes

Many HIV+ individuals develop swollen lymph nodes during initial infection and this can persist even during the clinically latent phase. Histological examination of the lymph nodes shows increasing disruption of the normal lymph node architecture, an influx of CD8 T cells and eventual loss of germinal centres.

There is a loss in CD4 T cell function as well as cell numbers in HIV infection

As the infection progresses HIV+ individuals show a loss of CD4 T cell function that cannot be explained simply on the basis of reduced CD4 T cell numbers and is somehow associated with the effects of the virus (see Figure 14.3). Transplantation recipients taking immunosuppressive drugs can show drops in CD4 T cell numbers similar to those seen in HIV+ individuals but the transplant recipients demonstrate much better CD4 T cell function. A number of HIV proteins, including gp41 and Vpr, have immunosuppressive properties which can inhibit T and B cell proliferation and the production of some cytokines such as 1L-12. One consequence of the loss of Th activity is the reduced ability to mount a delayed-type hypersensitivity reaction (Chapter 9) and as the infection progresses the ability to generate antibody responses is also lost.

Antibody abnormalities occur following HIV infection

Somewhat paradoxically, HIV+ individuals can demonstrate increased total levels of serum Ig despite the impaired ability to generate specific antibody responses. Again the basis for this is not clear but it is thought to reflect abnormalities in the normal regulation of immune responses caused by the virus. Possibly also related to generalised immune dysfunction, HIV+

individuals can show increased production of autoantibodies to antigens such as red blood cells, spermatozoa or myelin (a component of nerve sheaths) and they may suffer from flare-ups of allergies such as eczema.

14.4.2 The immune response to HIV varies among different individuals

Immediately following infection with HIV there is a period of rapid viral replication resulting in high levels of virus in the blood (viraemia) (Figure 14.3). HIV is an intracellular pathogen and as such would be expected to stimulate a strong CD8 cytotoxic T cell response and possibly a delayed-type hypersensitivity response, as well as the production of antibody. In fact strong antibody responses to the gp120 and gp41 envelope proteins, the p24 protein of the nucleocapsid (also known as gag) and some of the proteins that make up the reverse transcriptase (pol antigens) are seen soon after viral infection. There is also a powerful CD8 cytotoxic T cell against gp120, p24 pol and Tat antigens. The immune response against HIV results in a reduction in viraemia, that is, the level of virus in the blood. Different people vary in their ability to control the virus during the latent phase and have been put into three categories:

- **Progressors** have the highest levels of virus and in the absence of treatment would develop AIDS the most quickly.
- **Viraemic controllers** have lower viral levels and appear to control the virus much better than progressors. Without treatment they may eventually develop AIDS, although much later after infection than progressors.
- **Elite controllers** represent less than 1% of HIV infected people and are able to control the virus to extremely low or undetectable levels and most maintain relatively normal levels of blood CD4 T cells. Some of these have been infected for 25 years without treatment and have not developed AIDS.

The reason why these elite controllers maintain very low levels of virus is not fully understood but is due to genetic and other factors. There is obviously considerable interest in finding out the immunological basis for this 'elite' control for the purpose of vaccine design.

During the latent phase there is continuous destruction of the virus with the immune system eliminating up to 30% of the total viral load *every day*. However, the high replication rate of HIV means that the viral load stays the same although, as described above, viral load differs between progressors, viraemic controllers and elite controllers. There is also considerable destruction of CD4 T cells with up to 2×10^9 cells being destroyed every day. Most of these are replaced by the immune system so that the overall CD4 levels in the blood decline only very slowly. Therefore, although the latent phase is clinically latent it is a very dynamic situation with high viral destruction and replication and extensive CD4 T cell destruction and replacement. Two questions occur as a consequence of this picture: why, with the exception of elite controllers, does the immune response not clear the virus, and what is the cause of the extensive CD4 T cell destruction?

In terms of why the immune system does not clear the virus, HIV has a number of features that enable it to survive in the face of a powerful immunological onslaught:

- The very high replication rate of HIV is an important reason why the virus is not totally cleared.
- HIV can hide as a provirus where it is not detectable by the immune system.
- HIV has a very high mutation rate, so that antigens to which antibody and CD8 Tcs have been made mutate and are no longer recognised by the immune system. Antibodies and CD8 Tcs are made against the mutated antigen but this can mutate again and so the process of evading the immune response goes on and on.

The cause of CD4 T cell loss is thought to be a combination of direct killing by the virus and destruction of virally infected cells by the immune system (Figure 14.4). CD4 cells that are infected with the virus express viral antigens on their surface. These may be viral peptides presented by class I MHC or they may be soluble gp120 bound to CD4 on the T cells. HIV+ individuals can have large amounts of soluble gp120 in their blood and lymph, which will bind to CD4. Because of the expression of viral antigen, CD4 T cells can be killed in a number of ways (see Figure 14.4):

- **Antibody + complement.** Anti-gp120 binds to the gp120 bound to CD4, resulting in complement fixation and activation and cell lysis.
- **Antibody-dependent cell-mediated cytotoxicity.** Macrophages and NK cells have Fc receptors on their surface. This enables them to bind to the Fc portion of the anti-gp120 antibody bound to the CD4 T cell and kill the cell.
- **CD8 cytotoxic T cells** kill virally infected cells expressing HIV antigenic peptides in association with their class I MHC.

By generating these responses the immune system is actually doing what it is supposed to; it is generating effector mechanisms that eliminate virally infected cells and thereby the virus. Unfortunately the main type of infected cell is the CD4 T cell and therefore the virus kills the very cells needed to generate an immune response against it. Although the immune system makes a valiant effort to replace the CD4 T cells that are destroyed every day, the production of CD4 T cells never quite matches the destruction and there is a gradual drop in CD4 T cell numbers. Eventually the CD4 T cell numbers become too low to maintain normal immune responsiveness and immunosuppression ensues.

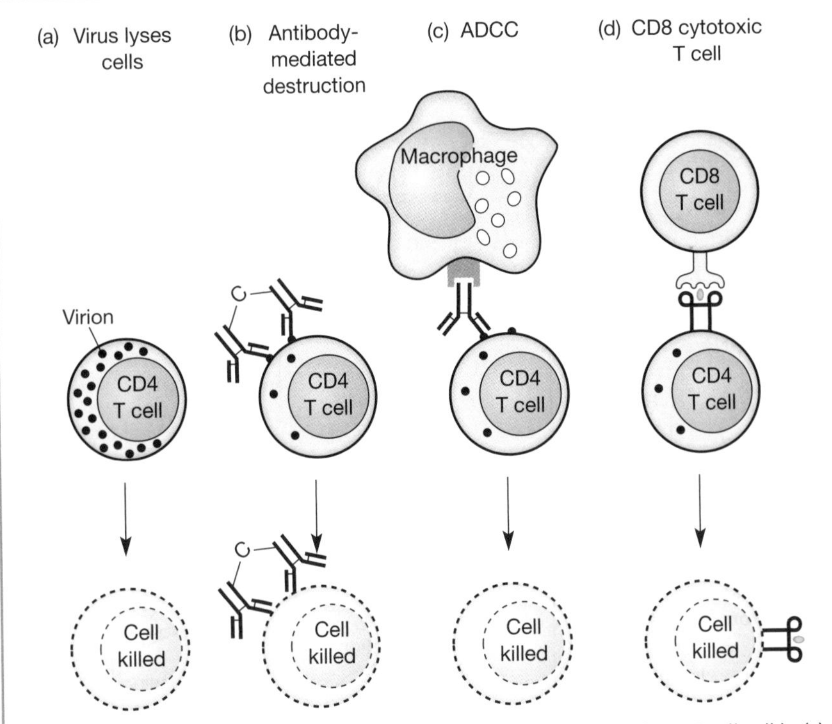

Figure 14.4 Killing of CD4 T cells. (a) HIV may directly cause lysis of CD4 T cells. (b), (c) Infected cells bearing HIV antigens may be killed by antibody and complement (b) or killed by antibody-dependent cell-mediated cytotoxicity (c). (d) HIV-infected cells presenting HIV peptides on their class I MHC molecules may be killed by CD8 Tcs.

14.5 Chemotherapy can prolong the life of HIV-infected people

The high mutation rate of HIV is not only a problem for the immune system. It is also a problem in trying to treat HIV+ individuals with antiviral drugs. The first anti-HIV drug was zidovudine (AZT), a reverse transcriptase inhibitor introduced in 1987. Although it inhibited replication of HIV, the virus very quickly mutated and became resistant to the drug.

There are now five types of anti-HIV drug in use clinically (Figure 14.5). There are two types of reverse transcriptase inhibitors, nucleoside analogue reverse transcriptase inhibitors which terminate DNA chain elongation and non-nucleoside reverse transcriptase inhibitors which inhibit the reverse transcriptase enzyme (Figure 14.6). HIV protease inhibitors inhibit the cleavage of the polyproteins (Figure 14.6). The most recently developed anti-HIV drugs are the entry blockers and integrase inhibitors (Figure 14.5). Because of the rapid development of resistance to single drugs, anti-HIV drugs are given as combinations of three or four different drugs in a therapy called highly active antiretroviral therapy, or HAART.

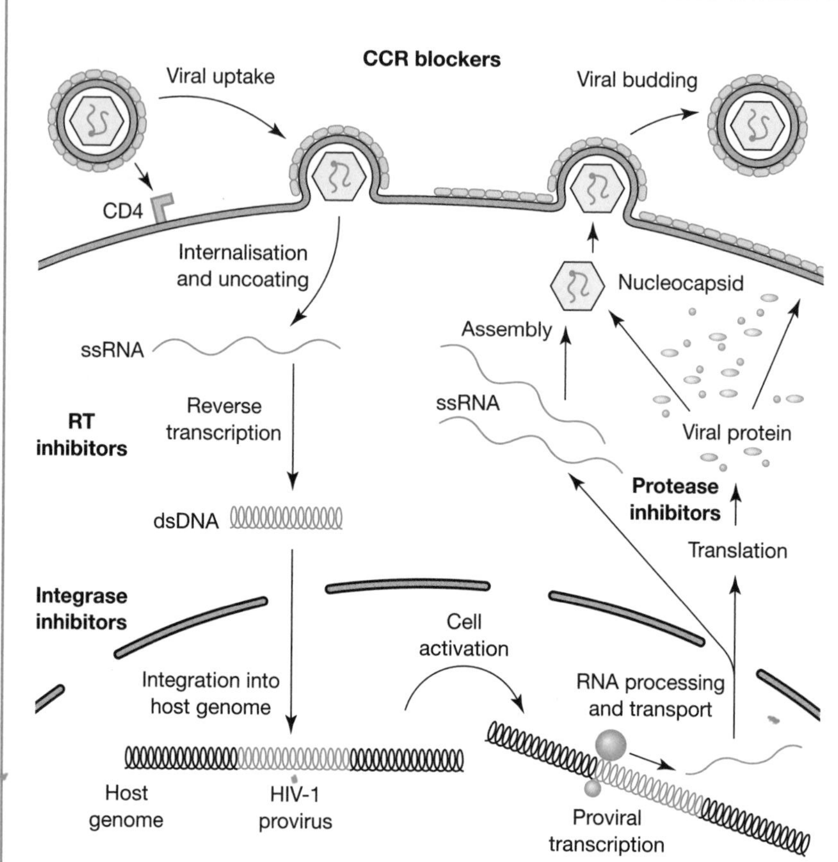

Figure 14.5 Anti-HIV drugs. CCR blockers block the binding of gp120 to CCR5. RT inhibitors inhibit reverse transcription. Integrase inhibitors inhibit the integration of viral DNA into the host geneome. Protease inhibitors inhibit the HIV protease.

HAART typically consists of two nucleoside reverse transcriptase inhibitors and one or two of the other types of anti-HIV drugs. It has revolutionised the outlook of people with HIV. It can be very effective at lowering levels of virus and raising CD4 T cell levels and results in significant reduction in mortality and morbidity so that many AIDS patients are able to leave the AIDS clinic and return home. However, there are problems with HAART. It is not a cure for AIDS so treatment is lifelong. There is considerable toxicity, especially to the liver, bone marrow and gut, which means some people cannot take the drugs. The regime for taking the drugs is complicated and intrusive to a normal lifestyle. Some of the drugs have to be taken with food and some without, and some drugs cannot be taken within a certain time period of each other. This contributed to poor compliance with the treatment. However this problem is being addressed by combining the drugs into single pills called fixed-dose combination pills so

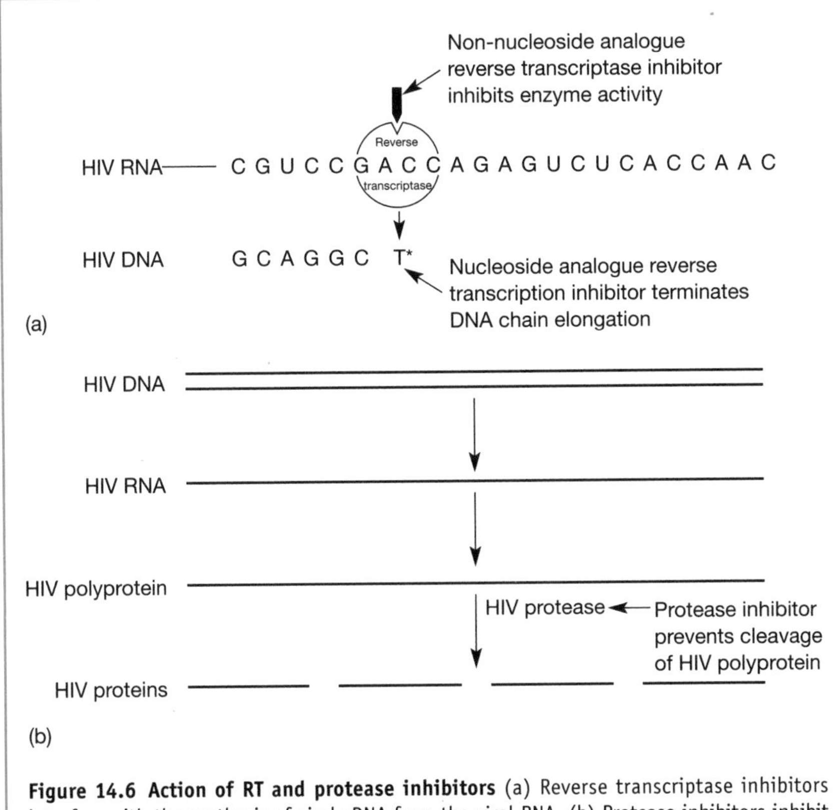

Figure 14.6 Action of RT and protease inhibitors (a) Reverse transcriptase inhibitors interfere with the synthesis of viral cDNA from the viral RNA. (b) Protease inhibitors inhibit the cleavage of viral polyproteins into viral proteins. Without the viral proteins replication cannot occur.

that only one or two pills need to be taken on a daily basis. Finally the drug treatment is very expensive – about $15 000 a year – so that it is not readily available in the countries that need it the most.

14.6 HIV has proven very difficult for vaccine development

Attempts to generate vaccines for HIV have fluctuated in intensity over the years for a variety of different reasons, both economic and scientific. Initially two type of vaccines were considered:

- **Prophylactic vaccines.** The aim of these vaccines was to provide protection to individuals who were not infected with the virus; in this respect the aim was the same as for vaccines against other infectious agents such as measles, mumps, polio, smallpox, etc.
- **Therapeutic vaccines.** Because AIDS initially was considered a disease of immunosuppression where HIV-infected people did not make a very

good immune response against the virus, it was thought it might be possible to boost the immune response of HIV-infected people so that they could mount a response and clear the virus. With the current appreciation that most HIV-infected people make a very strong response against the virus it is not so clear how much this response could be boosted. It may be that the best use of therapeutic vaccines will be to change the type of immune response to one which controls the virus.

The development of prophylactic vaccines against HIV has considerable difficulties of both an immunological and a logistical nature.

One of the main immunological issues concerning development of an HIV vaccine is which of the many different current approaches to vaccine development (see Chapter 10) to use, given the many subtypes and variants of the virus and its high mutation rate. Another immunological issue is what type of immune response to try to stimulate. Additionally there are no good small animal models of HIV.

The main logistical problems are in testing the vaccines. Traditionally vaccines were developed against acute diseases that often occurred in waves or epidemics (e.g. measles or mumps) and had a low mortality rate. It was therefore comparatively straightforward to vaccinate a section of the population and see whether these people were protected from the disease compared with non-vaccinated individuals. It was possible this way to get an answer to the effectiveness of the vaccine in a relatively short time. Because AIDS is a chronically progressive disease the conventional approach would take many years to give an answer – too long for most people to wait. A number of vaccines have been tested in clinical trials using DNA, recombinant and antigen/peptide approaches. Unfortunately, until recently none of the vaccines were successful but in 2009 one vaccine did show a 30% reduction of HIV infection. While 30% protection would not be considered acceptable for a conventional vaccine the trial did at least show that vaccines could provide protection against HIV infection and the challenge now is to produce a more effective vaccine.

14.7 Summary

- AIDS as a disease was first identified in 1981 in groups of high-risk individuals: homosexual men, intravenous drug users and haemophiliacs. Currently it is estimated that 30 million people are infected with the AIDS virus.
- The virus causing AIDS was isolated in 1983 and called the human immunodeficiency virus (HIV). A second strain, called HIV-2, was identified in 1986.

- HIV is a retrovirus, i.e. its genetic material consists of RNA. It infects CD4+ cells, mainly CD4 T cells but also monocytes and dendritic cells.
- Clinically HIV infection is characterised by the development of opportunistic infections, especially with the yeast *Pneumocystis carinii*, and the development of a rare tumour, Kaposi's sarcoma.
- The development of AIDS can take over 25 years from infection. Following infection and seroconversion there is a latent period that is largely aclinical.
- Although the latent period is aclinical there is an extensive amount of viral destruction and replication and also destruction and replacement of CD4 T cells. However, there is a gradual net loss of CD4 T cells, leading eventually to immunosuppression and the development of opportunistic infections, which are fatal without treatment.
- The most successful chemotherapy for HIV is highly effective antiretroviral therapy (HAART) with a combination of reverse transcriptase inhibitors and viral protease inhibitors. However, this therapy has significant side-effects, is difficult to comply with, and is very expensive.
- HIV vaccines are faced with the problem of the high mutation rate of the virus and logistical problems of testing the efficacy of a vaccine against a chronic infection. A recent vaccine trial has shown a small protective effect.

14.8 Questions

1) In the diagram of an HIV particle on the next page, match the arrows with items from the following list (N.B. not all labels in list will necessarily be used):

   (i) Envelope proteins
   (ii) gp120
   (iii) gp41
   (iv) Integrase
   (v) Lipid envelope
   (vi) Nucleocapsid
   (vii) p17
   (viii) p24
   (ix) Protease
   (x) Phosphatase
   (xi) Reverse transcriptase
   (xii) Ribonuclease
   (xiii) ssRNA
   (xiv) Serine protease

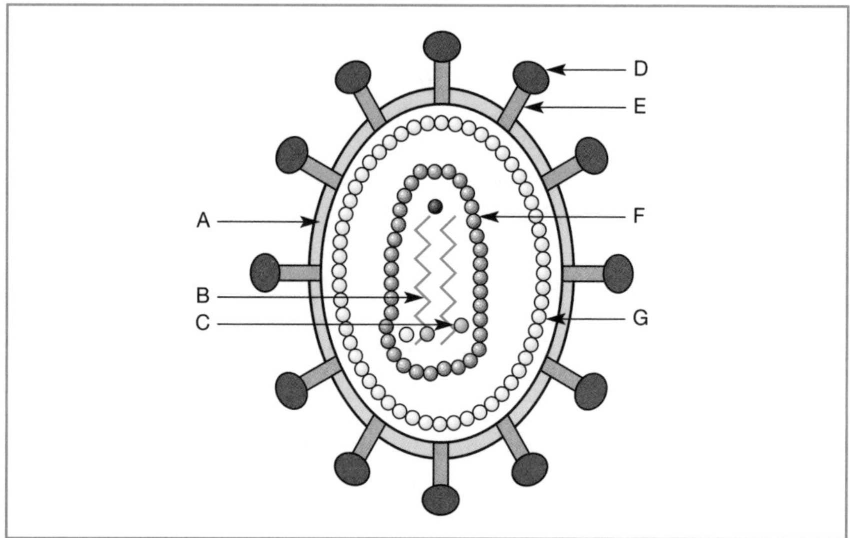

2) Draw a diagram showing the time course of HIV/AIDS in terms of (i) clinical stage, (ii) level of viraemia, (iii) CD4 T cell count, (iv) antibody response, (v) CD8 cytotoxic T cell response. Assume no treatment is given.

3) What are the features about HIV that make it so difficult for the immune system to eliminate it?

4) Describe what is meant by HAART.

5) What particular problems are associated with development of a vaccine against AIDS?

The answers to these questions can be found on page 342.

14.9 Further reading

1) O'Connell KA, Bailey JR, Blankson JN. (2009) Elucidating the elite: mechanisms of control in HIV-1 infection. *Trends in Pharmacological Sciences* 30: 631–637.

2) Sharp PM, Hahn BH. (2008) AIDS: Prehistory of HIV-1. *Nature* 455: 605–606.

3) AIDS Knowledge Database. **hivinsite.ucsf.edu/InSite?page=SM**

CHAPTER FIFTEEN

Manipulating the immune system: transplantation and tumours

Learning objectives

To understand contemporary approaches to manipulating the immune system in terms of transplantation and immunotherapy of tumours.

Key topics

- Transplantation
  - Types of transplant
  - Transplantation antigens
  - Immune responses to transplants
  - Preventing graft rejection
- Tumour immunology
  - Tumour antigens
  - Antibody-based tumour therapy
  - Tumour vaccines

15.1 Introduction

Part of the reason for understanding as much as possible about how the immune system functions is to aid in the development of treatments that can manipulate the immune system to prevent or treat diseases in which the immune system is involved. One method of manipulating the immune system, that of vaccination against infectious disease (described in Chapter 10), has prevented millions of deaths and has removed the need for billions of children to go through the misery and danger of many common childhood illnesses such as measles, mumps, polio and diphtheria. However, manipulation of the immune system does not stop at vaccination. In situations such as transplantation and cancer it is also desirable to alter the way the immune system behaves, either to prevent graft rejection or to harness the power of the immune system to attack tumour cells. This chapter will cover some aspects of immune manipulation in the fields of transplantation and cancer.

15.2 Transplantation: from kidneys to faces

Transplantation involves the transfer of tissue, cells or organs from one anatomical site to another. These two anatomical sites may be different sites on the same person, e.g. with a skin transplant for burns, or they may be equivalent anatomical sites in different people, e.g. a heart transplant from one individual to another. The first clinical kidney transplant was done in 1954 and today transplantation of a variety of tissues, cells and organs is now an accepted medical procedure for treating a wide variety of diseases and further treatments using transplants are being developed (see Table 15.1). The latest controversy has been that associated with face transplants.

15.2.1 Categories of transplant

As with so many other areas of immunology, transplantation has its own jargon. Terms have been used to describe the relationship between donor and recipient in transplantation:

- **Autograft.** A transplant of tissue from one site to another in the same individual is called an autograft. The most common instance is the transplantation of healthy skin to the burn area of burn victims to try to prevent infection and dehydration and promote wound healing.
- **Isograft.** An isograft is a transplant performed between genetically identical individuals, e.g. identical twins. Like autografts, isografts will normally survive without any immunosuppression.
- **Allograft.** Transplants between non-genetically identical individuals of the same species are called allografts. This is the most common category of transplant and the graft will be rejected in the absence of immunosuppressive therapy.
- **Xenograft.** This involves transplanting tissue from one species to another, e.g. baboon to human or pig to human. As might be expected, xenografts are subject to the strongest rejection responses and pose extra problems in preventing their rejection.

15.2.2 Transplantation antigens: the barriers to survival of transplants

Transplants between non-identical donors and recipients (allografts or xenografts) are rejected because the immune system recognises foreign antigens on the graft and does what it is supposed to: mount a specific immune response to eliminate the foreign organ. The antigens that stimulate graft rejection are called, unsurprisingly, transplantation antigens.

Table 15.1 Types of clinical transplant

Transplant	*Disease treated*
Kidney	Nephritis, diabetic complications
Heart	Heart failure
Lung/heart + lung	Cystic fibrosis
Pancreas/islets of Langerhans	Insulin-dependent diabetes mellitus
Liver	Congenital defects of the liver, hepatitis, cirrhosis
Cornea	Cataracts
Skin (usually autologous)	Burns
Bone marrow	Congenital haematopoietic defects (e.g. thalassaemia, severe combined immunodeficiency), leukaemia, lymphoma
Face	Animal attack, cancer

The transplantation antigens that stimulate rejection of allografts fall into two categories: major transplantation antigens and minor transplantation antigens.

Major transplantation antigens stimulate strong rejection responses

The major transplantation antigens are the class I and class II MHC molecules, HLA-A, B and C and HLA-DP, DQ and DR. As described in Chapter 4, MHC genes are highly polymorphic so that the donor and recipient will almost certainly have different alleles of at least some of the MHC loci. The cells of the graft will therefore have MHC molecules on their surface that are foreign to the recipient. Foreign (allogeneic) MHC antigens stimulate a particularly vigorous immune response (Figure 15.1). The reasons for this are not totally clear. Class I MHC alloantigens stimulate strong antibody and CD8 cytotoxic T cell responses. Class II MHC alloantigens stimulate CD4 T cells to become effector Th and also stimulate antibody responses. The Th help in the generation of antibody against class I and class II MHC, the generation of CD8 cyotoxic T cells, and the development of delayed-type hypersensitivity reactions (Figure 15.1).

Minor transplantation antigens also contribute to the rejection response

Proteins other than MHC are polymorphic, although to a much lesser degree than class I and II MHC. The graft donor and recipient will have different forms of many of these proteins which can act as antigens. Because they do not stimulate as strong a rejection response as MHC they are called minor histocompatibility antigens. Minor histocompatibility antigens are recognised by the recipient's T cells in association with self-MHC in the same way as any antigen from a pathogen (Figure 15.2). The

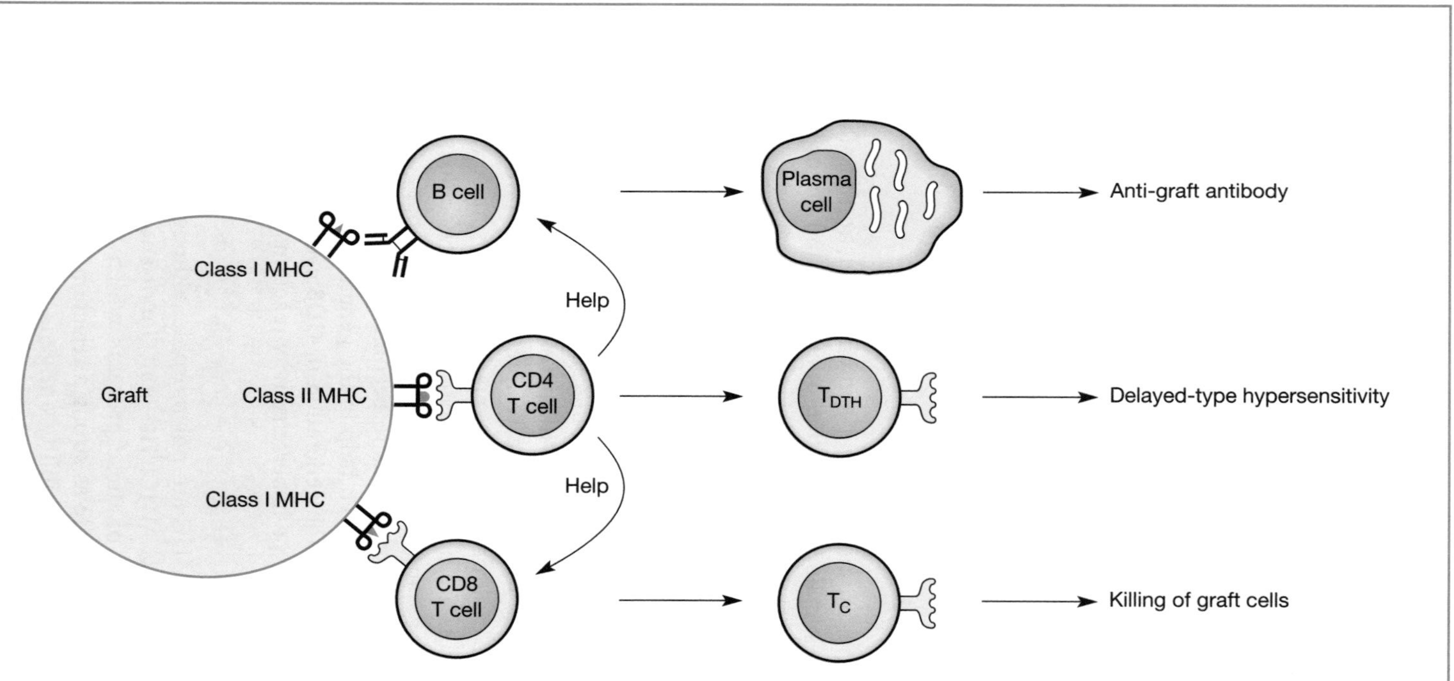

Figure 15.1 Immune response to foreign MHC. Foreign class I MHC on transplants stimulate strong antibody and CD8 Tc responses. The foreign class II MHC stimulates CD4 T cells, which can help in the generation of B cells or Tcs against class I MHC or can promote delayed-type hypersensitivity response within the graft. Foreign class II MHC can also stimulate B cells, resulting in antibody formation against the class II MHC.

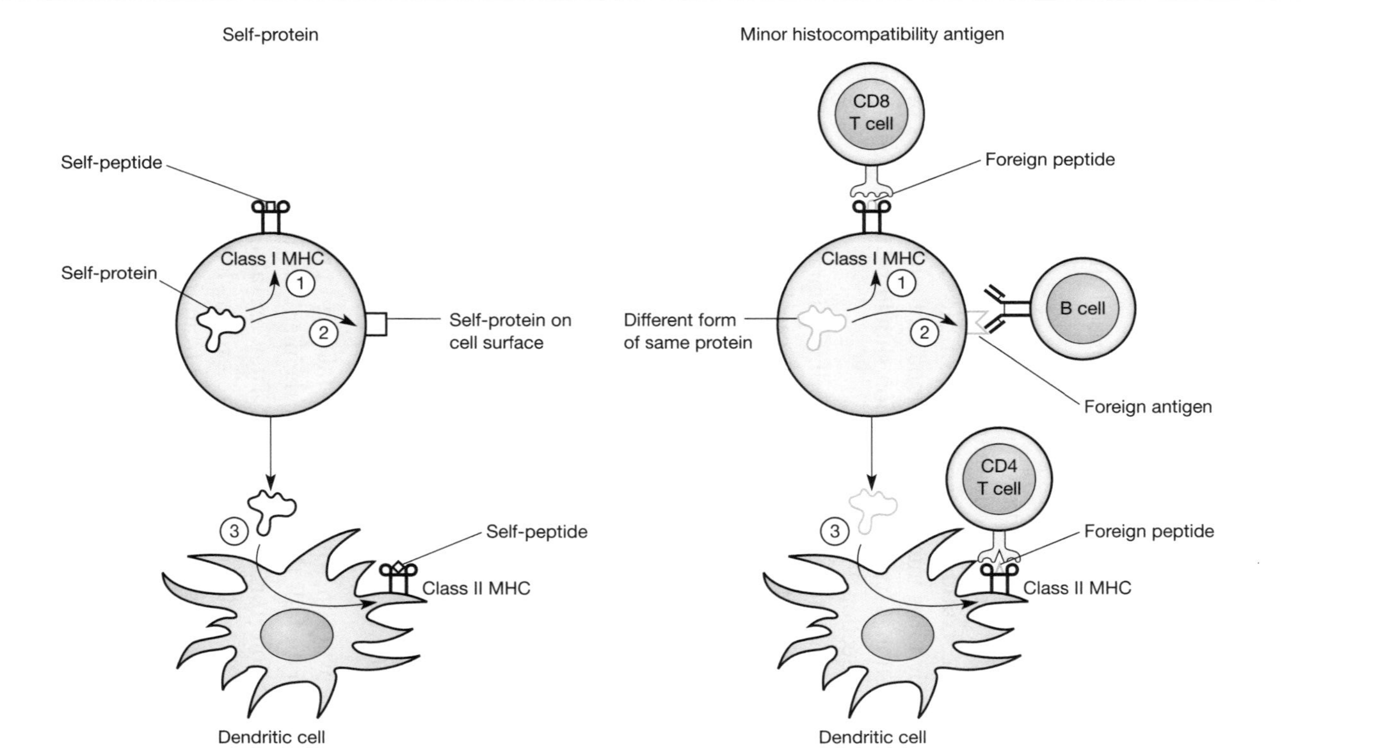

Figure 15.2 Minor transplantation antigens. Any protein for which the gene is polymorphic can potentially act as a minor transplantation antigen. Surface proteins can stimulate antibody production and internal proteins can be presented by class I MHC to simulate CD8 T cell responses. Protein that is taken up by dendritic cells can be processed and presented on class II MHC to stimulate CD4 T cells, which provide help for the antibody and CD8 T cell responses.

self-MHC can be on the recipient's cells or, if the graft and recipient are matched for any MHC molecules, on the graft cells – in this case self- and graft MHC are shared. There are potentially hundreds of minor transplantation antigens although only about eighty have been identified. Most stimulate only T cells although antibody is also made to antigens other than MHC on a graft.

15.2.3 Transplants can be the target of different types of immune response

Graft rejection responses have been categorised into three main types: hyperacute, acute and chronic.

- **Hyperacute rejection** occurs if the recipient has pre-existing antibody against the graft. The antibody binds to antigens on the graft, resulting in complement fixation and activation of the clotting pathways. The graft essentially turns into a massive blood clot within minutes or hours of transplantation. Hyperacute rejection should be prevented by cross-matching (see Section 15.2.4).
- **Acute rejection** is the equivalent of a primary immune response against the graft. All kinds of effector mechanism can be generated against the graft: antibody, cytotoxic CD8 T cells and delayed-type hypersensitivity responses. Which of these occurs depends on a number of factors, including the degree of matching. They act against the graft in the same way as they act against pathogens, as described in Chapters 8 and 9. Without immunosuppression the graft would be rejected in 7–20 days.
- **Chronic rejection** is a process that usually starts after the graft has been in place for more than a year. One of the main pathological changes seen is thickening of the graft's arterial walls, which can lead to occlusion of the artery and ischaemia. Eventually a large enough part of the graft is without an adequate blood supply and the graft stops functioning.

15.2.4 A number of approaches are used to prevent graft rejection

There are three ways in which steps are taken to prevent graft rejection (Figure 15.3). First of all the donor and recipient are blood typed so that they are compatible for blood groups. They are also cross-matched to check that the recipient does not have pre-existing antibody to the donor. Cross-matching involves taking some serum from the recipient and adding it to leukocytes from the donor together with some complement. If the recipient has anti-donor antibodies they will bind to the donor leukocytes and fix complement, resulting in lysis of the leukocytes which can be visualised using fluorescent dyes.

The second step to improving successful transplantation is matching the recipient and donor for HLA by tissue typing. HLA typing was described in Chapter 4. Unless the donor is a family member there will only be a small

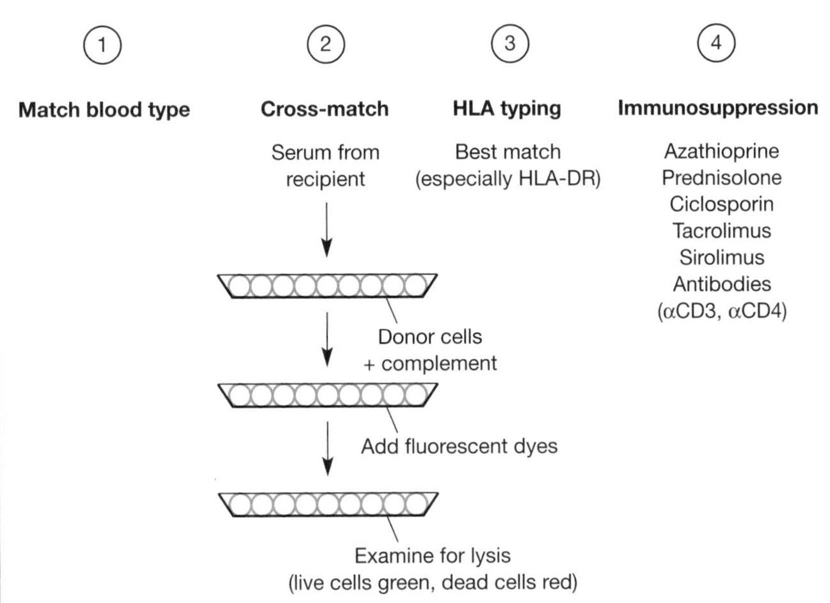

Figure 15.3 Steps in preventing transplant rejection. To prevent hyper-acute rejection caused by pre-existing antibody against donor antigens, the recipient and donor are matched for blood type and also cross-matched. Cross-matching involves incubating serum from the recipient with white blood cells from the prospective donor in the presence of complement. If the recipient has antibodies against donor antigens, the antibodies will bind to the donor cells and fix complement, resulting in lysis of the cells. The lysis can be detected by adding fluorescent dyes, which stain dead cells red while live cells are seen as green under a microscope. Potential donors and recipient are HLA-typed (see Chapter 4) and the donor with the best match is chosen for the transplant. Finally immunosuppressive agents are given to the recipient to try to prevent the anti-graft immune response.

chance of matching the donor and recipient at all six HLA loci, especially taking into account that MHC molecules are co-expressed and therefore people may well express two different alleles at one or more of the MHC loci. However, the more HLA matches there are between donor and recipient, the better the chance of a successful transplant. Matching at *HLA-DR* (Figure 15.4) has been shown to have the most beneficial effect but matching at *HLA-A* and *HLA-B* is also of benefit.

The third step in preventing graft rejection, and the one that made clinical transplantation feasible, is the use of immunosuppression. Immunosuppression involves the use of drugs or antibodies to inhibit immune responses against the graft. Although the drugs are very effective at inhibiting the response to the graft they also inhibit the response to pathogens. Transplant recipients are consequently at increased risk of infection and also of the development of some types of tumours.

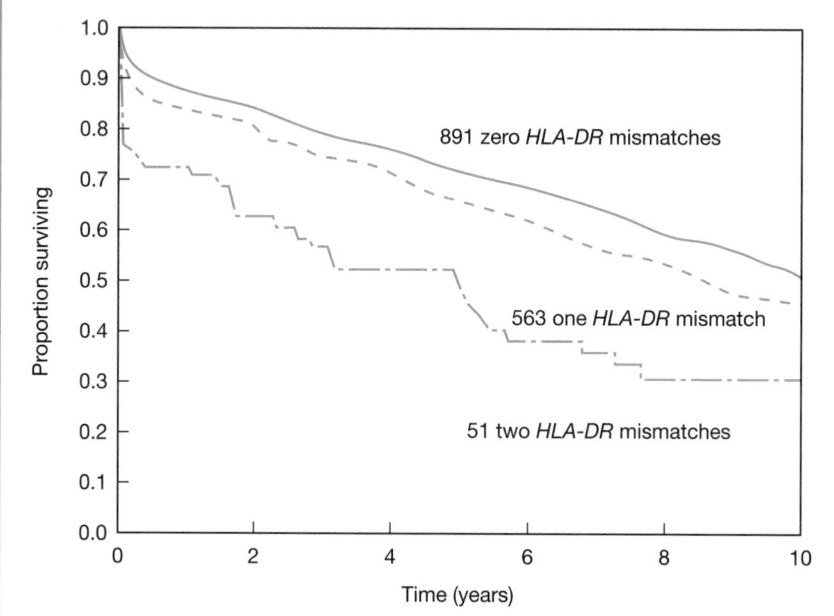

Figure 15.4 The effect of *HLA-DR* matching on graft survival. Transplants where donor and recipient are matched at both *HLA-DR* alleles show better survival than those with only one or no matches. The data are on kidney transplants performed in the transplant centre of Manchester Royal Infirmary from 1985 to 1999 (courtesy of Dr Phillip A. Dyer).

15.2.5 Clinical immunosuppression uses many type of drugs and antibodies

The first class of immunosuppressive drugs were the anti-mitotic agents. They were able to inhibit the proliferation of lymphocytes required during a specific immune response. A commonly used anti-mitotic drug in transplantation is azathioprine. A problem with anti-mitotic drugs is that other tissues containing rapidly dividing cells, such as bone marrow and gut epithelium, are also affected.

Corticosteroids affect the transcription of many genes in lymphocytes and other cells. A consequence of this is that steroids affect lymphocyte migration, resulting in fewer cells entering the graft. They also inhibit macrophage and monocyte function and have general anti-inflammatory effects. The most commonly used steroid in transplantation is prednisone.

Up until the 1980s the immunosuppressive regimen of choice was a combination of azathioprine and prednisone. This could be quite effective and in some centres one-year kidney graft survival rates exceeded 80%. In the 1980s the first of a new breed of immunosuppressive drugs was introduced into clinical transplantation with dramatic effects on graft survival rates. This was **Ciclosporin,** a fungal product, which, like the later members tacrolimus and sirolimus, specifically inhibits CD4 T cell function. The drugs differ in the molecular details of their action but a major effect they have in common is to inhibit the **IL-2 pathway.** Ciclosporin and tacrolimus inhibit the production of IL-2, and sirolimus inhibits signalling through the IL-2 receptor. Because IL-2 plays a vital role in the proliferation of CD4 T cells, inhibition of the pathway provides a powerful immunosuppressive effect.

Finally, antibodies can be used in transplantation to inhibit T cell function. These include polyclonal antithymocyte globulins from the serum of sheep or goats immunised with human thymocytes, or monoclonal antibodies (see Chapter 10) to T cell antigens such as CD3 or CD4. These antibodies act by causing the destruction of T cells or interfering with their function. They cause profound immunosuppression and therefore are used primarily in the period immediately after transplantation and to treat acute rejection episodes where the conventional immunosuppression is failing to prevent the rejection response.

15.3 Using the immune system against tumours

For over one hundred years there was considerable controversy about the role of the immune system in cancer and particularly whether individuals made immune responses against tumour cells. Recent investigations have shown that T cells and antibody specific for tumour antigens can be found in people with cancer. The presence of antibody can sometimes be detected long before clinical diagnosis and a test for cancer of the prostate has been licensed on the basis of detecting anti-prostate cancer antibodies.

Tumour antigens can be of many different types (see Table 15.2). Some viruses are associated with tumours and the immune response can target viral antigens. This is the case for the vaccine against the human papilloma virus which is being used to protect against cervical cancer. Other tumour antigens are the result of mutations in the genes of the cancer cells. It is thought that on average six mutations are required for a cell to become truly tumorigenic. Sometimes when genes mutate the mutation alters the protein produced so that it is recognised as foreign (Figure 15.5). Other antigens are the so-called onco-foetal antigens. These are proteins which are normally only expressed during foetal life but not in adult cells. However, during the development of tumours some of these proteins are re-expressed.

Tumour cells can evade and subvert the immune system

If tumours do express antigens that can be recognised by the immune system the big question is how do cancer cells avoid destruction and develop into full-blown tumours? It appears that tumours use two main mechanisms to avoid immune destruction; these are **escape** and **subversion**. Tumour cells are constantly mutating. When tumour cells mutate it is possible that the mutated cells are no longer recognised by the immune system. Obviously these cells have a selective advantage and so will preferentially grow – this is known as immune escape. One way tumours avoid

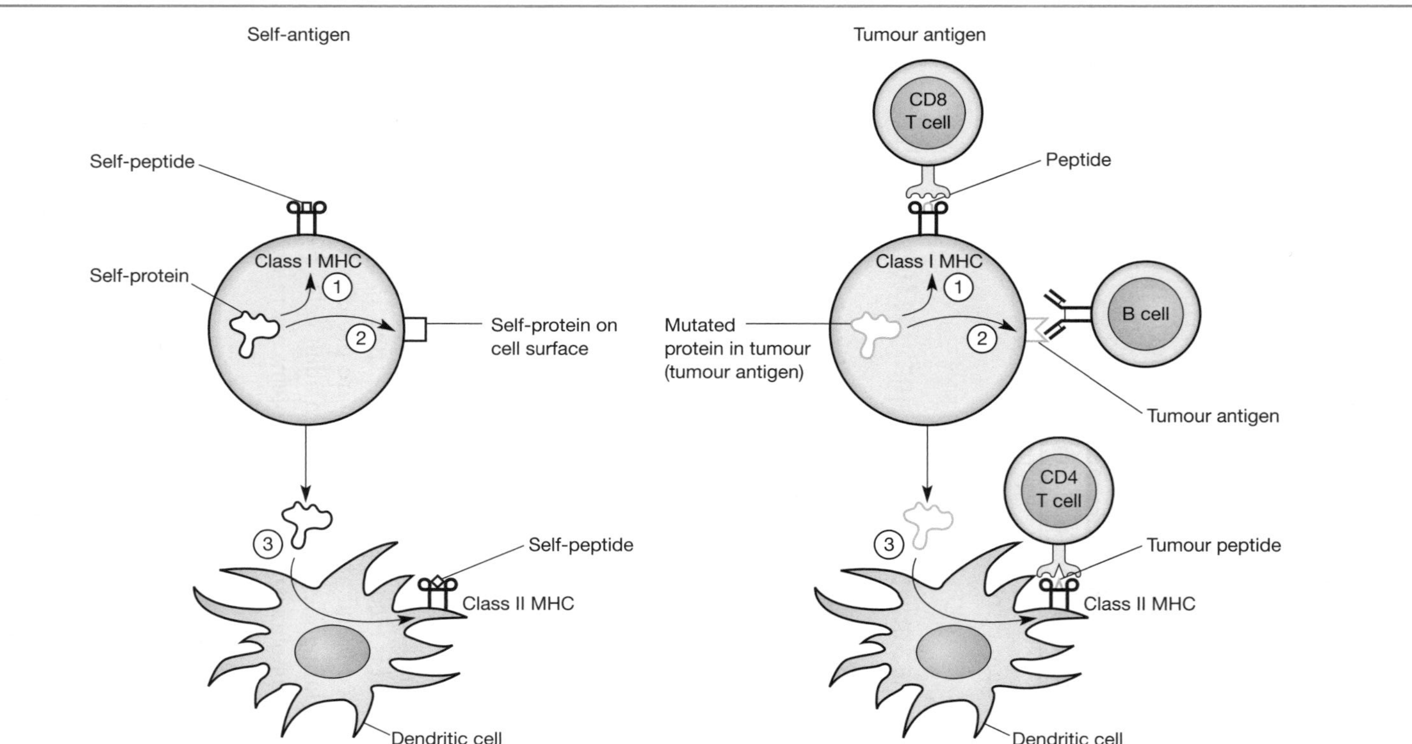

Figure 15.5 Tumour antigens. Mutation of a gene in a tumour cell can lead to a different form of the protein, which is recognised as foreign by the immune system. Surface proteins can stimulate antibody production and internal proteins can be presented by class I MHC to stimulate CD8 T cell responses. Protein that is taken up by dendritic cells can be processed and presented on class II MHC to stimulate CD4 T cells, which provide help for the antibody and CD8 T cell responses.

Table 15.2 Some examples of tumour antigens

Antigen	*Type of antigen*	*Expression in tumour cells*	*Expression in normal cells*
Antibody antigens			
Carcinoembryonic antigen (CEA)	Foetal	Colon + others	Yes
MUC-1	Aberrantly glycosylated mucin	Breast	Yes
p53	Mutated form of protein	Many	No
T cell antigens			
MAGE-1, MAGE-3	Foetal	Melanoma, breast	Testis
Ras	Mutated form of protein	Many	No
Tyrosinase	Differentiated cell product	Melanoma	Yes
Human papilloma virus E6 and E7 proteins	Viral gene product	Cervical carcinoma	Yes

detection by CD8 cytotoxic T cells is to down-regulate HLA class I expression on the cell surface so avoiding recognition by the TCR.

The second mechanism by which tumours avoid destruction is by producing factors that are immunosuppressive. Tumour cells can produce a variety of these factors such as transforming growth factor-β (TGF-β) and prostaglandin-E_2 that are able to inhibit immune responses. Initially the immunosuppression is local to the tumour so the individuals still make normal responses against microbial pathogens. However, as the tumour progresses, the immunosuppression becomes systemic so that patients are at increased risk of infection. Tumours are also good at inducing Tregs which can inhibit the immune response against the tumour.

With the increased knowledge about the interactions between the immune system and developing tumours there is significant interest in harnessing the immune response to provide protection against cancer. There are two main approaches that utilise the immune system to try to destroy tumour cells. These are the use of antibodies against tumour cells and the development of tumour vaccines.

15.3.1 Antibodies can act against tumours in many ways

Monoclonal antibodies (see Chapter 10) that are specific for tumour antigens have been developed and can be used as therapeutic agents in patients. Two variations have been considered. Unmodified antibodies can have anti-tumour effects in a number of ways:

- **Killing of tumour cells.** Antibodies can bind to the tumour cells and recruit other components of the immune system to kill the tumour cells. This could be through complement fixation and lysis or the involvement of antibody-dependent cell-mediated cytotoxicity (Figure 15.6).

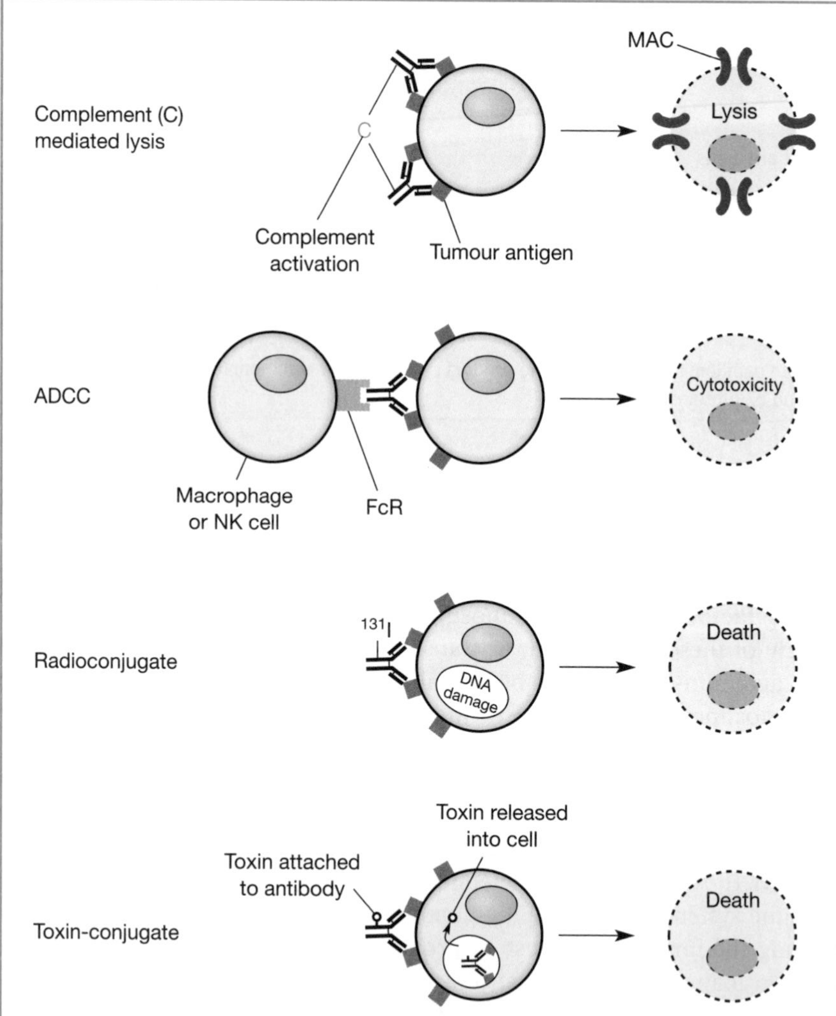

Figure 15.6 Anti-tumour therapy with antibodies. Anti-tumour cell antibody can bind to the tumour cell and either fix complement, resulting in tumour cell lysis, or promote ADCC. Radioactive isotopes or toxins can be conjugated to the anti-tumour antibody, which then specifically targets the radiation or toxin to the tumour cell.

- **Modulation of the immune system.** Antibodies can be used to try to boost the response against tumours. Tumours appear to be very good at inducing Tregs and so one approach is to deplete Tregs by giving antibodies against CD25 which is on the surface of Tregs.
- **Blockage of growth-factor receptors.** The growth of some tumour cells is promoted or dependent on growth factors, such as epidermal growth factor, binding to receptors on the tumour cells. Antibodies against these receptors can bind to the receptor and block the binding of the growth factor.
- **Inhibition of vascularisation of tumours.** As solid tumours grow they need to generate new blood vessels to supply oxygen and nutrients to the tumour. This is known as neo-vascularisation and is controlled in part by various growth factors. One mAb is specific for vascular endothelial growth factor (VEGF) and blocks its action thereby inhibiting neovascularisation. It is licensed for use in cancer of the colon, breast and lung.

A second approach with monoclonal antibodies is to attach them to toxic agents, which may be radioactive isotopes or toxins (Figure 15.6). These agents are too toxic to use directly but the theory is that the antibody will bind to antigens on the tumour cell and be internalised, leading to delivery of the toxic agent specifically to the tumour cells. Specific agents that have been considered for use include ^{131}I, bacterial toxins such as diphtheria toxin, and plant toxins including ricin, a very powerful toxin obtained from the castor bean. A number of these immunotoxins are now licensed for use.

The monoclonal antibodies that have been licensed for clinical use are shown in Table 15.3 and more are in clinical trials.

Table 15.3 Monoclonal antibodies licensed as anti-tumour agents

Antibody	Target antigen	Tumour type
Rituximab	CD20	Non-Hodgkin lymphoma
Cetuximab	EGF-R	Colorectal cancer
Trastuzimab	Her2 (EGF-like receptor)	Breast cancer
Alemtuzumab	CD52	Chronic lymphocytic leukaemia
Bevacizumab	VEGF	Colorectal cancer (metastatic)
^{131}I-tositumomab (Radio-Ab)	CD20	Non-Hodgkin lymphoma
Gemtuzumab-Ozogamicin (Toxin-Ab)	CD33	Acute myeloid leukaemia

EGF-R, epidermal-growth-factor receptor
VEGF, vascular endothelial growth factor

15.3.2 Tumour vaccines – stimulating immune responses to tumours

Tumour vaccines represent an approach to try to stimulate immune responses to tumours in the same way that microbial vaccines are used to stimulate responses against pathogens. The types of vaccines being developed are similar in approach to vaccines against infectious disease with some additional, innovative methods being developed in tumour vaccines. These include the following.

Peptide and DNA-based vaccines

These have formed the basis of vaccines involving the injection of antigens or peptides with adjuvants (Figure 15.7), or the use of DNA vaccines. In addition to using adjuvants, other agents, especially cytokines such as interleukin-2, granulocyte/macrophage colony-stimulating factor (GM-CSF) and interleukin-12, have been incorporated into tumour vaccines to try to boost the immune response. Many of these vaccines have been tested in clinical trials, especially for melanoma and breast cancer patients, using a number of antigens/peptides including the MAGE family and MUC-1 (Table 15.2).

DNA-based tumour vaccines use the same approach as with infectious agents (see Section 10.2.3) involving injection of DNA-encoding tumour antigens or peptides intra-muscularly or dermally via gene guns. It has been shown that the bacterial DNA acts as an adjuvant because the unmethylated CpG motifs are recognised by TLRs on dendritic cells and stimulate dendritic cells to become good stimulators of T cells.

Tumour cells themselves can be used as vaccines

One of the problems with the antigen/peptide approach is that for many tumours no tumour-specific antigens have been identified and even some of the known tumour antigens are not expressed by tumours in all individuals. Another approach has been to use irradiated tumour cells or tumour cell lysates as vaccines. These have the advantage that both unique tumour and unknown shared antigens will have the potential to stimulate an immune response. Whole vaccines can be based on either cells derived from the tumours of the patient themselves or on tumour cell lines, although the latter, being derived from other individuals' tumours, will not stimulate responses against the patient's unique tumour antigens. Generally they are also given with an adjuvant. Whole vaccines have been tried in clinical trials for melanoma and pancreatic cancer, among others.

Tumour cell vaccines can be improved by transfecting tumour cells with specific genes

One of the findings from experimental and clinical trials with tumour cells as vaccines was that on their own they did not stimulate a very good immune response. An alternative approach to the use of adjuvants in

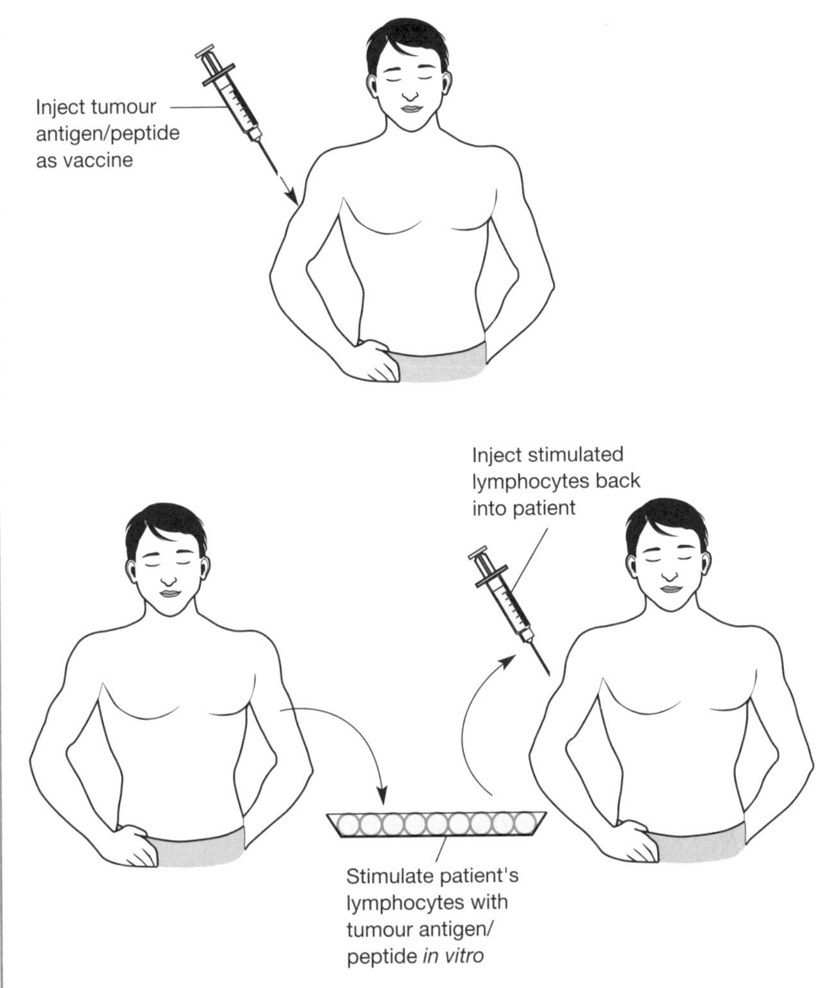

Figure 15.7 Tumour vaccines. Known tumour antigens can be used as vaccines to try to stimulate immune responses against the antigen, which then destroy the tumour cells. The antigen/peptide is either injected *in vivo* to stimulate a response (top) or the patient's cells are incubated with antigen and injected back into the patient (bottom).

whole vaccines is to transfect tumour cells with genes coding for products that would boost the immune response. This approach has the advantage that it does not require knowledge of the particular tumour antigens but instead involves manipulation of the tumour cells so that they directly or indirectly stimulate immune responses. A typical approach is to take out some tumour cells and transfect them with genes that will make the tumour immunogenic or stimulate an inflammatory response (Figure 15.8). Tumour cells have been transfected with co-stimulatory molecules (CD80 or CD86) so that they can directly stimulate CD4 T cells. In experimental

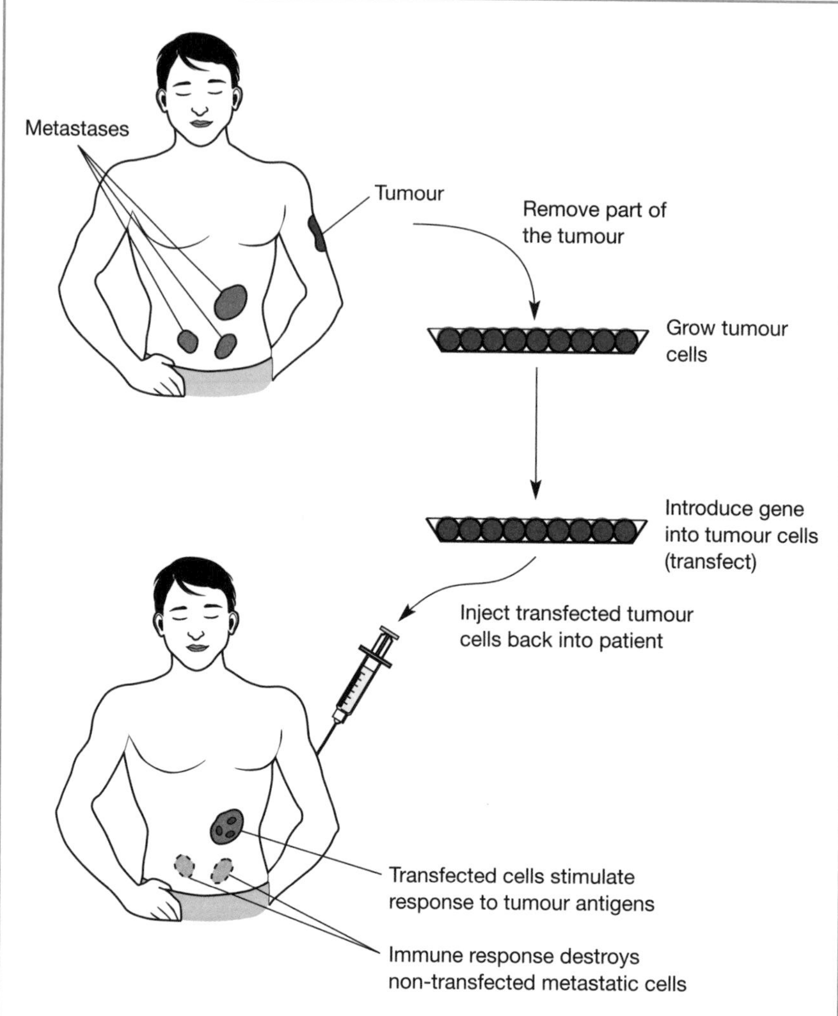

Figure 15.8 Transfected tumour cell vaccines. Part of the tumour is removed and grown in tissue culture. The tumour cells are transfected with a gene coding a particular protein that will make the tumour cells more immunogenic (e.g. GM-CSF or CD80). The transfected tumour cells are injected back into the patient in the hope that they will stimulate an immune response against the transfected cells that will also destroy non-transfected tumour cells.

situations the transfected tumour cells stimulated a response that protected against both transfected and non-transfected tumours. The most successful experimental approach has been to transfect tumour cells with granulocyte macrophage colony-stimulating factor (GM-CSF) or IL-12. When injected back into animals the transfected tumour cells secrete the cytokine to boost the response. GM-CSF is able to stimulate dendritic cells so that they are better at processing tumour antigens and carrying them to draining lymph nodes, where they present antigen on their class II MHC and activate tumour-specific CD4 T cells. They are also good at presenting antigenic peptide on class I MHC and activating CD8 T cells to become Tcs. These GM-CSF transfected tumour cells have also been shown to provide protection against both transfected and non-transfected tumours. Similarly IL-12 transfected tumour cells stimulate good responses, presumably through the ability of IL-12 to boost Th1 development. It is vital that the transfected tumours also protect against non-transfected tumours because in the clinical situation you will be taking only a small proportion of the tumour cells for transfection and the transfected cells must be able to stimulate an anti-tumour response that is effective against non-transfected tumour cells.

Dendritic cells can be used as a basis for vaccines

The success of GM-CSF-transfected tumour cells through the activation of dendritic cells led to the idea of basing vaccines on dendritic cells. This was made feasible by the development of techniques for producing large numbers of dendritic cells from blood. Peripheral blood monocytes can be cultured in tissue culture with GM-CSF and IL-4 which causes the cells to differentiate into dendritic cells. These dendritic cells can then be incubated with tumour cell lysates or tumour-specific antigens/peptides so that they take up antigen and present it on their class I and class II MHC molecules. The dendritic cells are then injected back into the patients in the hope they will stimulate tumour-specific T cells. This approach has been tested in clinical trials for melanoma and gastro-intestinal cancers.

Cytokines, antigen-specific T cells and natural killer cells can be used against tumours

Cytokines have been given systemically to try to boost immune responses, with limited success for some tumours. Although not strictly a vaccine, cytokines are being given with vaccines to increase immunogenecity (see above). Another use of cytokines, particularly IL-2, is to enable tumour-specific T cells to be grown *in vitro*. Lymphocytes extracted from tumours, so-called tumour-infiltrating lymphocytes (TILs), can be grown in tissue culture with IL-2 alone to expand the number of T cells that have recently been activated with tumour antigen *in vivo*. Alternatively TILs or blood-derived T cells can be cultured with tumour antigen/peptide and antigen-presenting cells and IL-2 so that tumour antigen-specific T cells are stimulated by antigen and grow under the influence of IL-2. The cells can then be injected back into the patient to exert anti-tumour activity.

A final use of cytokines is in improving natural killer cell activity. Natural killer cells from blood can be grown in tissue culture and proliferate in the presence of high concentrations of IL-2 to become what are known as lymphokine-activated killer (LAK) cells. These LAK cells have enhanced tumour-killing activity and can be injected back into patients in the hope they will kill tumour cells.

It is evident from these examples that many innovative approaches are being used to induce or increase anti-tumour immune responses. Although they have been used in clinical trials, results have not been spectacular. Part of the problem is that the early clinical trials were in terminally ill patients with widespread untreatable tumours and the trials were not controlled. Results of the trials in terms of partial or complete tumour remissions were compared with historical remission rates for the particular tumours. However, results in some of the trials were encouraging enough for the approaches to be taken forward to larger controlled trials. It will probably be important to use vaccines in situations where the tumour burden is much smaller, either earlier on in the disease or following surgery or chemotherapy to reduce the tumour burden and using vaccines against the residual tumour. This, combined with ever-increasing knowledge of the interaction between the immune system and tumours, will hopefully lead to the development of effective tumour vaccines that can result in the elimination of established tumours.

15.4 Summary

- The most common form of transplant is an allograft, which is between two non-genetically related individuals of the same species. There is also considerable interest in xenografting tissue from other species to humans.
- Allografts stimulate strong immune responses against foreign MHC and non-MHC antigens. These can result in the generation of antibody, CD8 T cells and delayed-hypersensitivity responses against the graft.
- Graft rejection is prevented by immunosuppressive drugs, which are effective but can leave the recipient at increased risk of infection and the development of certain tumours.
- Immunotherapy of tumours involves two main approaches: the use of anti-tumour antibodies to destroy tumour cells and the development of tumour vaccines.
- Anti-tumour antibodies can be given in unmodified form or attached to radioactive or toxic agents. A number of these have been licensed for clinical use.
- Because tumours do not stimulate very good immune responses, two main types of tumour vaccine are being tested to try to stimulate anti-tumour immune responses. One type involves immunisation with identified tumour antigens to stimulate responses against the antigen. The other type involves transfecting tumour cells with genes that will make tumour cells stimulate an immune response that will be directed against both transfected and non-transfected tumour cells.

15.5 Questions

1) What are the organs/tissues that are used in clinical transplantation?
2) What is meant by the terms 'major transplantation antigens' and 'minor transplantation antigens'?
3) Describe the different types of rejection that can occur with an allograft and explain the measures used to try to prevent these.
4) What sort of antigens can act as tumour antigens?
5) Explain the difference between active and passive tumour immunotherapy.

The answers to these questions can be found on page 343.

15.6 Further reading

1) Chinen J, Buckley RH. (2010) Transplantation immunology: solid organ and bone marrow. *Journal of Allergy and Clinical Immunology* 125:S324–335.
2) Schuster M, Nechansky A, Loibner H, Kircheis R. (2006) *Cancer Immunotherapy Biotechnology* 1:138–147.

Answers

Chapter 1

1) Bacteria, viruses, yeast/fungi, parasites (protozoa and worms).
2) C
3) (i) D (ii) B (iii) C (iv) A (v) E (vi) F
4) Endotoxins are structural components of pathogens, e.g. lipopolysaccharride (LPS) is part of bacterial cell wall. Exotoxins are secreted toxins, e.g. cholerotoxin, diphtheria toxin.
5) Because it breaches the strong physical barrier formed by skin which only a few pathogens can penetrate.

Chapter 2

1) Monocytes are a type of white blood cell that can leave the bloodstream, enter tissues and differentiate into macrophages.
2) E
3) B
4) Endocrine – the hormone is synthesised by one gland and travels in the bloodstream to act on cells at a distant site. Paracrine – the hormone is secreted and acts on cells in close proximity. Autocrine – the hormone is secreted and binds to receptors on cell that secreted hormone.
5) (i), (iii), (iv), (ii).
6) Adhesion molecules allow leukocytes to bind to endothelium – initially via selectins. Chemokines increase the strength of binding of integrins to Ig superfamily members. Cytokines (especially TNFα) increase the expression of adhesion molecules on endothelial cells. Chemokines also direct leukocyte movement toward the site of infection within tissues.

7) (i) Redness is caused by vasodilation and swelling is caused by oedema due to increased vascular permeability allowing plasma fluid to enter the tissue at the site of inflammation. (ii) Vasodilation increases blood flow to the area and increased vascular permeability makes it easier for soluble factors and cells to enter the tissue to combat the pathogens causing the inflammation.

8)

Arrow	**Label**
1	A
2	D
3	C
4	B
5	E

Chapter 3

1) B cell, CD4 T cell, CD8 T cell

2)

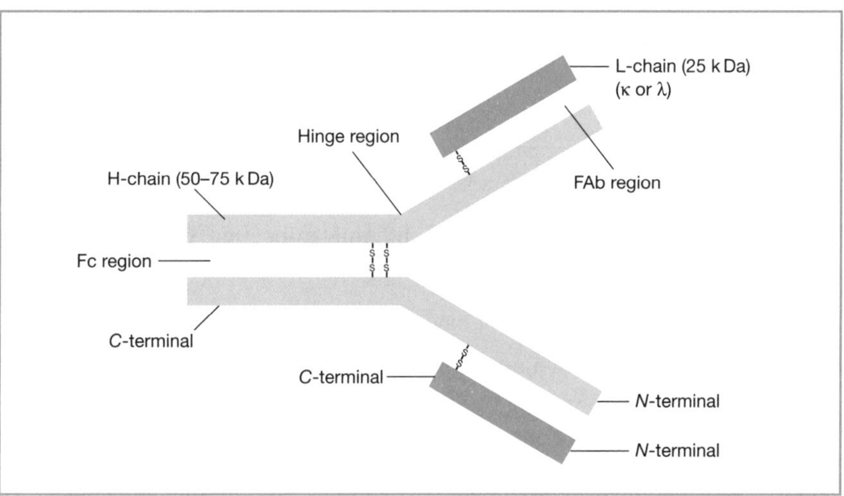

3) C

4) An antigen is the whole macromolecule (most commonly protein), parts of which are recognised by antibody or T cells. An antigenic epitope is the small part of an antigen that is recognised by antibody or a T cell in conjuction with MHC.

5) B

6) Affinity is the strength of binding of one binding site of an antibody to one antigenic epitope. Avidity is the combined strength of binding of all the antigen binding sites of an antibody molecule to antigenic epitopes.

Chapter 4

1) Because they finish their production in the thymus and are therefore thymus (T) dependent.

2) CD4 T cells and CD8 T cells.

3)

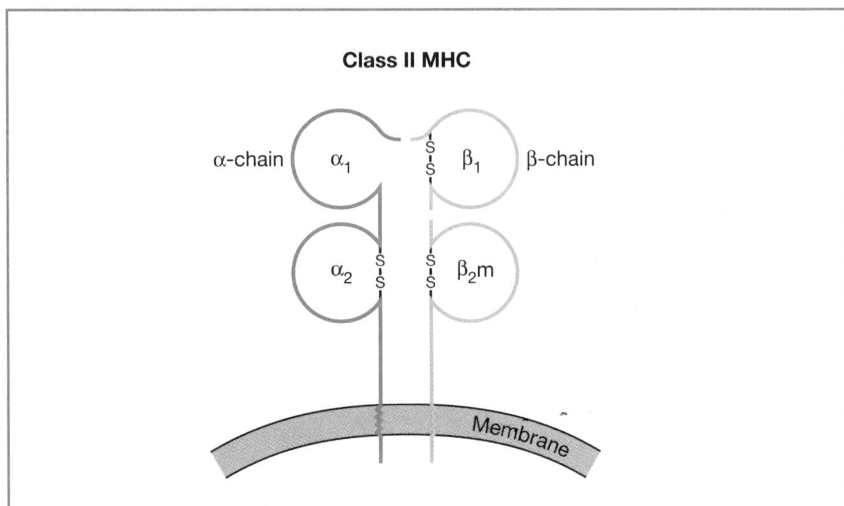

4) A) T cell receptor for antigen (TCR). B) Antigen-peptide. C) Class I MHC.

   It is a CD8 T cell.

5) (i) It is very extensive with loci having hundreds of alleles.

   (ii) It means that within a population there will always be some individuals who will be able to present antigenic peptides from any pathogen.

6) Exogenous antigens are endocytosed, the endosome fuses with a lysosome and the antigen will be degraded by lysosomal enzymes. The endolysosome fuses with a compartment of peptide loading (CPL) and the lysosomal enzymes break down the invariant chain bound to the class II MHC molecule and allows peptide from the antigen to bind to the MHC.

   Endogenous antigens are produced in the cytoplasm and broken down in proteasomes. The antigenic peptides produced in the proteasomes are transported into the Golgi by TAP transporters and the petides can bind to newly synthesised class I MHC molecules.

Chapter 5

1) C

2) During differentiation in the <u>bone marrow</u> developing B cells rearrange their <u>immunoglobulin/antibody</u> genes. The heavy chain genes rearrange <u>D</u> to <u>J</u> and <u>V</u> to <u>D</u> while the light chain genes rearrange <u>V</u> to <u>J</u>. In a similar

fashion, developing T cells in the thymus rearrange their TCR genes. The β-chain genes rearrange D to J and V to D while the α-chain genes rearrange V to J. The purpose of this gene rearrangement is to generate a large number of antibody and TCRs with specificity for a huge number of different antigens.

3) Number of α-chains: $46 \times 49 \times 100 = 225400$

   Number of β-chains $57 \times 2 \times 100 \times 13 \times 100 = 14820000$

   Number of combinations of α-chains and β-chains $= 3.34 \times 10^{12}$

4) Allelic exclusion stops B and T cells rearranging the second copy of their receptor genes if the first rearrangement is successful. This stops lymphocytes expressing more than one receptor for antigen: if a lymphocyte expressed 2 different receptors with specificities for 2 different antigens this would make control of responses to one antigen much more difficult.

5) Lymphocytes are produced with random specificities for antigen. If a lymphocyte doesn't see its antigen in a specific time it is better to replace it with another cell with a different antigen specificity which may see its antigen.

Chapter 6

1) A) adenoids, B) salivary glands, C) lymphatic vessel, D) bone marrow, E) thymus, F) lymph node, G) lymphatic vessel, H) spleen, I) mucosal lymph follicle.

2) There is no heart to pump fluid round lymphatic vessels, fluid is moved by muscle movement and valves. Lymphatic vessels contain mainly lymphocytes and dendritic cells with no red blood cells.

3) It is the recirculation of lymphocytes through blood and lymphoid tissue and is important to maximise the chance of a lymphocyte meeting its antigen.

4) (i) Spleen

   (ii) Gut lymphoid tissue, mesenteric lymph nodes.

   (iii) Lymph node.

   (iv) Draining lymph nodes (tracheobronchial)

5) A

Chapter 7

1) (iii), (v) and (ix), (viii), (vi), (xi), (i), (ii) and (iv), (vii), (x)

2)

Cytokine	Function
IL-2	A
IL-4	B
IL-5	B
IL-6	C
IFNγ	B
TGFβ	B

3) Rapid movement of cells in lymphoid tissues promotes interaction of cells, chemokines attract cells to each other and adhesion molecules promote the stable interaction of cells with each other.

4) (i) F, (ii) D, (iii) E, (iv) C, (v) B, (vi) A.

5) B cells switch to IgA and differentiate into plasma cells in mucosal lymphoid tissue, including mesenteric lymph nodes. Plasma cells go to the lamina propria and secrete IgA. IgA is taken up by epithelial cells and secretory piece is added. IgA is secreted onto the mucosal surface.

Chapter 8

1)

Arrow	**Label**
A	(v)
B	(i)
C	(ii)
D	(iii)
E	(iv)

2) Lysosome-independent mechanisms: oxygen radicals (superoxide anion [O_2^-], hydrogen peroxide [H_2O_2], singlet oxygen [$^1O_2$] and free hydroxyl radicals [$\cdot OH$]) and NO.

   Lysosomal-dependent mechanisms: Generation of chlorine products (hypochlorous acid [HOCl], hypochlorite [OCl^-] and chlorine [Cl_2], proteolytic enzymes, defensins.

3) Classical, alternative and lectin.

4) A) normal, B) C–, C) C–Ab–.

5) Formation of MAC and lysis. Opsonisation. Inflammation. Chemotaxis.

Chapter 9

1)

Label	**Cytokine**
C	IL-6 + TGFβ
B	IL-4
D	IFNγ
A	IL-12
F	IL-17
E	IL-4

2) Granule exocytosis involving mainly perforin and granzymes resulting in apoptosis of target cell.

   Engagement of Fas on target cell by Fas-L on Tc triggering Fas apoptotic pathways in target cell.

3) Viruses can only replicate in live host cells so killing host cell would stop viral replication.

4) (i) Tc
   (ii) Ab + complement
   (iii) Th17
   (iv) Th2 and IgE production.
5) To activate macrophages so that they are better able to kill intracellular pathogens.

Chapter 10

1) The secondary antibody response is faster, bigger and better than the primary response.

   It is faster because there are more memory B and CD4 T cells. Memory B cells have already undergone affinity maturation and class switch.

   It is bigger because there are more memory B cells.

   It is better because the memory B cells have already undergone affinity maturation.

2) Attenuated vaccines stimulate good antibody and Tc responses. There are dangers of reversion to virulence and they can cause disease in immunodeficient recipients.

   Killed vaccines stimulate good antibody responses but do not stimulate Tcs. There is a danger to personnel growing and handling virulent pathogens. There is the possibility of incomplete killing/inactivation.

3) An adjuvant is an agent that boosts the immune response to an antigen.

4)

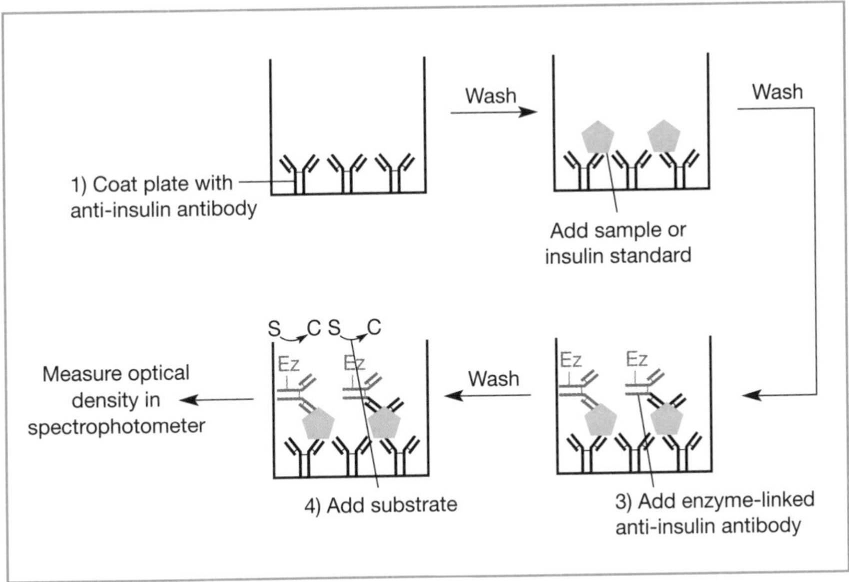
1) Coat plate with
anti-insulin antibody
Wash
Add sample or
insulin standard
Wash
3) Add enzyme-linked
anti-insulin antibody
Ez
Wash
S C S C
Ez
4) Add substrate
Measure optical
density in
spectrophotometer

5) Polyclonal antibodies are produced following immunisation of a person or animal with antigens. They contain a mixture of different antibodies against various antigenic epitopes on the antigens.

   Monoclonal antibody is produced by culturing hybridoma cells and purifying antibody from the culture supernantant. All the monoclonal antibody produced by one hybridoma line is identical and reacts against a single antigenic epitope.

Chapter 11

1) If the self-antigen is soluble the B cell is anergised. If the self-antigen is cell-bound the B cell undergoes apoptosis.
2) Positive selection is where thymocytes specific for self-MHC are rescued from cell death whereas thymocytes specific for foreign-MHC die by apoptosis.

   Negative selection is where developing thymocytes with specificity for self-MHC + self antigen-peptide are induced to undergo clonal deletion.

   Positive selection is important because it gets rid of thymocytes that are specific for MHC foreign to the individual. The TCR of these cell would never bind to antigenic peptide/MHC because they would never see the foreign MHC for which they are specific. They are therefore useless and would 'clog up' the immune system if they were not eliminated.

   Negative selection is important because it gets rid of thymocytes that are specific for self-antigen/self MHC. If they were allowed to develop into T cells they would have the potential to react against self-antigens causing autoimmunity.
3) Central tolerance occurs in the thymus and results in the deletion of self-reactive thymocytes or their conversion into Tregs thus helping to prevent autoreactivity. Peripheral tolerance is induced in the peripheral immune sytem after the T cells have left the thymus and occurs through anergy or induction of Tregs. Both are necessary because not all self-antigens are expressed in the thymus and peripheral tolerance deals with self-antigens not expressed in the thymus.
4) nTregs are induced in the thymus by contact of self-antigen specific developing CD4 T cells with self-antigen/MHC on medullary thymic epithelial cells (mTECs). iTregs develop from naïve peripheral CD4 T cells that are induced by recognition of self-antigen/MHC.
5) If the TCR on the CD4 T cell recognises antigen/MHC in the absence of co-stimulatory signals it becomes anergic. This helps prevent autoimmunity because many parenchymal (tissue-specific cells) can express MHC and present self-antigenic peptide but do not express co-stimulation molecule ligands and therefore induce anergy in self-reactive T cells.

Chapter 12

1)

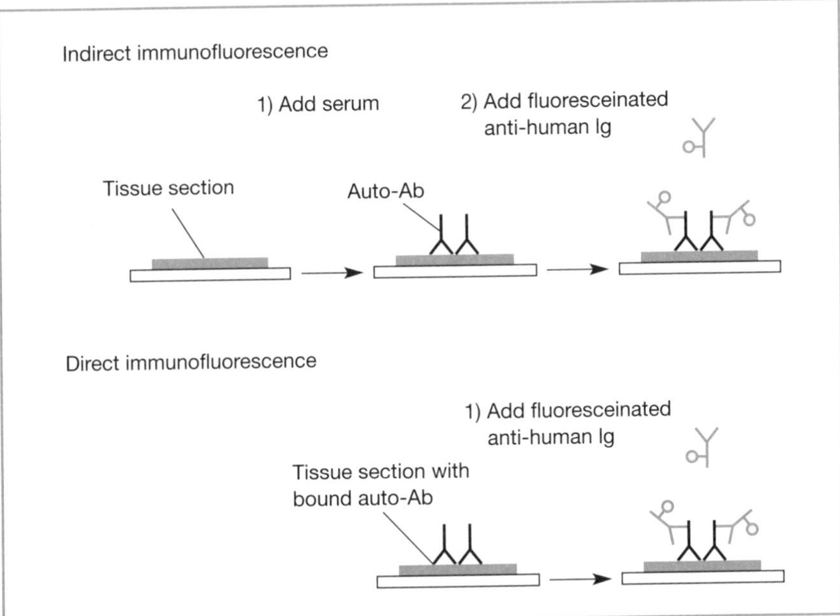

2) (i) Activation of complement pathway leading to lysis (e.g. autoimmune haemolytic anaemia), (ii) Inducing opsonisation of target cell (e.g. autoimmune thrombocytopenia, (iii) Blockage of function (e.g. myasthenia gravis, pernicious anaemia), (iv) stimulation of receptors (e.g. Grave's disease), (v) Formation of immune complexes leading to activation of complement and inflammation (e.g. SLE).

3) Many genes (multigenic) especially class II MHC, and environmental factors such as infectious agents, toxins/chemicals and possibly diet.

4) A monogenic disease is where a mutation in a single gene causes disease (e.g. cystic fibrosis). Multigenic diseases are where many genes make contributions to susceptibility and it is different alleles of the genes that affect susceptibility, not a mutation in a single gene.

5) Molecular mimicry is where an antigenic epitope on a foreign antigen (e.g. derived from a microbe) resembles an epitope on a self-antigen. Molecular mimicry can contribute to the development of autoimmunity in 2 main ways. Antibodies produced against the foreign epitope can react against the self-epitope (e.g. with *Streptococcus* and heart-myosin in rheumatic fever). Foreign epitopes could lead to activation of self-reactive T cells to become effector T cells which bind to self-epitopes and cause autoimmunity (only demonstrated experimentally so far).

Chapter 13

1) (i) C, (ii) D (iii) B, (iv) E, (v) A.
2) The following are COMMON sources of allergens (True/False)
   - (i) T
   - (ii) T
   - (iii) F
   - (iv) T
   - (v) F
   - (vi) T
   - (vii) T
   - (viii) T
   - (ix) F
   - (x) T
3) The seasonality suggests that Fred is allergic to some pollen/flower/grass that appears in June. You could test by skin testing for allergens.
4) Desensitisation by giving slowly increasing amounts of allergen. Anti-IgE. Anti-cytokine antibodies (e.g. against IL-5).
5) In type II hypersensitivity the antibody is directed against antigens on the affected tissue/organ (e.g. anti-Rhesus IgG in haemolytic disease of the newborn). In type III the hypersensitivity is caused by immune complexes where the antigen is not on the affected tissue (e.g. pigeon fancier's lung).

Chapter 14

1) A) (v)
   B) (xiii)
   C) (xi)
   D) (ii) or (i)
   E) (iii) or (i)
   F) (viii) or (vi)
   G) (vii) or (vi)

2)

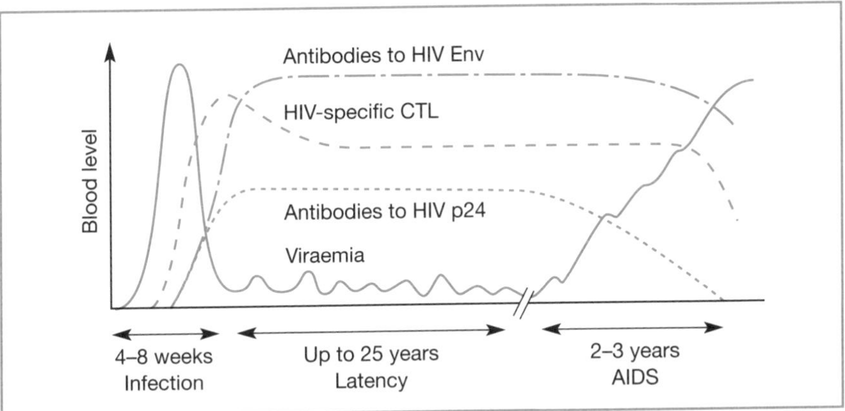

3) It has a very fast replication rate and also a high mutation rate so it mutates genes for antigen so that antibody/T cells can no longer recognise antigen/ antigenic epitope.

4) HAART stands for 'highly active anti-retroviral therapy' It is a combination of anti-HIV drugs – typically two reverse-transcriptasae inhibitors and a protease inhibitor although other combinations are used.

5) There are many different types and sub-types of the virus. It is not known what is the best type of immune response to stimulate or which target antigens are best. The best type of vaccine is not known. HIV has a high mutation rate. Because of the chronic nature of the disease it takes a long time for vaccine trials to give answers about whether vaccines are protective.

Chapter 15

1) Kidney, heart, lung, pancreas/Islet of Langerhans/liver, cornea, skin, bone marrow, face.

2) Major transplantation antigens are foreign class I (HLA-A,B,C) and class II (HLA-DP, DQ, DR) MHC molecules and stimulate strong rejection responses. Minor transplantation antigens are any protein which has different alleles giving different structural forms of the protein. They stimulate much weaker rejection responses.

3) Hyperacute rejection occurs within minutes to hours of the transplant operation and is caused by the recipient having pre-existing antibodies against donor antigens. It can be prevented by checking that the recipient does not have antibodies to the donor (cross-matching). Acute rejection occurs days to weeks after the transplant and is due to specific immune responses against the graft. It can be prevented/reduced by a combination of blood-group matching, matching donor and recipient at HLA (especially DR, A and B) and immunosuppression. Chronic rejection occurs months to years after the transplant and is characterised by thickening of the arterial walls leading to narrowing and eventual occlusion of the arteries resulting in ischaemia.

4) Viral antigens (e.g. Papilloma virus and cervical cancer), oncofoetal antigens (e.g. CEA), differentially glycosylated products (e.g. MUC-1), proteins from mutated forms of the gene (e.g. Ras), differentiated cell products (e.g. tyrosinase).

5) Antibodies can be used to harness the immune system to kill tumour cells e.g. via ADCC. They can also be used to modulate the immune system, to block/inhibit growth factor receptors (e.g. Herceptin) and to inhibit neovascularisation (e.g. ant VEGF). Antibodies can be conjugated to radioisotopes or toxins to increase their toxicity.

Glossary

Acute phase proteins Proteins produced by the liver in response to systemic inflammation. Include C-reactive protein and serum amyloid.

Acute rejection Primary immune response against a transplant.

Adhesion molecules Molecules expressed on the surface of cells to allow cell–cell adhesion.

Adjuvant A substance that boosts an immune response against an antigen.

Affinity The strength of binding of a single antigen binding site of an antibody to an antigenic epitope.

Affinity maturation The increase in antibody affinity during a germinal centre reaction due to somatic hypermutation.

AIDS (acquired immunodeficiency syndrome) Disease caused by the human immunodeficiency virus (HIV) leading to loss of immune function.

Allele A particular form of a gene that is polymorphic, i.e. has more than one variant.

Alleleic exclusion The stopping of Ig or TCR genes from rearranging the second copy of the gene if rearrangement of the first copy is successful.

Allergy Disease caused by production of IgE; includes asthma, hayfever and eczema.

Allograft A graft between two non-genetically identical members of the same species.

Anaphylaxis A severe, potentially life threatening, form of systemic allergic response.

Anergy A state of non-responsiveness induced in cells; most commonly seen in immunological tolerance.

Antibody A 4-chain protein produced by plasma cells that binds to target molecules (antigens) and mediates protection against infection.

Antibody class The type of antibody determined by the Fc part of the Ig heavy chain; the different classes of antibody are IgA, IgD, IgE, IgG and IgM.

Antibody-dependent cell-mediated cytotoxicity (ADCC) Killing of target cells through killer cells (macrophages and NK cells) binding to the Fc part of antibody bound to the target cell.

Antigen A molecule of which part is recognised by antibody or the T cell receptor.

Antigen-presenting cell A cell such as a dendritic cell or macrophage which presents antigen to a B cell or T cell.

Antigen processing The breaking down of proteins so that peptides derived from them can be presented by MHC molecules to T cells.

Apoptosis A programmed form of cell death which does not result in release of cell contents or inflammation.

Asthma A type of allergy affecting the lungs resulting in wheezing and shortness of breath.

Atopy A state of immunological reactivity such as IgE associated with allergy but not necessarily accompanied by clinical symptoms.

Attenuated vaccine A live vaccine where a virulent pathogen is grown under conditions where it loses its virulence but can be used as a vaccine.

Autoantibody An antibody against a self-antigen of a person or animal.

Autocrine Hormone action where hormones are produced by one cell and act on the cell that produced it.

Autograft A graft from one part of a person to another part of the same person – most commonly of skin for treatment of burns.

Autoimmune disease A disease involving the production and involvement of antibody and/or T cells against self-antigens.

Autoimmunity Disease caused by the immune system attacking self-antigens; includes rheumatoid arthritis and multiple sclerosis.

Avidity The overall strength of binding of antibody to its target antigen determined by the sum of strength of binding of individual antibody binding sites.

β2-microglobulin A 12 Kda protein that associates with class I MHC α-chains to form the HLA class I product.

Body cavity A fluid-filled space within the body in which organs may be situated.

Budding Process by which virus particles leave cells by being shed from the cell surface, usually without killing the host cells.

Cathelicidins Antimicrobial peptides found in skin and other secretions.

CD antigen CD is cluster of differentiation. A system for naming molecules on the surface of cells based on antibody recognition.

Cell lysis Bursting open of cells leading to their death and release of contents.

Cell migration The movement of cells throughout the body, especially through blood and tissues.

Chemokines Cytokines that have chemotactic activity, the ability to control cell movement.

Chemotaxis The movement of cells in a directed manner, usually following an increasing concentration gradient of a chemotactic agent.

Chemotherapy The use of drugs for clinical treatment, for example in cancer or transplantation.

Chronic rejection The slow rejection of transplants, usually occurring years after the transplant operation.

Cilia Hair-like structures lining the respiratory tract.

Class switch The process by which B lymphocytes change the class of antibody they express on their cell surface, from IgM and IgD to IgG, IgA or IgE.

Clonal deletion The process in which lymphocytes specific for an antigen (usually a self-antigen) undergo cell death by apoptosis on recognising the antigen.

Clonal ignorance A mechanism of immunological tolerance based on the inability of lymphocytes to recognise sequestered antigen, e.g. lens protein in the eye.

Clonal selection The process by which lymphocytes specific for a particular antigen are stimulated to respond to the antigen.

Collectins Proteins that can bind sugars on microbial surfaces and promote the elimination of microbes.

Colony-stimulating factors Cytokines that were originally identified by their ability to stimulate the production of particular cell types in bone marrow cultures.

Combinatorial diversity The creation of different antigen binding specificities in Ig or TCR by the random combination of different V, D and J regions.

Commensal organism Microbes, usually bacterium but sometimes yeast, that lives in various anatomical locations, for example the gut, and benefits the host.

Compartment of peptide loading An intracellular vesicle where antigen peptide binds to class II MHC to be transported to the cell surface.

Complement A system of over twenty blood proteins with various biological activities.

Complement pathways Three pathways by which the complement cascade is activated; called the classical, alternative and lectin pathways.

Complementarity determining region (CDR) The parts of the variable region of Ig or TCR that determines the antigen specificity of the Ig or TCR.

Concordance The incidence of a disease in different groups, for example identical twins or non-identical siblings, more or less closely related to each other.

Cross-matching In transplantation – the testing of a potential recipient's serum for antibodies against potential donor cells.

Cross reactivity The situation where Ig or TCR binds to more than one antigen.

Cytokines Small proteins that act like hormones to enable communication between cells of the immune system.

Cytotoxic T cells (Tcs) CD8 T cells that have the ability to kill other cells, for example virally infected or tumour cells.

Damage-associated molecular patterns (DAMPs) Molecules that are exposed after damage to host cells that can be recognised by PRRs.

Defensins Antimicrobial peptides found in skin and other secretions.

Delayed-type hypersensitivity (DTH) A type of immune response involving the recruitment and activation of monocytes by CD4 T cells at a site of infection or other antigen.

Dendritic cell A type of bone marrow-derived cell found in most tissues that is very good at activating T cells.

D-gene Gene segments found in the genes for Ig heavy chain (Dh) or the TCR β-chain (Dβ): D stands for diversity.

Differentiation The process by which cells change gene expression and function, for example during the production of blood cells in the bone marrow.

DNA vaccine Thte use of DNA constructs encoding antigen from a pathogen to stimulate immunity to the pathogen.

Eczema A red, itchy skin condition caused by an allergic reaction.

Elephantiasis Massive swelling of the breasts, testes or legs caused by blockage of lymphatic vessels in worm infections.

ELISA Enzyme-linked immunosorbant assay – an assay based on the use of antibodies to measure a variety of substances.

Endocrine Hormone action where hormones are produced by one gland and travel through the bloodstream to act on distant cells.

Endotoxins Toxins that are components of microbes, e.g. cell walls, and not secreted.

Epithelial cell Cells forming the outer layers of skin and mucosa of the respiratory, digestive and genito-urinary tracts.

Epitope Structural parts of antigens that are recognised by antibody or the TCR.

Exotoxins Toxins secreted by pathogens, e.g. cholera.

Extravasation Movement of cells across the vascular endothelium thereby entering tissue from blood.

Fab region The part of the antibody molecule that is involved in antigen recognition.

Fas A death receptor on cells which activates apoptosis when bound by its ligand.

Fc receptor A receptor expressed on various cell types that can bind the Fc region of the antibody molecule.

Fc-region The part of the antibody molecule that determines its biological function; stands for 'Fragment-crystallisable'.

Fluorescence microscopy The use of fluorescent dyes to visualise objects using a UV-microscope.

Follicular dendritic cell (FDC) A specialised cell found in germinal centres that presents antigen to B cells.

γ/δ T cell A type of T cell using a receptor for antigen made up of a γ-chain and a δ-chain.

Gene rearrangement The process by which Ig or TCR genes recombine during the development of B or T lymphocytes.

Germinal centre Structure within lyphoid tissue where plasma cells and memory B cells are produced.

Gram stain A stain used to identify bacteria based on the structure of their cell walls; bacteria can be classified as Gram positive or Gram negative.

Granulolysin A protein found in the granules of cytotoxic T cells that can cause the death of other cells.

Granzyme A proteolytic enzyme found in the granules of cytotoxic T cells that can stimulate apoptosis of target cells.

HAART Highly active antiretroviral therapy – a combination of drugs used to treat HIV-infected people.

Haematopoeisis The process of the production of red and white blood cells in the bone marrow.

Haemophiliac Individual with defect in blood clotting system.

Heavy chain The larger protein chains of the antibody molecule.

Helminth Parasitic worm, e.g. *Ascaris*, the cause of roundworm.

Helper T cells (Ths) CD4 T cells that co-operate with other cells in immune responses.

Herd immunity A situation where a high enough proportion of a population is immune to an infectious pathogen so that the pathogen cannot exist within the population.

High endothelial venule A specialised blood vessel in the paracortex of lymph nodes that has cuboidal endothelial cells and is the site where lymphocytes enter lymph nodes from blood.

Histamine An inflammatory mediator secreted by mast cells that promotes vasodilation and increased vascular permeability.

HIV Human immunodeficiency virus – the cause of AIDS.

Human leukocyte antigen (HLA) The name of the human major histocompatibility complex (MHC).

Hygiene hypothesis The idea that reduced exposure of children to infectious organisms has resulted in an increase in allergy and autoimmune disease.

Hyperacute rejection Very rapid (minutes to hours) rejection of transplant due to pre-formed antibodies in the recipient against graft antigens.

Hypersensitivity Harmful immune responses against usually harmless antigens, e.g. in allergy.

Hyphal form of yeast A form of fungus that grows as filamentous, branched threads.

Immunity A state of protection against a pathogen induced through previous exposure to the pathogen or vaccination.

Immunoglobulin superfamily Large family of proteins characterised by possessing 110 amino acid immunoglobulin domains.

Immunological synapse Specialised area of contact between cells of the immune system where antigen receptors, adhesion molecules and co-stimulatory receptors are concentrated.

Immunotoxin A conjugate of antibody and a toxin designed to increase the killing ability of the antibody.

Infectious organism A bacterium, virus, yeast or parasite capable of infecting individuals.

Inflammatory response The response to infection or tissue damage involving redness and swelling and the influx of cells and proteins.

Innate immune system Cells and tissues that are pre-existing before infection and provide an immediate response to infection.

Integrase An enzyme that promotes the integration of viral DNA into the host's DNA.

Interferons A family of cytokines with which can inhibit viral replication.

Interleukins One of the families of cytokines initially named because they acted as hormones enabling communication between white blood cells (leukocytes).

Invariant chain A protein that is added to class II MHC to stop self-peptides from binding to the newly formed class II MHC.

Isograft A graft between two genetically identical members of the same species; in humans between identical twins.

J-chain A protein that joins the subunits of IgM and IgA.

J-gene Gene segments found in the genes for Ig heavy chain (Jh) or the TCR β-chain (Jβ): J stands for joining.

Junctional diversity Novel nucleotide sequences formed at the junctions of V, D and J during the rearrangement of Ig and TCR genes.

Kinins A group of plasma peptides with various functions, especially in inflammation.

Light chain The smaller protein chains of the antibody molecule.

Lymph node Small bean-shaped structures containing lymphocytes and other cells where adaptive immune responses are generated.

Lymphatic vessel Vessel of the lymphatic system with similarities to blood vessels but also significant differences.

Lymphocytes White blood cells that compose the specific immune system. Main types are B and T cells.

Lymphopoeisis The production of lymphocytes in bone marrow and thymus.

Lysosome An intracellular vesicle containing degradative enzymes that can act against microbes.

Lysozyme An enzyme present in sweat, tears, saliva and other secretions that can break down bacterial cell walls.

Lytic virus A virus that can cause infected cells to burst open (lyse) releasing virus to infect other cells.

Macrophage Type of cell found in all tissues and organs that is able to engulf microbes and kill them.

Major histocompatibility complex A 3.6 Mb region of DNA containing over 200 genes coding for three classes of product; Class I MHC and class II MHC present antigen-peptides to T cells.

Marginal zone Region of the spleen between the red and white pulp, containing many macrophages and B cells.

Mast cells Cells found in most tissues involved in the inflammatory response.

Membrane attack complex A complex of C5–9 complement components that form pores in cell membranes resulting in lysis of the cells.

Memory In immunological terms memory is generated after a first exposure to antigen and is able to generate faster, bigger and better responses to the antigen.

Memory B cell Memory B cells are generated during immune responses and are able to respond more quickly to antigen than naïve B cells.

Memory T cell Memory T cells are generated during immune responses and are able to respond more quickly to antigen than naïve T cells.

MHC-restricted recognition Recognition of the MHC/antigenic peptide by the T cell receptor.

Mitogen A molecule (usually a protein) that can make cells proliferate; in lymphocytes the proliferation is independent of the antigen specificity of the cell.

Monoclonal antibody An antibody produced by hybridoma cells which are formed by fusion of myeloma cells and B cells. All antibody produced by one hybridoma cell line is identical.

Monocyte A type of white blood cell that is able to enter tissues and become a macrophage.

Mucosa Tissue lining some organs and body cavities including digestive, respiratory and genito-urinary tracts.

Mucosal associated lymphoid tissue (MALT) Structures of the immune system, especially lymphoid nodules, associated with mucosa of the digestive, respiratory and genito-urinary systems.

Mucus Stcky secretion of cells lining the gastro-intestinal, respiratory and genito-urinary tracts.

Natural killer cell A type of lymphocyte that can kill infected cells and produce various cytokines.

Negative selection The process by which developing T cells which recognise self-antigen in the thymus are stimulated to die by apoptosis.

Neutralisation The ability of antibody to block the activity of toxins or prevent attachment of microbes to cells or tissue.

Neutrophil Type of white blood cell involved in inflammatory response. Also known as polymorphonuclear leukocyte (PMN).

Non-lytic virus A virus that can replicate in infected cells without causing lysis of the cells.

Opsonins Proteins that can bind to molecules on microbes and receptors on leukocytes to facilitate recognition of pathogens.

Oxygen radicals Chemically very active forms of oxygen with unpaired electrons which can damage microbial molecules; examples include hydroxyl ions and superoxide.

Paracrine Hormone action where hormones are produced by one cell and act on nearby cells.

Pathogen An infectious organism capable of causing disease, e.g. influenza virus, tuberculosis bacterium.

Pathogen-associated molecular patterns (PAMPS) Molecules on pathogens that can be recognised by pattern recognition receptors, e.g. LPS, viral RNA, bacterial DNA.

Pattern recognition receptors (PRRs) Receptors on cells of the innate immune system and others used to recognise molecules of microbes.

Perforin A protein in the granules of cytotoxic T cells and natural killer cells that can assemble to form pores in the membranes of target cells.

Peritoneum Membrane lining the abdominal cavity.

Peyer's patches Organised lymphoid structures located along the small intestine.

Phagocytosis The uptake and ingestion of microbes or damaged tissues or cells by phagocytes.

Plasma cells Antibody-secreting cells derived from B lymphocytes.

Polymorphism The existence of different forms of a gene.

Positive selection The process by which developing T cells in the thymus that are specific for self-MHC are rescued from cell death.

Primary lymphoid tissue Lymphoid tissue where lymphocytes are produced – primarily bone marrow and the thymus.

Prostaglandins A group of small biologically active lipid molecules derived from arachidonic acid.

Protease An enzyme that can break down proteins.

Proteasome An intracellular cylindrical structure that breaks down proteins.

Protozoa Single-celled parasites, e.g. *Plasmodium*, the cause of malaria.

Provirus A stage of the viral life cycle where viral DNA is integrated into the host DNA.

Radio-antibody An antibody that is linked to a radioactive isotope.

Recombinant vector vaccine A vaccine where DNA coding for pathogen antigens is incorporated into the DNA of another virus or bacterium.

Reverse transcriptase An enzyme that transcribes RNA into DNA.

Rhinitis Hay fever.

Salivary gland Exocrine glands responsible for the production of saliva.

Secondary lymphoid tissue Lymphoid tissue where immune responses are generated, consisting primarily of lymph nodes, spleen and mucosa-associated lymphoid tissue (MALT).

Secretory piece A protein that is added to secetory IgA that helps protect the IgA from degradation by proteolyic enzymes.

Specific immune system Cells and tissues that are involved in generation new immune responses after infection.

Spleen A 'fist-shaped' organ in the abdomen that is responsible for removing old red blood cells and mounting immune responses against blood-borne antigen.

Subunit vaccine A vaccine consisting of a peptide or carbohydrate derived from a pathogen.

Superantigen Toxin that causes excessive stimulation of CD4 T cells, e.g. in toxic shock syndrome.

T lymphocytes Lymphocytes produced in the thymus. The two main types are CD4 and CD8 T cells.

TAP transporter A structure in the membrane of the rough-endoplasmic reticulum (RER) that transports antigenic peptides from the cytoplasm into the RER.

TCR (T cell receptor) A protein dimer used by T cells to recognise antigen-peptide presented by MHC.

T-dependent antigen An antigen that needs the involvement of CD4 T cells to help B cells become antibody-producing plasma cells.

Thymus A bi-lobed structure above the heart that is the site of production of T lymphocytes.

T-independent antigen An antigen that can stimulate antibody production without the need for CD4 T cells.

Tissue-typing The process of identifying which alleles of the class I and II MHC HLA genes a person possesses.

Tolerance In immunological terms, the induction of immunological unresponsiveness to an antigen.

Toxoid An inactivated form of toxin used as vaccine to stimulate antibody that can neutralise the native toxin.

Transfection The introduction of foreign DNA into a cell.

Transplantation Transfer of organs, tissues or cells, usually from one individual to another.

Transplantation antigens Antigens that stimulate rejection of grafts.

Treg A CD4 T cell that is able to inhibit the activity of other cells of the immune system.

Tropism The types of cell a particular virus can infect.

Tumour Uncontrolled growth of cells.

Tumour antigens Antigens on tumours that can stimulate immune reponses.

Tumour vaccine Agent designed to stimulate or boost an immune response against tumours.

Tumour-necrosis factor A cytokine that is important in inflammatory responses.

Vascular permeability The 'leakiness' of blood vessel walls to molecules of different sizes.

Vasodilation An increase in the diameter of blood vessels.

Western blot A technique where proteins by separated by electrophoresis, transferred to a membrane and probed with antibodies.

Wheal and flare The swelling (wheal) and redness (flare) associated with an inflammatory response in the skin.

Xenograft A transplant between different species.

Zoonoosis Infection that has jumped from one species to another, e.g. HIV from chimpanzee to man.

Index

Note: Figures, Tables and Boxes are indicated by *italic page numbers*; Glossary entries by **emboldened numbers**. Plate numbers are listed at end of each relevant index entry.